Second International Symposium on Civil Engineering and Environmental Research

IOP Conference Series: Earth and Environmental Science
Volume 1184

Banjarmasin (Online), Indonesia
23-25 October 2022

Editors:

Irfan Prasetia
Ade Yuniati Pratiwi

ISBN: 978-1-7138-7446-1
ISSN: 1755-1307

Printed from e-media with permission by:

Curran Associates, Inc.
57 Morehouse Lane
Red Hook, NY 12571

proceedings.com

Some format issues inherent in the e-media version may also appear in this print version.

Institute of Physics
Dirac House, Temple Back
Bristol BS1 6BE UK

Phone: 44 1 17 929 7481
Fax: 44 1 17 920 0979

techtracking@iop.org

Additional copies of this publication are available from:

Curran Associates, Inc.
57 Morehouse Lane
Red Hook, NY 12571 USA
Phone: 845-758-0400
Fax: 845-758-2633
Email: curran@proceedings.com
Web: www.proceedings.com

TABLE OF CONTENTS

GEOTECHNICAL ENGINEERING

WATER AND ENVIRONMENTAL ENGINEERING

CONSTRUCTION MANAGEMENT

TRANSPORTATION MANAGEMENT AND ENGINEERING

STRUCTURAL ENGINEERING

ENVIRONMENTAL ENGINEERING

PREFACE

As a university located in the South Borneo region, Lambung Mangkurat University (ULM) has academic and moral responsibilities to conserve Borneo's "lungs of the World". Especially for the Graduate School of Civil Engineering ULM, a series of national seminars has been held annually. The Graduate school of Civil Engineering has already organized the Annual National Seminar since 2014 on several themes and sub-themes regarding sustainable development. Furthermore, since 2021, the Graduate School of Civil Engineering has been organizing the International Symposium on Civil Engineering and Environmental Research (ISCEER) to improve the quality of the seminars and reach a broader range of speakers and participants also planned to be held annually. In 2022 we already held the 2nd ISCEER. The theme for the 2nd ISCEER is "Research and Technology Updates in Civil and Environmental Engineering to Support Sustainable Development". The symposium was held on October 24 - 25, 2022.

In the symposium, we discussed the latest findings regarding Research and Technology Updates in Civil and Environmental Engineering to Support Sustainable Development, including but not limited to the following:

a. Structural and material engineering
b. Water engineering
c. Environmental engineering
d. Geotechnical engineering
e. Transportation management, and
f. Construction management

This symposium aims to produce documents on points of view and ideas generated by cross-disciplinary speakers who have great concern for Civil Engineering and Environmental Research in terms of sustainable development in general and specifically in green construction. This activity is expected to encourage the passion for researching green technology for sustainable development. As a result, this is moving towards sustainable development publications published in national and international journals. It becomes suggestions for policy-makers in the context of sustainable development, which are always close to people's lives and their environments.

Apart from that, in the current pandemic situation, the condition of the Coronavirus Disease 2019 (Covid -19) pandemic has implications for many aspects of life. Therefore, same as the 1st ISCEER, the 2nd ISCEER was held online using Zoom Platform. All of the Keynote speakers, Invited Speakers, and Regular Speakers from different countries presented their latest research finding through video conference. In addition to the speakers, participants followed the event online through Zoom Platform or watched live streaming of the activities through YouTube.

The committee organizes online activities in Banjarbaru, South Kalimantan, Indonesia. Even though it was held online, it was carefully prepared using suitable multimedia devices so that the impression of an offline event was still felt, especially during the opening ceremony. This symposium was attended by more than 200 people consisting of academics, activists, researchers, and relevant stakeholders related to green technology for sustainable development from all over the world. The conference was divided into

IOP Conf. Series: Earth and Environmental Science 1184 (2023) 011001 doi:10.1088/1755-1315/1184/1/011001

several sessions. The presentation is held on a panel (consisting of 3 presenters) with 10 presentation times for each presenter, followed by 15 minutes of Q & A.

On this occasion, the committee of the 2nd ISCEER would like to express our deep appreciation to our co-host the Cebu Technological University, Philippines, and our sponsors, Public Works and Construction Development Offices of South Kalimantan Province and others, for their supports and participation in realizing this event through enormous contributions from their sponsorships. Finally, we hope this special event can contribute many benefits to developing knowledge, especially for sustainable development in Civil and Environmental Engineering.

See you again at the 3rd ISCEER
Best regards, ISCEER Committee Graduate School of Civil Engineering ULM

Dr. Eng. Irfan Prasetia, S.T., M.T.
Head of Graduate School of Civil Engineering ULM, Editor of conference materials, Lambung Mangkurat University

Ade Yuniati Pratiwi, S.T., M.Sc., Ph.D.
Chairwoman of the Organizing Committee, Editor of conference materials, Lambung Mangkurat University

Advisory Board

Prof. Dr. Ahmad, S.E., M.Si.	:	Rector of ULM
Dr. H. Aminuddin Prahatama Putra, M.Pd	:	Vice-Rector for Academic Affairs of ULM
Prof. Dr. Ir. H. Yudi Firmanul Arifin, M.Sc	:	Head of Center for Wetland Innovation and Technology, ULM
Dr. Bani Noor Muchamad, S.T., M.T.	:	Dean of Faculty of Engineering, ULM
Prof. Meilana Dharma Putra, S.T., M.Sc., Ph.D	:	Vice Dean for Academic Affairs of the Faculty of Engineering, ULM

Organizing Committees

Chief Director		
Dr. Eng. Irfan Prasetia, S.T., M.T.	:	Graduate School of Civil Engineering, ULM
Chief Executive		
Ade Yuniati Pratiwi, S.T., M.Sc., Ph.D.	:	Graduate School of Civil Engineering, ULM
Secretary		
Endah Widiastuti, S.T., M.T	:	Department of Civil Engineering, ULM
Hendriyani, S.E.	:	Department of Civil Engineering, ULM
Event Division		
Eddy Nasrullah, S.T., M.T.	:	Department of Civil Engineering, ULM

Papers / Publications Division

Prof. Dr. Iphan F. Radam, S.T., M.T.	:	Graduate School of Civil Engineering, ULM
Aqli Mursadin, S.T., M.Sc., Ph.D.	:	Graduate School of Civil Engineering, ULM
Dr. Nursiah Chairunnisa, S.T., M.Eng.	:	Graduate School of Civil Engineering, ULM
Dr. Ing Yulian F. Arifin, S.T., M.T.	:	Graduate School of Civil Engineering, ULM
Ir. Ratni Nurwidayati, M.T., M.Eng.Sc.	:	Department of Civil Engineering, ULM
Ir. Retna Hapsari Kartadipura, M.T.	:	Department of Civil Engineering, ULM
Candra Yuliana, S.T., M.T.	:	Department of Civil Engineering, ULM
Arya Rizki Darmawan, S.T., M.T.	:	Department of Civil Engineering, ULM
Abdul Karim, S.T., M.T.	:	Department of Civil Engineering, ULM

Public Relations Division

Nova Widayanti, S.T., M.T	:	Department of Civil Engineering, ULM

Peer Review Statement

All papers published in this volume have been reviewed through processes administered by the Editors. Reviews were conducted by expert referees to the professional and scientific standards expected of a proceedings journal published by IOP Publishing.

- **Type of peer review:** Single Anonymous
- **Conference submission management system:** Morressier
- **Number of submissions received:** 38
- **Number of submissions sent for review:** 38
- **Number of submissions accepted:** 33
- **Acceptance Rate (Submissions Accepted / Submissions Received × 100):** 86.8
- **Average number of reviews per paper:** 2
- **Total number of reviewers involved:** 16
- **Contact person for queries:**
  Name: Abdul Karim
  Email: abdulkarim@ulm.ac.id
  Affiliation: Lambung Mangkurat University

The Stability Analysis of the Bored Pile Foundation as Liquefaction Hazard Mitigation at the University of Tadulako Reconstruction Project

H R Pasin[1,2], S Ismanti[1,*], and H C Hardiyatmo[1].

[1] Department of Civil and Enviromental Engineering, Faculty of Engineering, Universitas Gadjah Mada Yogyakarta, Indonesia
[2] Directorate General of Human Settlement, Ministry of Public Works and Housing, Republic of Indonesia.

*Corresponding Author's e-mail address: sito.ismanti@ugm.ac.id

Abstract. A Pile foundation is one of the most important structural elements. Pile foundations are used to support buildings when the hard soil layer is located very deep to keep the structure stable and standing when loaded by internal and external forces. The University of Tadulako is located in a moderate liquefaction-prone zone. In planning for rebuilding post-earthquake, where this area suffered massive damage, it was built on loose sandy soil with shallow groundwater levels. As a result, it has a possibility of liquefaction hazards in this area. This study aims to analyze the stability of the bored pile foundation as liquefaction hazard mitigation under service and extreme conditions. The analysis is carried out using two approaches, analytical method and computer-aided numerical simulation using *RSPile* and *GEO5 2019*. The type of foundation plan that will be used is the bored pile foundation with a diameter of 0.6 meters with a depth of 10 to 13 meters. The result shows the foundation remains stable under service conditions, but under extreme conditions, the foundation collapsed for type P1.B, and recommended redesign becomes type P1.A. However, after redesigning, the foundation becomes inefficient, and further research is needed on efficient foundation design.

1. Introduction

A pile foundation is one of the most important structural elements. Pile foundations are used to support buildings when the hard soil layer is located very deep to keep the structure stable and standing when loaded by internal and external forces [1]. Therefore, it is necessary to carry out a stability analysis of the foundation to keep the building structure safe. Because if engineers do not calculate them, the foundation will fail and collapse, as happened in the case of the 1964 Niigata earthquake, where an apartment building collapsed due to a loss of soil-bearing capacity to resist the structure keeping stand and stable, which can be seen in **Figure 1**. Ishihara and Koga (1981) explained that the incident was caused by the influence of liquefaction that occurred beneath the building structure [2]. Liquefaction is a phenomenon of losing ground strength due to strong ground shaking in loose sediment that is inundated by water [3]. Liquefaction can cause failure in the bearing capacity, excessive settlement, or slope movement [4].

The University of Tadulako Palu is located at a high risk of geological disaster hazards, covering a high earthquake-prone zone and a moderate liquefaction-prone zone [5]. On September 28, 2018, there was a strong earthquake of 7.5 Mw which caused liquefaction in several places surrounding Palu City, Central Sulawesi, as seen in **Figure 2** [6,7]. In planning for rebuilding after the 2018 Palu earthquake,

IOP Conf. Series: Earth and Environmental Science 1184 (2023) 012001 doi:10.1088/1755-1315/1184/1/012001

where this area suffered massive damage, it was built on loose sandy soil with shallow groundwater levels. As a result, it has a possibility of liquefaction hazards in this area. Thus, mitigation efforts are needed to reduce the impact of liquefaction occurs.

Figure 1. Buildings that collapsed due to loss of soil-bearing capacity when the 1964 Niigata Earthquake occurred [2]

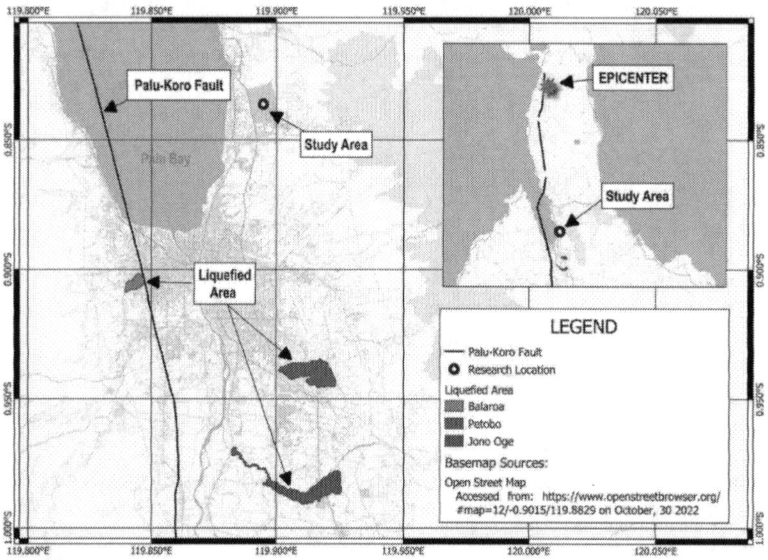

Figure 2. Study area and liquefied area during the 2018 palu earthquake [7].

This study aims to study liquefaction potential and mitigation strategies around the building reconstruction work project at the University of Tadulako based on a literature review, such as the soil investigation report, project planning document, and previous research in the study area. Analyzing stability of the bored pile foundation on the bearing capacity using the analytical method by O'Neil and Reese (1989) [8]. For pile displacement and internal force using numerical methods by computer-aided numerical simulation, namely *Rocscience RSPile version 3.005 64-bit* and *Finesoftware GEO5 2019*. Verification was carried out to find out whether the pile foundation was stable or not against the load working using applicable standards, Indonesian Geotechnic Code, SNI 8460 2017 [9], and the suggestion and recommendations if the foundation was not stable.

2. Research Method

2.1. Liquefaction Potential in Study area

As explained above, the University of Tadulako is located close to the earthquake source and not far from the coastal Palu bay. The soil in the study area is dominated by loose sandy soil. Based on Sassa and Takagawa (2019) [10] and survey documentation belonging to the Ministry of Public Works and Housing (2018), as seen in **Figure 3**. The following is evidence of the impact caused by liquefaction around the study area, namely: (1) surficial exposure of sand boils in the coastal bay documented by Sassa and Takagawa [10], (2) differential settlement, tilting column, and wall in the Law College building, (3) cracks and differential settlement direction in Auditorium of the Faculty of Teacher Training and Education, (4) tilting column in the Faculty of Teacher Training and Education College building, (5) tilting wall and differential settlement in Dean of the Faculty of Teacher Training and Education building, (6) differential settlement, cracks, and collapse of the in the wall, tilting column in Rectorate building, (7) ground deformation and differential settlement in the Physical Education, Health and Recreation Study building, and (8) differential settlement, and tilting column in the integrated laboratory of the Faculty of Teacher Training and Education.

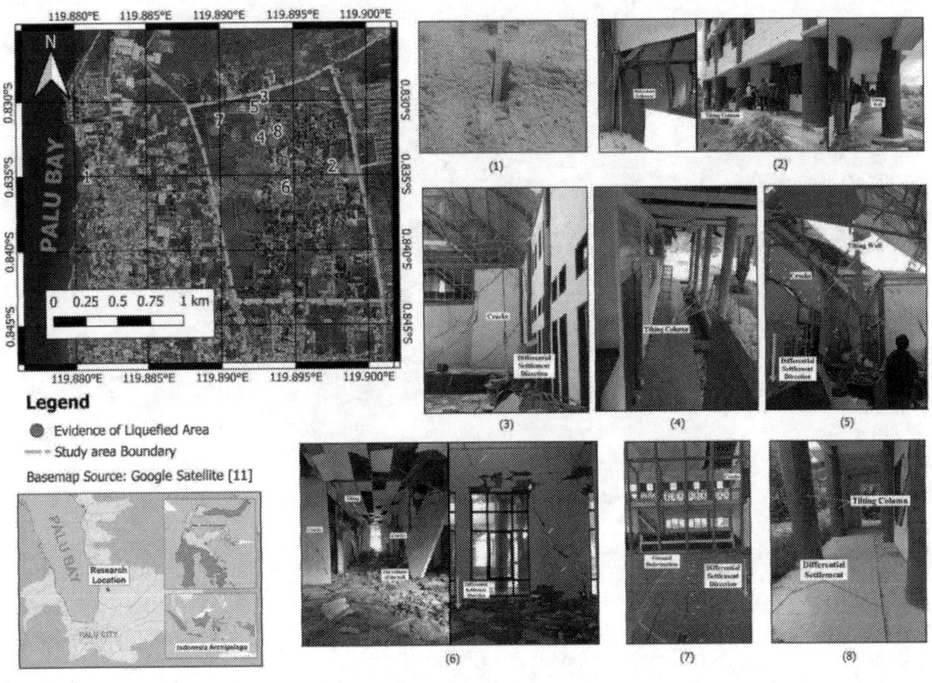

Figure 3. Evidence of liquefied area, ground settlement, and building collapse in the surrounding study area (Modified from [10,11] and survey documentation belonging to the Ministry of Public Works and Housing, 2018).

IOP Conf. Series: Earth and Environmental Science 1184 (2023) 012001 doi:10.1088/1755-1315/1184/1/012001

Previous studies have investigated liquefaction hazards surrounding the study area by Pasin *et al.* regarding the evaluation of liquefaction potential [12] and the settlement due to liquefaction, using SPT data from as many as 14 points scattered in the study area [13]. Liquefaction has a significant effect on the borehole BH-05, with the depth prone to soil liquefaction 5 to 9 meters, with a very high category of liquefaction potential [12]. For liquefaction-induced settlement, the largest settlement value was obtained at drill point DP-05 of 10.234 centimeters with a medium category of settlement due to liquefaction [13]. At these two points, the Dean's Building of the Faculty of Teacher Training and Education will be reconstructed on liquefiable soils as seen in **Figure 4.** It needs special attention on this building, and it is necessary to take preventive measures to reduce the impact caused. This is a challenge for Engineers because the planned buildings stand on soil that has liquefaction potential. Pasin *et al.* (2022) recommend for mitigation is to place deep foundations under layers that are vulnerable to potential liquefaction hazards [12]. Hence, it is necessary to analyze the foundation's stability against liquefiable soil and select appropriate liquefaction mitigation alternatives if the previously planned mitigation is not safe if liquefaction occurs in this building.

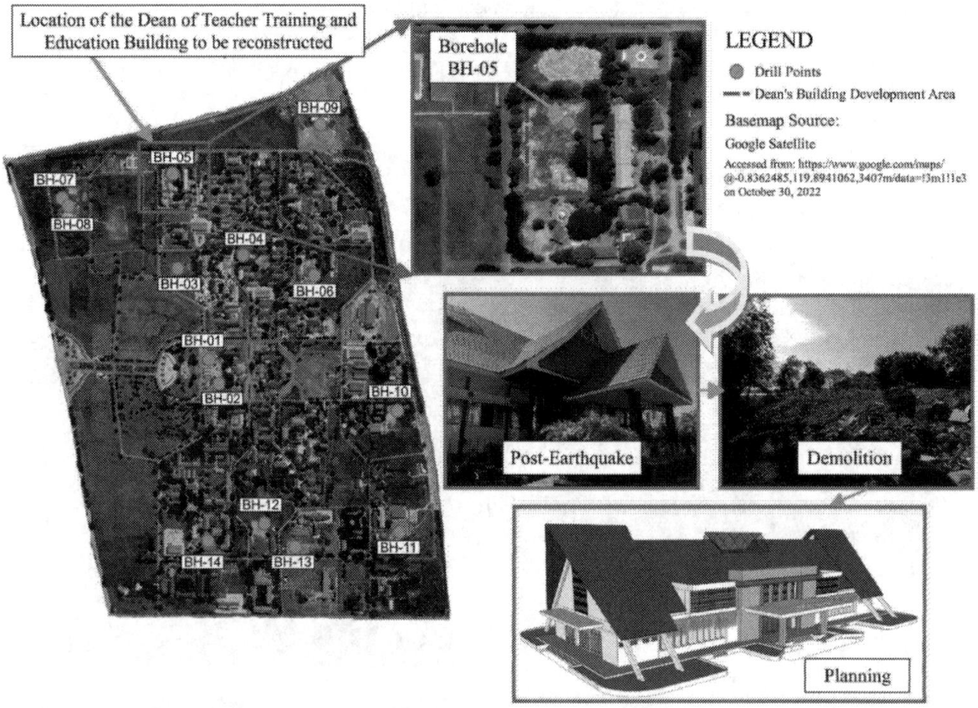

Figure 4. The location of the Dean of the Teacher Training and Education Building is to be reconstructed and developed [11].

2.2. Mitigation Strategies

There are many methods and strategies to select suitable mitigation efforts to resist the potential liquefaction hazard. According to Idriss and Boulanger [14], mitigation methods and techniques include taking no action and accepting potential damage and risk, stopping the development, choosing a different site, soil reinforcement or lowering the standards, or modifying the liquefaction-resistant structure. According to Towhata [15], there is a flow chart in mitigation planning against liquefaction hazards, as presented in **Figure 5.**

IOP Conf. Series: Earth and Environmental Science 1184 (2023) 012001 doi:10.1088/1755-1315/1184/1/012001

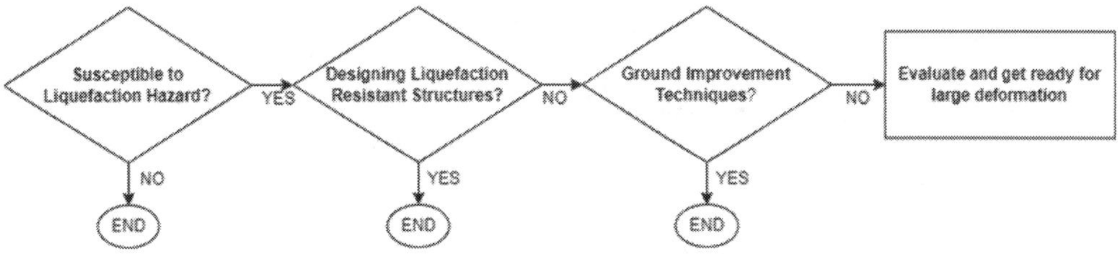

Figure 5. The flowchart in mitigation planning against liquefaction hazard [15].

Figure 5 shows the initial steps for mitigation planning of a liquefaction disaster. At first, it is necessary to see whether, during the preliminary investigation survey, the study area's high potential liquefaction is completed. If there is no liquefaction potential, the problem related to liquefaction is completed. If the study area has a high liquefaction potential, it is necessary to evaluate the strength of the structure. Can the structure resist the liquefaction hazard? If the structure cannot resist the liquefaction potential hazard, soil improvement is carried out. Then it is evaluated that the soil improvements that have been carried out have been able to resist the liquefaction potential hazard. If soil improvement efforts and structural strengthening measures such as strengthening pile foundations must be further detailed concerning how much of an acceptable deformation risk is there and if the worst condition occurs? Is it going to be a serious problem or not? Looking for new solutions and ideas related to how to focus on reducing residual deformation of the liquefied soil layer so that the facility can be maintained or restored quickly to avoid adverse negative effects such as economic losses, infrastructure, and loss of life.

For this research, we focused on designing liquefaction-resistant structures, such as deep foundations, especially bored pile foundations, to penetrate the liquefied soil layer and reach the hard soil layer [12]. Selection of a bored pile as a liquefaction mitigation hazard in this case rather than a driven pile foundation because the University of Tadulako is an education function and is currently under reconstruction post-earthquake, therefore using this foundation is more suitable than a driven pile because selecting a driven pile can cause a sound disturbance, vibration, and ground deformation that can cause damage to the surrounding buildings [1]. Besides that, soil conditions in the study area are dominated by sandy soils, where sand soil drilling work can result in density disturbances and is prone to landslides. So that the selection of an appropriate implementation method for field conditions is the wet method. According to Hardiyatmo [1], the wet method is suitable for drilling past the water table added with bentonite solution (clay solution) to prevent landslides in the borehole. According to Day (2012), In the case of the Niigata earthquake that caused liquefaction of the sand deposits in the low-lying areas, applying a deep foundation as a mitigation effort to prevent the movement of the foundation during an earthquake is one of the most effective and precise action, where concrete buildings using deep foundations up to a depth of 20 meters were not damaged [16].

2.3. Bored Pile Configuration and Layout

Two types of bored pile foundations are used in this dean's building plan, namely, type P1.A, and type P1.B. Both are designed as a single pile with the same diameter of 0.6 meters. The length for each type of pile is 13 meters for P1.A, and 10 meters for P1.B. Both piles have the same pile cap configuration, namely for a pile cap thickness of 0.8 meters and a pile cap cross-sectional area of 1 m^2. The strength of the material planned for each type of bored pile is the concrete strength of 25 MPa, and the yield strength of the reinforcement (rebar) is 420 MPa. The reinforcement configuration is also the same as the diameter of the reinforcement of 16 mm, a number of longitudinal bars of 10 pieces, the diameter of the shear reinforcement are 16 mm, and the distance between the shear bars is 200 mm. The configuration for each type of bored pile foundation is presented in **Table 1.** The detailing and layout of the bored pile foundation in the Dean's Building of the Faculty of the Teacher Training and Education College can be seen in **Figure 6** and **Figure 7**.

IOP Conf. Series: Earth and Environmental Science 1184 (2023) 012001 doi:10.1088/1755-1315/1184/1/012001

Table 1. Configuration of bored pile foundation

Bored Pile Foundation	P1.A	P1.B
Single or Group Pile	Single Pile	Single Pile
Pile Diameter (m)	0.60	0.60
Length (m)	13.00	10.00
The thickness of the Pile Cap (m)	0.80	0.80
Cross-Sectional Area (m2)	1.00	1.00
Concrete Strength f'c (MPa)	25.00	25.00
Rebar Strength fy (MPa)	420.00	420.00
Rebar Diameter (mm)	16.00	16.00
Number of Longitudinal Bars	10.00	10.00
Shear Reinforcement Diameter (mm)	16.00	16.00
Spacing of Shear Bars (mm)	200.00	200.00

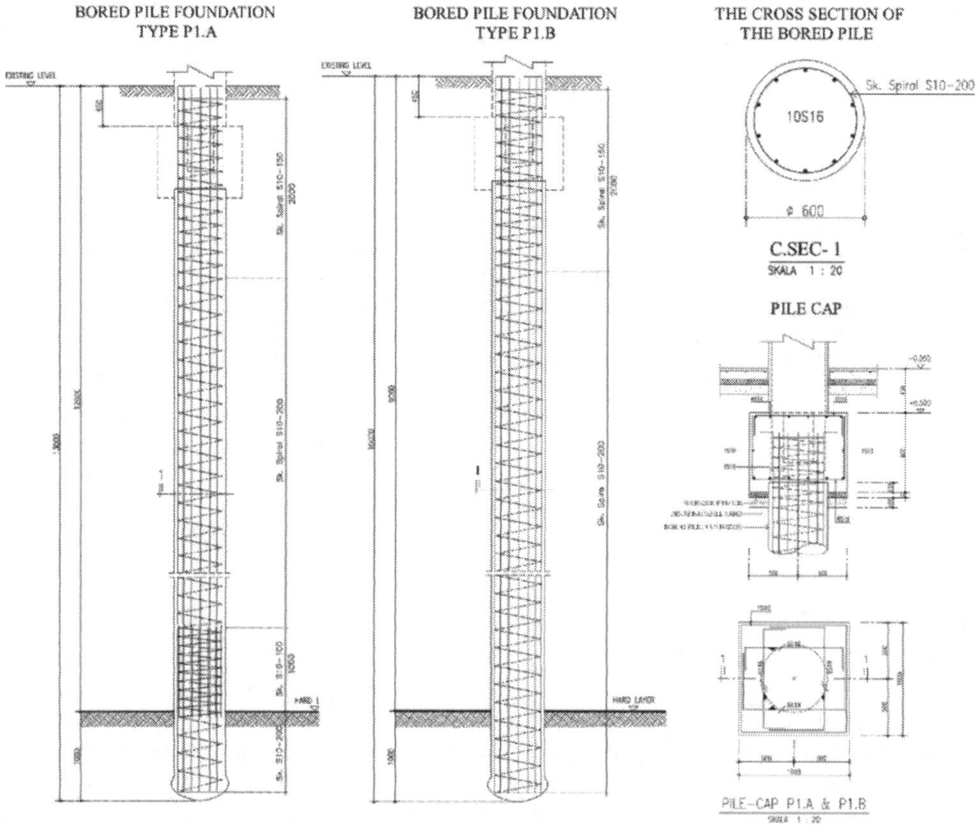

Figure 6. Detailing of the bored pile foundation and pile cap

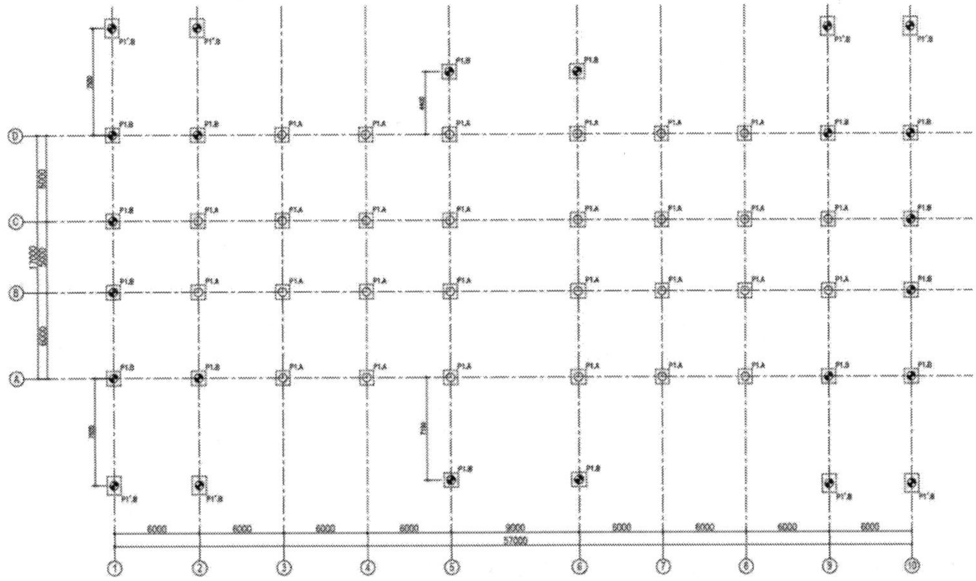

Figure 7. Layout of the bored pile foundation in the Dean's Building of the Faculty of Teacher Training and Education College.

2.4. Stability Analysis of Bored Pile Foundation

2.4.1. Bearing Capacity using Analytical Method

In this study, the stability analysis of the bored pile foundation is used by the O'Neil and Reese (1989) method [8] to calculate bearing capacity, which has considered the weight of the installed bored pile with water pressure effects.

$$Q_u = Q_b + Q_s - W_p' \quad (1)$$

where Q_u is the ultimate bearing capacity in kilo newton, Q_b is end resistance of the bored pile in kilo newton, and W_p' is the weight of installed bored pile with water pressure effects in kilo newton.

$$Q_b = A_b f_b \quad (2)$$

$$Ab = \frac{1}{4} \pi D^2 \quad (3)$$

$$f_b = 0.60 \sigma\tau N_{60} \leq 4500 \text{ kPa} \quad (4)$$

$$N_{60} = C_E x C_B x C_R x C_S x N_{SPT} \quad (5)$$

where Ab is the bored pile base area, f_b is end resistance per unit area in kilo newton per square meter, D is the diameter of the bored pile in meters, and N_{60} is blow count for an energy ratio of 60%.

$$Q_s = A_s f_s \quad (6)$$

$$f_s = \beta P_o' \quad (7)$$

$$\beta = K \, tg \, \delta \quad (8)$$

where A_s is the blanket area of pile in square meter, f_s is friction resistance per unit area in kilo newton per square meter. P_o' is overburden pressure in the middle soil layer. δ is friction angle between the ground and the pile in kilo newton per square meter.

$$W_p' = W_p - U \quad (9)$$

$$W_p = \gamma_c V \quad (10)$$

$$U = \gamma_w V \quad (11)$$

where W_p is the weight of the bored pile in kilo newton, U is water pressure in kilo newton, and γ_c is the specific gravity of concrete in kilo newton per cubic meter, where the value is 24 kN/m^2, γ_w is the specific gravity of water, where the value is 9.81 kN/m^2, V is the volume of bored pile.

$$Q_a = \frac{Q_u}{SF} \quad (12)$$

$$P_u < Q_a \ldots \text{ Safe} \quad (13)$$

where Q_a is the allowable bearing capacity in kilo newton, SF is the safety factor based on Indonesian Geotechnics Code SNI 8460: 2017 [9], where the value is 2.5, P_u is the axial force due to factored load.

2.4.2. Negative Skin Friction

Negative skin friction is a downward pulling force acting on the pile blanket due to the settlement of the liquefied soil layer, so that this friction causes an additional load acting on the pile foundation. There are several equations that can be used to calculate negative skin friction for cohesive soil layers using Johannessen and Bjerrum (1965) [1], namely as follows:

$$Q_{nsf} = pL\alpha\, c_u \quad (14)$$

where Q_{nsf} is negative skin friction in kilo newton, p is perimeter in meter, L is pile length in meter, α is adhesion factor, c_u is undrained cohesion in kilo newton per square meter. And for negative skin equation for non-cohesive soil layers, it can be seen in which are as follows:

$$Q_{nsf} = A_s\, c_a \quad (15)$$

$$c_a = P_o' K \tan\varphi \quad (16)$$

$$K = 1 - \sin\varphi \quad (17)$$

where P_o' is average effective overburden pressure in kilo newton per square meter, K is lateral pressure coefficient, and the friction angle between the pile and the ground in degree (°).

2.4.3. Pile Displacement and Internal Force

Analysis of pile displacements and internal forces on the bored pile foundation using computer-aided numerical simulation methods, such as *RSPile*, and *GEO5*. Why is this analysis done numerically? because lateral load pile foundation in soil needs to be analyzed using a nonlinear pile model (finite element method) in which pile capacity as a function of pile deflection is provided (elastic subsoil - p-y curves method) [17] [18] and difficult to solve by analytical method because it requires a count of iterations until there is a balanced condition between the soil reaction and the stresses on the pile [1]. A numerical method is a method for solving mathematical problems by means of arithmetic operations with an approximate value [19].

The *RSPile* is a software developed by Rocscience to analyze general piles numerically. *GEO5* is a software developed by Finesoftware to analyze geotechnical problems numerically. Each software is carried out in 2 stages, namely: stage 1 for service conditions, and stage 2 for extreme conditions. The difference between the two stages is the loading conditions used. Many parameters must be considered before carrying out the analytical procedure using the respective software. These parameters include soil properties, soil thickness, groundwater, material specification, pile configuration, geometry, pattern, and loading. Input parameter values for the sandy soil used can be seen in **Table 2.** In the clay layer, the lateral capacity analysis using the RSPile model requires input parameters in the form of average undrained shear strength and strain factor, using the values recommended by Reese et al. (1975) [17] for clay soil conditions above the groundwater table as presented in **Table 3** and for clay soil conditions below the water table, use the recommended parameter values based on Matlock (1970) [17] as presented in **Table 4.**

Table 2. Lateral soil coefficient values for water-saturated sand and sand above the water table based on Reese et al. (1974) [17]

Condition	Relative Density	Loose	Medium	Dense
Saturated sand	k_{py} (kN/m^3)	5400	16300	34000
Sand on the surface of the water	k_{py} (kN/m^3)	6800	24400	61000

Table 3. Value of ε_{50} for normally consolidated clay by Reese dkk. (1975) [17]

Consistency	*Su* (kPa)*	ε_{50}
Soft	< 48	0,020
Medium	48 - 96	0,010
Dense	96 – 192	0,005

Table 4. Value of ε 50 for overconsolidated clay by Matlock (1970) [17]

Su (kPa)	50 - 100	100 - 200	200 - 400
ε_{50}	0,007	0,005	0,004

Modeling using GEO5 in lateral capacity analysis requires input of soil parameters which include soil specific gravity, water saturated soil specific gravity, internal internal friction angle, cohesion value, passion ratio, deformation modulus, and soil lateral reaction modulus. The value of the deformation modulus can be seen in **Table 5**, and soil lateral reaction modulus values **Table 6.** In the analysis of the lateral capacity of the liquefied soil layers using GEO5 software, it is represented by reducing the value of the lateral soil reaction modulus (k) using the value recommended by Brandenberg et al, (2005). The reduction factor for the modulus of soil reaction under liquefaction conditions can be seen in **Table 7.**

Table 5. Correlation value between soil consistency and deformation modulus, Meyerhoff (1956) [20]

Soil Type	Deformation Modulus, *Edef*		*Possion's Ratio*
	MN/m2	lb/in2	
Loose sand	10,35 – 24,15	1500 – 3500	0,20 – 0,40
Medium sand	17,25 – 27,60	2500 – 4000	0,20 – 0,40
Dense sand	34,50 – 55,20	5000 – 8000	0,30 – 0,45
Silty sand	10,35 – 17,25	1500 – 2500	0,20 – 0,40
Sand and gravel	69,00 – 172,50	10000 – 25000	0,15 – 0,35
Soft clay	2,07 – 5,18	300 – 750	
Medium clay	5,18 – 10,35	750 – 1500	0,20 - 0,50
Stiff clay	10,35 – 24,15	1500 - 3500	

IOP Conf. Series: Earth and Environmental Science 1184 (2023) 012001 doi:10.1088/1755-1315/1184/1/012001

Table 6. The range of soil lateral reaction modulus values is based on Bowles (1997) [18]

Soil type	k (MN/m^3)
Dense sandy gravel	220 - 400
Medium dense gravel	155 – 300
Medium graded sand	110 – 180
Fine sand	80 – 100
Saturated stiff clay	30 – 110
Plastic clay	40 – 140
Saturated plastic clay	10 -80
Soft clay	2 – 40

Table 7. Reduction factor for soil reaction modulus in liquefaction conditions [21]

$(N_1)_{60cs}$	m_p
< 8	0 – 0,1
8 – 16	0,1 – 0,2
16 – 24	0,2 – 0,3
> 24	0,3 – 0,5

2.4.4. Load Working on Pile Foundation

Load is a force generated by the building's own weight (internal) and external forces working on the structure such as building utilization, and environmental influences around the building. Based on the Indonesian Minimum Design Load Code, Load is a force or other action due to the weight of all building materials, occupants, and belongings, environmental effects, differences in movement, and restraint forces due to changes in dimensions [22]. The load working on the bored pile foundation is determined by the analysis of the building structure that has been given a load factor and combined load before the load is transferred to the pile foundation. Load factor is a factor that calculates the deviation between the actual load and the nominal load, namely the uncertainty in the analysis that converts the load into a load effect and considers the possibility of more than one extreme load occurring simultaneously [22]. The load working on the bored pile foundation used in this analysis is presented in **Table 8.**

Table 8. The load working on bored pile foundation

No.	Load Combination	Description
1	1.0 DL + 1.0 LL	Due to Normal Load
2	1.0 DL	Due to Dead Load
3	1.0 LL	Due to Live Load
4	1.0 EX	Due to Static Earthquake Direction X
5	1.0 EY	Due to Static Earthquake Direction Y
6	1.0 SPECX	Due to Dynamic Earthquake Direction X
7	1.0 SPECY	Due to Dynamic Earthquake Direction Y
8	COMB1 (1.40 DL)	Due to Combination Load 1
9	COMB2 (1.20 DL + 1.60 LL)	Due to Combination Load 2
10	COMB3 (1.48 DL + 1.00 LL + 1.30 SPECX + 0.39 SPECY)	Due to Combination Load 3
11	COMB4 (1.48 DL + 1.00 LL + 1.30 SPECX - 0.39 SPECY)	Due to Combination Load 4
12	COMB5 (1.48 DL + 1.00 LL - 1.30 SPECX + 0.39 SPECY)	Due to Combination Load 5
13	COMB6 (1.48 DL + 1.00 LL - 1.30 SPECX - 0.39 SPECY)	Due to Combination Load 6
14	COMB7 (1.48 DL + 1.00 LL + 0.39 SPECX + 1.30 SPECY)	Due to Combination Load 7
15	COMB8 (1.48 DL + 1.00 LL + 0.39 SPECX - 1.30 SPECY)	Due to Combination Load 8
16	COMB9 (1.48 DL + 1.00 LL - 0.39 SPECX - 1.30 SPECY)	Due to Combination Load 9
17	COMB10 (1.48 DL + 1.00 LL -0.39 SPECX - 1.30 SPECY)	Due to Combination Load 10
18	COMB11 (0.62 DL + 1.30 SPECX + 0.39 SPECY)	Due to Combination Load 11
19	COMB12 (0.62 DL + 1.30 SPECX - 0.39 SPECY)	Due to Combination Load 12
20	COMB13 (0.62 DL - 1.30 SPECX + 0.39 SPECY)	Due to Combination Load 13
21	COMB14 (0.62 DL - 1.30 SPECX - 0.39 SPECY)	Due to Combination Load 14
22	COMB15 (0.62 DL + 0.39 SPECX + 1.30 SPECY)	Due to Combination Load 15
23	COMB16 (0.62 DL + 0.39 SPECX - 1.30 SPECY)	Due to Combination Load 16
24	COMB17 (0.62 DL - 0.39 SPECX + 1.30 SPECY)	Due to Combination Load 17
25	COMB18 (0.62 DL - 0.39 SPECX - 1.30 SPECY)	Due to Combination Load 18

2.4.5. Verification Strategies

As mentioned earlier, the verification strategy is an examination of the stability of the bored pile foundation against extreme loads that work on a structure by dividing a safety factor or checking the threshold value based on design tolerance in applicable standards and verifying whether the planned structure is safe or not and giving advice, suggestions, and recommendations if the foundation is unstable. **Table 9** shows the allowable requirements and thresholds value based on applicable standards.

Table 9. Allowable Requirement and threshold value based on Applicable Standard

No.	Allowable Requirement	Threshold Value	Applicable Standard
1.	Safety Factor for Static Condition	2.5	
2.	Safety Factor for Earthquake Condition	1.6	Indonesian Geotechnics Code,
3.	Lateral Displacement	2.5 cm	SNI 8460: 2017
4.	Total Settlement	15.0 cm	

3. Results and Discussion

The stability analysis of bored pile foundations is calculated on soil that has very high liquefaction potential and has the most impact on buildings to be built as liquefaction hazard mitigation in the Tadulako University Teaching and Education Faculty Dean Building. Based on the analysis of liquefaction potential and liquefaction-induced settlement obtained from previous studies, the BH-05 drill point has a very high vulnerability to liquefaction potential. This is indicated by an LPI value of 16.996 and settlement due to liquefaction of 10.234 cm. In the analysis of the safety factor for the potential, it is obtained that the depth that is vulnerable to liquefaction is located at a depth of 5 to 9 meters. This is indicated by the value of the SF factor of safety less than 1 (SF < 1), and marked with a red label. The results of liquefaction potential and reduction due to liquefaction were obtained from the liquefaction hazard analysis at the research location can be seen in **Table 10.**

Table 10. The results of the analysis of the calculation of the safety factor, LPI, and liquefaction-induced settlement value for the BH-05 Dean of FKIP drill point [12, 13]

Depth (m)	N	Soil Type	FC (%)	σ_{vc} (kPa)	σ_{vc}' (kPa)	C_N	$(N1)_{60}$	$(N1)_{60\text{-}cs}$	CSR	CRR	SF	LPI	Settlement (cm)
1	1	Sandy Silt	60,72	17,82	17,82	1,58	1,52	n.a.	0,35	2,00	n.a.	0,000	0,000
3	3	Sandy Silt	60,72	53,45	53,45	1,18	3,86	n.a.	0,35	2,00	n.a.	0,000	0,000
5	15	Silty Sand	25,28	99,44	79,82	1,10	20,04	25,13	0,42	0,30	0,720	2,098	2,361
7	14	Silty Sand	25,28	145,43	106,19	0,98	16,66	21,76	0,45	0,23	0,503	7,428	6,108
9	13	Silty Sand	25,28	191,42	132,56	0,89	14,76	19,85	0,46	0,20	0,424	13,828	10,234
11	35	Silty Sand	25,28	237,41	158,93	0,88	39,63	44,73	0,47	1,73	3,725	16,996	10,234
13	52	Silty Sand	25,28	283,40	185,30	0,85	56,74	61,83	0,46	1,64	3,567	16,996	10,234
15	55	Silty Sand	25,28	329,39	211,67	0,82	57,95	63,04	0,45	1,56	3,460	16,996	10,234
17	60	Silty Sand	25,28	375,39	238,05	0,80	61,29	66,38	0,44	1,49	3,384	16,996	10,234
19	60	Silty Sand	25,28	421,38	264,42	0,78	59,62	64,71	0,43	1,43	3,331	16,996	10,234
20	60	Silty Sand	25,28	444,37	277,60	0,77	58,86	63,96	0,42	1,40	3,310	16,996	10,234
											Category	**Very High**	**Medium**

The analysis was performed under the following condition: (1) service condition is a normal condition before liquefaction occurs and does not affect the soil. Loading in this condition only includes dead load and live load. Service loads must be identified for each serviceability condition being investigated [22]. (2) extreme conditions are a condition when during or after an earthquake occur, and liquefaction is considered to compute and affects the soil completely. Loading in this condition includes static and dynamic earthquake load in all directions. seismic loads must be calculated in addition to combined loads [20]. Both conditions are based on maximum loading result. The difference between the two conditions is that in seismic conditions, skin friction on the bored pile foundation was considered not to have the capacity for this resistance, and negative skin friction was calculated to become total loading, as presented in **Table 11.**

Table 11. The result of the maximum loading acting on bored pile foundations

Pile type	Conditions	Negative Skin Friction (Qnsf) [kN]	Axial Load (Pu) [kN]	*Pile Cap* Weight (W) [kN]	Total Axial Load (Pu_{total}) [kN]	Lateral Load (FX) [kN]	Lateral Load (FY) [kN]	Bending Moment Mx [kNm]	Bending Moment My [kNm]
1	*2*	*3*	*4*	*5*	*6*	*7*	*8*	*9*	*10*
P1.A	Service	0,00	792,65	13,78	806,42	37,28	-19,62	-13,73	-27,47
	Extreme	1657,37	901,54	13,78	2572,68	123,61	14,72	10,79	68,67
P1.B	Service	0,00	370,82	13,78	384,59	0,00	20,60	14,72	1,96
	Extreme	1657,37	548,38	13,78	2219,52	93,20	33,35	34,34	67,69

Table 12, **Table 13**, **Table 14**, and **Table 15** shows the result of the stability analysis of the bored pile foundation based on the initial design which covers bearing capacity, lateral displacement, and settlement on pile, axial force, shear force, and bending moment. From the verification result on the bored pile foundation type P1.B, the safety factor of the bearing capacity and settlement on pile does not meet standard against total axial load in extreme (liquefaction) conditions, with each value obtained, 0.54 for the safety factor, and 36.1 cm for settlement on pile. Hence, it is recommended to redesign the bored pile foundation type from initial design to alternative design. So that the bearing capacity, and pile strength increase in extreme conditions or when liquefaction occurs.

At this stage the alternative design of the foundation uses a design that complies with the initial design stage with an alternative foundation design used, namely that originally P1.B with a 10-meter pole length was changed to P1.A with a 13-meter-long pile, then re-evaluated the stability of the foundation when the pile length is increased.

For lateral displacement, settlement on pile, and internal force based on the initial design using *RSPile* in all types of bored pile foundations, conditions, and directions compared to the allowable requirement, the Indonesian Geotechnics Code SNI 8460: 2017 [9], the bored pile foundation is meet the standards.

Table 12. The result of the bearing capacity based on the initial design

Pile Type	Condition	End Resistance (Q_b) [kN]	Friction Resistance (Q_s) [kN]	Bored Pile Weight Effective (Wp') [kN]	Ultimate Bearing Capacity ($Q_{ult\text{-}netto}$) (11+ 12+13) [kN]	Safety Factor (14 / 6)	Verification
1	*2*	*11*	*12*	*13*	*14*	*15*	*16*
P1.A	Service	1226,79	4184.03	52,16	5358.67	6,64	OK
	Extreme	1226,79	1918.89	52,16	3093.53	1,20	OK
P1.B	Service	676,46	2836.96	40,12	3473,29	9,03	OK
	Extreme	676,46	571,81	40,12	1208,15	0,54	Not OK

Table 13. The result of the analysis of lateral displacement and settlement of bored pile foundations based on the initial design using *RSPile*

Pile Type	Condition	Lateral Displacement		Settlement on Pile (cm)	SNI 8460:2017		Verification		
		X	Y		Allowable Lateral Displacement (cm)	Allowable Settlement on Pile (cm)	Lateral Displacement X	Lateral Displacement Y	Settlement on Pile
P1.A	Service	0,11	-0,04	-0,11			OK	OK	OK
	Extreme	2.11	0,02	-0,45	2,5	15	OK	OK	OK
P1.B	Service	0,00	-0,04	-0,17			OK	OK	OK
	Extreme	2,35	0,07	-36,1			OK	OK	Not OK

Analysis of the internal force using the *RSPile* software has limitations. We cannot verify whether the stability analysis of the bored pile results obtained is safe or not. As we can see in **Table 14**, analysis of the internal force of the bored pile foundation only includes axial force, shear force, and bending moment, without verification statement. Thus, in this study, we used *GEO5* Software to analyze the resistance of the internal force based on the strength of the material of the bored pile foundation.

Table 14. The result of the Internal Force based on the initial design using *RSPile*

Pile Type	Condition	Axial Force [kN]	Shear Force [kN]		Bending Moment [kNm]	
			X	Y	X" Z"	Y" Z"
P1.A	Service	806,00	19,60	37,20	-13,70	27,50
	Extreme	2570,00	-14,70	123,00	10,80	-263,00
P1.B	Service	384,00	20,50	-1,09	-14,70	-1,96
	Extreme	2220,00	-33,10	217,00	34,30	-297,00

Table 15 shows the result of the internal force using *GEO5* software. All internal forces, such as axial force, shear force, and bending moment, that work in the bored pile foundations remain meet the threshold of the material strength capacity of the bored pile foundation.

Table 15. The result of the Internal Force based on the initial design using *GEO5*

Pile Type	Condition	Axial Force [kN]			Shear Force [kN]			Bending Moment [kNm]		
		N_{Ed}	N_{Rd}	Verification	V_{Ed}	V_{Rd}	Verification	M_{Ed}	M_{Rd}	Verification
P1.A	Service	806,42	2785,18	OK	42,13	793,06	OK	97,74	337,58	OK
	Extreme	2572,68	3697,19	OK	124,48	793,06	OK	172,54	247,96	OK
P1.B	Service	384,59	2832,34	OK	20,60	793,06	OK	45,43	334,56	OK
	Extreme	2219,52	3872,48	OK	98,99	793,06	OK	127,55	222,55	OK

Table 16. The results of stability of the bored piled foundation at the research location

Stages of Analysis	Initial Design P1.A		Initial Design P1.B		Alternative Design	
Condition	**Service**	**Extreme**	**Service**	**Extreme**	**Service**	**Extreme**
Maximum Loading, *Pu* (kN)	806,42	2572,68	384,59	2219,52	384,59	2219,52
Ultimate Bearing Capacity, *Qu* (kN)	5358,67	3093,53	3473,2	1208,15	5358.6	3093.5
Safety Factor, *SF*	6,64	1,20	9,03	0,54	13,93	1,39
Verification on Safety Factor Value	**OK**	**OK**	**OK**	**Not OK**	**OK**	**OK**
Lateral Displacement X, δ_X (cm)	0,11	2.11	0,00	2,35	0,002	1,38
Lateral Displacement Y, δ_Y (cm)	-0,04	0,02	-0,04	0,07	0,038	0,069
Verification on Lateral Displacement	**OK**	**OK**	**OK**	**OK**	**OK**	**OK**
Settlement on Pile, δ_Z (cm)	-0,11	-0,45	-0,17	-36,1	-0,0887	-2,04
Verification on Settlement	**OK**	**OK**	**OK**	**Not OK**	**OK**	**OK**
Ultimate Axial Force, N_{ED} (kN)	806,42	2572,68	384,59	2219,52	384.59	2219.52
Axial Force Capacity, N_{RD} (kN)	2785,18	3697,19	2832,34	3872,48	2748.49	3786.82
Verification on Axial Force	**OK**	**OK**	**OK**	**OK**	**OK**	**OK**
Ultimate Shear Force, V_{ED} (kN)	42,13	124,48	20,60	98,99	20,6	98,99
Shear Force Capacity, V_{RD} (kN)	793,06	793,06	793,06	793,06	793,06	793,06
Verification of Shear Force	**OK**	**OK**	**OK**	**OK**	**OK**	**OK**
Bending Moment, M_{ED} (kNm)	97,74	172,54	45,43	127,55	47,56	137,89
Bending Moment Capacity, M_{RD} (kNm)	337,58	247,96	334,56	222,55	339,9	235,27
Verification on Bending Moment	**OK**	**OK**	**OK**	**OK**	**OK**	**OK**

Based on the results of the recapitulation of the stability analysis of the bored pile foundations shown in **Table 16**, at the initial design analysis stage there were 2 (two) design conditions that did not meet the permit requirements required in **Table 9**, namely the safety factor, and settlement on pile under extreme conditions. This is because in extreme conditions, liquefaction is calculated on the soil at the study site, resulting in the pile foundation receiving excess loading due to negative skin friction, and the absorption of the bearing capacity which is considered to have no frictional resistance to the liquefied soil. As an alternative, in this study an alternative design was carried out with the aim of increasing the bearing capacity of the foundation in order to increase the value of the safety factor and reduce the settlement of piles that did not meet the permit requirements. The results obtained at the design alternative analysis stage show an increase in the safety factor and the resistance of settlement on pile under extreme conditions, along with an increase in the carrying capacity due to the extension of the pile. In alternative design, the safety factor value obtained of 1.39 during extreme conditions. But during service conditions, the design is inefficient or overdesign because the safety factor value obtained after redesign becomes 13.93. The difference in the safety factor is so large that it can be said that considerable investments are required to build a safe structure when a liquefaction disaster occurs. Accurately predicting natural disaster phenomena is very challenging, such as the 2011 Tohoku earthquake of magnitude M 9.0 that caused the accident at the nuclear reactor of the Fukushima Daiichi, where the design of this plant is based on the worst condition of the 1995 Kobe earthquake of magnitude M 7.5 [23]. Considering that in this study the foundation was designed under extreme conditions and this catastrophic event is a rare occurrence. Therefore, further research and special considerations are needed on how large deformation is acceptable under liquefaction hazards completely occurs so that the planned structure is safe, efficient, and economical.

4. Conclusion

Based on the results of the analysis of the stability of the bored pile foundation as mitigation of liquefaction hazard in the University of Tadulako Reconstruction Project, it can be concluded that liquefaction greatly contributes to the stability of the foundation, this can be seen in the decrease in the bearing capacity of the foundation value, and increased loading acting on the foundation in service, and extreme conditions. This results in a significant decrease in the value of the safety factor of the pile bearing capacity against loading, which at the initial design analysis stage P1.A, the safety factor value of 6.64 was obtained in service conditions, and there was a decrease in extreme conditions to 1.2. At the initial design analysis stage P1.B, the safety factor value of 9.03 was obtained in service conditions, and there was a decrease in extreme conditions to 0,54. At the alternative design analysis stage, the safety factor value of 13,93 was obtained in service conditions, and there was a decrease in extreme conditions to 1,39.

5. Acknowledgement

The authors would like to express their sincere gratitude for the support given by the Ministry of Public Works and Housing, the Republic of Indonesia, and Universitas Gadjah Mada for funding this research fully, and the Settlement Infrastructure Agency for Central Sulawesi Region for providing the necessary data for this research.

6. References

[1] Hardiyatmo H C, 2020 *Analisis dan Perancangan Fondasi II* 5th ed. Yogyakarta: Gadjah Mada University Press.

[2] Ishihara K and Koga Y, 1981 Case Studies of Liquefaction in the 1964 Niigata Earthquake *Japanese Society of Soil Mechanics and Foundation Engineering.* Vol.21, No.3.

[3] USGS, 2019 *Frequently Asked Questions: What is Liquefaction?* [online] Available at: https://www.usgs.gov/faqs/what-liquefaction#faq [Accessed: October 27, 2022].

[4] Andiani, 2020 Penggunaan Informasi Geologi untuk Penataan Ruang Pasca Bencana Palu, Sigi, dan Donggala [PowerPoint slides] Retrieved from https://kkp.go.id/an-component/media/upload-gambar-pendukung/DitJaskel/publikasi%20materi/pvmbg-22/Talkshow%20PVG%20pasca%20Bencana%20Palu%201.pdf.

[5] Central Sulawesi Province, 2019 Peta Zona Ruang Rawan Bencana Palu dan Sekitarnya (Alternatif 1).

[6] The United States Geological Survey USGS, 2018 M7.5 Palu, Indonesia Earthquake of September 28, 2018. Poster.

[7] Open Street Map, 2022 The Location of Study Area: Palu City Map 0° 54' 05.4" S 119° 52' 58.4" E *[online] Available at: https://www.openstreetbrowser.org/#map=12/-0.9015/119.8828* [Accessed 30 October 2022]

[8] O'Neil M W and Reese L C, 1989 New Design Method for Drilled Shafts from Common Soil and Rock Test. *Foundation Engineering: Current Principles and Practices*, pp.1026-1039.

[9] BSN, 2017 SNI 8460: 2017 – Persyaratan Perancangan Geoteknik.

[10] Sassa S and Takagawa T, 2019 Liquefied gravity flow-induced tsunami: first evidence and comparison from the 2018: Indonesia Sulawesi earthquake and tsunami disasters. *Landslides* 16. 195-200.

[11] Google Satellite, 2022 The University of Tadulako Map 0° 49' 51.5" S 119° 53' 33.6"E *[online] Available at: https://www.google.com/maps/place/0%C2%B049'51.5%22S+119%C2%B053'33.6%22E/@-0.8363964,119.888567,2589m/data=!3m1!1e3!4m5!3m4!1s0x0:0x151e75e024ffe33e!8m2!3d-0.8309778!4d119.8926564* [Accessed 30 October 2022].

[12] Pasin H R Ismanti S and Hardiyatmo H C, 2022 Evaluation of Liquefaction Potential Index in the Educational Infrastructure Reconstruction Project at the University of Tadulako, Palu City, Central Sulawesi *5th International Conference on Earthquake Engineering and Disaster Mitigation.*

[13] Pasin H R Ismanti S and Hardiyatmo H C, 2022 The Effects of Palu-Koro Fault on Predicting Ground Motion for Estimating Liquefaction-Induced Settlement of the Educational Infrastructure Reconstruction Project at the University of Tadulako *8th International Conference on Science and Technology.*

[14] Idriss I M, and Boulanger R W, 2008 Soil Liquefaction During Earthquakes *Earthquake Engineering Research Institute*, the University of California at Davis, USA.

[15] Towhata I, 2008 *Geotechnical Earthquake Engineering* Berlin: Springer.

[16] Day R W, 2012 Geotechnical Earthquake Engineering Handbook 2nd Edition with the 2012 International Building Code New York: McGraw-Hill.

[17] Rocscience Inc, 2022 *RSPile Laterally Loaded Piles Theory Manual.*

[18] GEO5, 2018 *Analysis of the horizontal bearing capacity of a single pile* Engineering manual No.16 Update:07/2018.

[19] Triatmodjo B, 2018 *Metode Numerik* Revisi ed. Yogyakarta: Beta Offset.

[20] Karlinasari R, 2017 Analysis of Lateral Deformation of Mini Pile Around the Vacuum Consolidation Area *Proceeding of PILE 2017*

[21] Brandenberg S J, 2005 Behavior of Pile Foundations in Liquefied and Laterally Speading Ground.

[22] BSN, 2020 SNI 1727: 2020 – Beban desain minimum dan kriteria terkait untuk bangunan gedung dan struktur lain.

[23] Acton J M and Hibbs M, 2012 Why Fukushima Was Preventable *Carnegie Endowment for International Peace.*

IOP Conf. Series: Earth and Environmental Science 1184 (2023) 012002 doi:10.1088/1755-1315/1184/1/012002

Equipment for Testing the Tensile Strength of Natural Fibers: Design and Implementation

Y F Arifin[1, 2, 3], R Siswanto[4], M Arsyad[2, 3], Novi Shintia[5], M H Ridha[3], M R Ramadhani[3]

[1]Wetland Based Material Research Center, University of Lambung Mangkurat
[2]Civil Engineering Master's Program, University of Lambung Mangkurat, Indonesia
[3]Civil Engineering Study Program, University of Lambung Mangkurat, Indonesia
[4]Mechanical Engineering Study Program, University of Lambung Mangkurat
[5]Banjarmasin State Polytechnic, Indonesia

y.arifin@ulm.ac.id

Abstract. The use of fibers, especially natural fibers, as construction materials has begun to develop. Natural fibers are also used for soil stabilization, where there are two mechanisms for increasing soil strength: tensile strength and fiber-soil friction. The purpose of this research is to design and build simple, inexpensive tools that can be used to test the tensile strength of a natural fiber. The tool consists of three main parts, namely the pull tool, loadcell, and display, and a supporting component, namely the computer. Prior to use, two calibrations were performed with respect to load cell accuracy and the vertical deformation of the tool. The equipment was used directly to measure the tensile strength of one of the natural fibers from oil palm empty fruit bunches. The relationship between stress and strain, fiber diameter, and fiber strength is discussed and compared with behaviors reported from the existing literature. The tensile strength of oil palm empty fruit bunches was found to be 107 MPa on average, with a strain at break of 6%. Fiber behavior, independent of diameter, was discovered to include elastic and plastic regions in the stress-strain relationship.

1. Introduction

Tensile strength, in addition to frictional resistance between the soil and the fiber, is an important component in its use as a strengthening material or soil stabilization. Many fiber tensile strengths have been reported in the literature, including the tensile strength of oil palm empty fruit bunch (OPEFB) fibers. Among the tensile strengths of OPEFB fibers reported in the literature, they are sorted from the highest to the lowest, as shown in Table 1. The tensile strength of OPEFB fiber varies greatly from the maximum value of 52-283 MPa. A very high value for the tensile strength of a material. Consequently, its application in composite materials or as soil reinforcement is highly promising.

Besides tensile strength, strain is also an important parameter. Compared to synthetic fibers, the strain at the fiber break is very large, for example, as shown in Table 1, namely 9.7–22.1%. The large variation of the two parameters, both tensile strength and strain, is caused by several factors, including the age of the plant, the age of the fiber after extraction, the condition of the fiber surface (surface damage, damaged cell wall, or has undergone treatment), measuring length, and fluctuations in gripping pressure during testing [1].

Tabel 1. Tensile strength and maximum strain of OPEFB fiber

No	Tensile strength (MPa)	Maximum strain (%)	References
1	283	15,4	[2]
2	248	14	[3]
3	137.6	22,1	[4]
4	130	9,7	[5]
5	110		[6]
6	91	14	[7]
7	80		[8]
8	71	11	[9]
9	52	10	[10]

On the other hand, fiber-reinforced soil's shear strength is made up of two parts: the shear strength of the soil matrix and the tensile stress in the fibers [11]. Also, the bond between the soil and the fiber in the pull-out mechanism and the tensile strength of the fiber itself cause the fiber to add to the increase in shear strength [12]. It was reported that soil samples mixed with fiber became more ductile, and strength and strain at collapse were also higher [11]. Fiber length was also found to have an effect on the soil-fiber mixture. The surface friction between the fiber and the soil is higher when the fiber is long [11]. This friction plays an important role in increasing the shear strength of soft soils mixed with fibers [13]. However, friction does not exist if the tensile strength of the fiber is low because the fiber breaks before friction begins. Therefore, it is important to keep testing the fibers to be utilized, particularly for reinforcing or stabilizing clay, which may undergo a change after a specific period of time in the soil. This research aims to develop a tensile testing instrument with modifications to the fiber clamp and implement it to test OPEFB fibers.

2. Material Used

The OPEFBs used were a by-product of one of the palm oil factories in Kec. Angsana, Kab. Tanah Bumbu, South Kalimantan Province. Fiber was taken from empty oil palm fruit bunches, which were air dried, and the fiber was extracted. Fiber was not treated before use. A total of 20 strands of fiber with an initial length of approximately 100 mm were selected. The fiber diameter was measured with a micrometer. Measurements were made at both ends and in the middle, and then the average was taken. Diameter is a very important factor in calculating tensile strength, which is the tensile force divided by the cross-sectional area of the fiber. In the tensile test, only fibers with diameters close to those required (i.e. 0.3 mm) were used.

3. Methods and Procedures

3.1. Equipment description

Figure 1 shows a sketch of the tool and its components. The developed tensile test equipment consists of a load frame, digital force gauge, driving motor, and computer. Generally, for fiber placement, a method as suggested by ASTM [14] is used. However, this method is difficult to apply to the OPEFB fiber test, which has a very small diameter between 0.2 and 0.6 mm. Therefore, the fiber binding tool was a chuck drill, which is designed to be able to hold material with a diameter of at least 0.2 mm. The shape of the tool is as shown in Figure 1(a). The clamp on the chuck drill used is made of stainless steel, which allows it to damage both the surface and the fiber itself. This was considered by ignoring broken fibers at the base or pinched parts. The tensile strength recorded was for the one whose fibers broke in the middle, as shown in Figure 1(b). This shows that the pinching that occurs in the fiber does not affect its tensile strength.

3.2. Calibrations

Another important procedure in fiber tensile testing is instrument calibration. In this tool, the calibration includes two important components: the calibration of the digital force gauge and the vertical deformation of the tool.

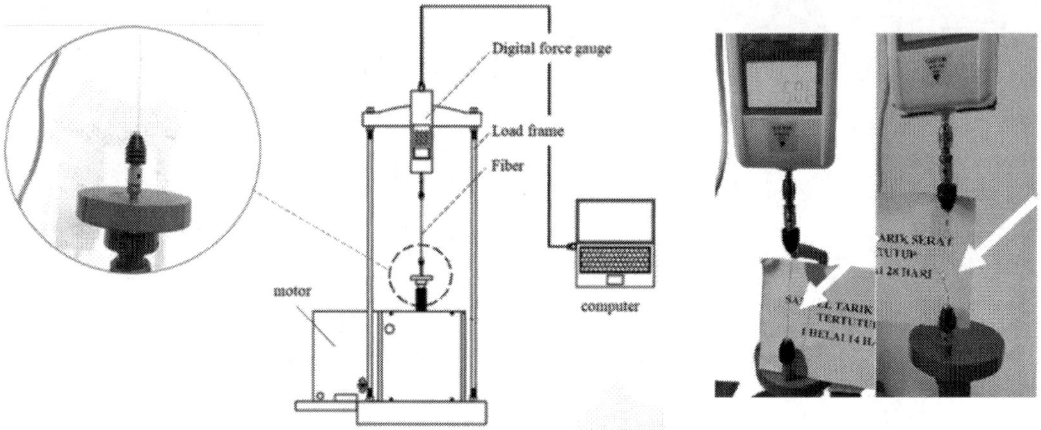

(a) (b)

Figure 1. (a) Sketch of tensile test equipment and its components, and (b) Considered fiber breaking mechanism

3.2.1. Digital force gauge

Even though it has been calibrated throughout the manufacturing process, the digital force gauge must be verified. The force measurement calibration was carried out by preparing predetermined loads starting at 500, 1000, 2000, 3000, and 5000 g. The load was then hung on a measuring device, and the measured force was recorded. Figure 2(a) shows the relationship between the weight of the load and the force measured on the force meter. As seen in Figure 2(a), the measured force and load are in agreement, where the measured force is respectively 5, 10, 20, 30, and 50 N. There is no measured force deviation up to a predetermined load.

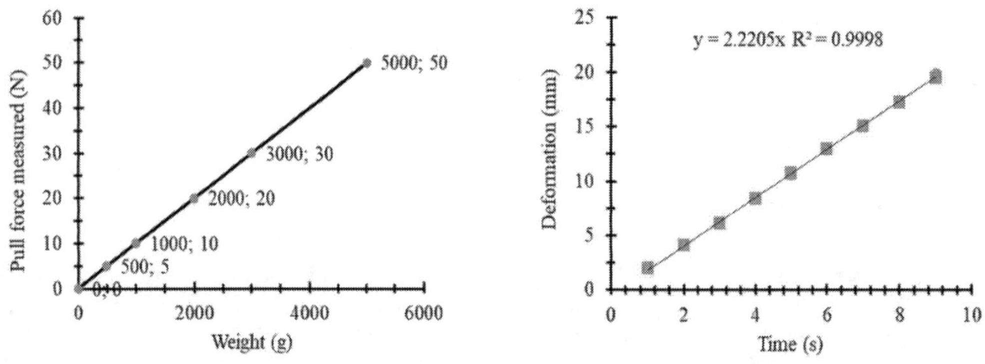

(a) (b)

Figure 2. (a) Weight and force relationship in the digital force calibration, and (b) Time and vertical deformation relationship

3.2.2. Vertical deformation calibration

One of the important components in measuring the tensile strength of a fiber is its deformation. Fiber deformation can be measured using a dial gauge mounted on the sample base plate. However, measuring forces simultaneously becomes difficult. Thus, calibration was carried out to obtain the vertical movement of the tool at a certain time. Figure 2(b) shows the relationship between time and vertical deformation. Equation 1 was derived from the data and can be used to compute the vertical deformation, which was then calculated as the fiber length extension. The strain was calculated as a ratio of the fiber's extended length to its initial length.

$$y = 2,2205x \tag{1}$$

where y is time and x is vertical deformation. The deformation is determined by the time interval of the force gauge.

4. Results and Discussions

4.1. Tensile strength and strain of OPEFB fiber

Before being tested, the sample diameter was measured using a micrometer, and the results were obtained as shown in Figure 3(a). From the 20 samples, the average fiber diameter was 0.28 mm, with the smallest fiber being 0.24 mm and the largest being 0.35 mm. From these data, it was also found that the standard deviation of the tested diameter was 0.03 mm.

Figure 3(b) shows typical results of the OPEFB fiber tensile test. This curve is similar to that reported by Omar et al. [15], where there are three areas resulting from the tensile test of OPEFB fibers, namely elastic, plastic, and collapsed areas. The fiber begins to be pulled into the elastic area at a strain of 0.06–0.08%, resulting in a tensile strength of 60–70 MPa. This strain was higher than reported by Ramlee et al. [16] (i.e., 0.03%) and Omar et al. [17] (i.e., 0.04%). More than that, continue with the plastic region until the 5–6% strain reaches a tensile strength of 97–107 MPa.

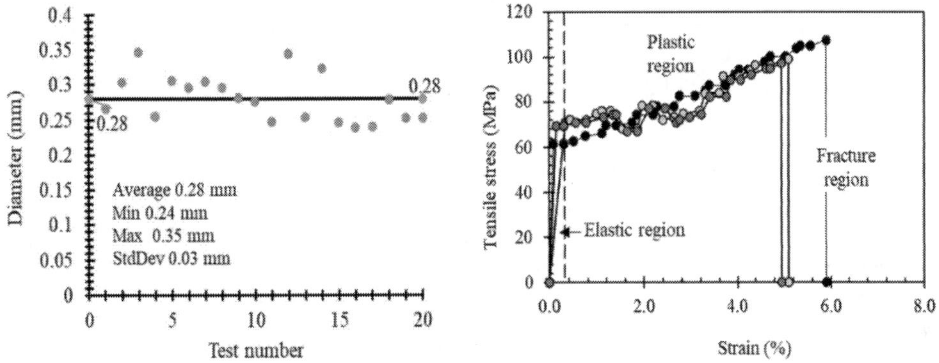

(a) (b)

Figure 3. (a) Samples diameter, and (b) Tensile stress and strain relationship

The tensile strength of OPEFB fiber samples from the 20 fibers tested is shown in Figure 4(a). From these data, the average fiber tensile strength was 106.79 MPa, with the smallest value being 61.186 MPa and the largest being 188.89 MPa. When compared with the tensile strength of the fibers shown in Table 1, the tensile strength of OPEFB taken from the palm oil mill in Kec. Angsana, Kab. Tanah Bumbu, South Kalimantan Province, is among the fiber tensile strengths reported by Danso et al. [18] and Abdullah et al. [7]. Maximum fiber strains ranged from 2.99% to 16.2%, with a standard deviation of 3.36%, as shown in Figure 4(b). The average fiber strain was 8.26%. When compared with the magnitude of the strain that occurs in Table 1, this strain is the lowest. Based on the very varied data on tensile strength and strain at failure, tests of both are very important to determine their role in a composite or in soil stabilization.

There are two components that act when the soil-fibre sample is subjected to stress (i.e., friction between the fiber surface and the soil and the tensile strength of the fiber) [13]. The first stage involves friction between the fibers and the soil, which causes the strength and strain to increase [11]. Tensile strength contributes to maintaining friction between the fiber surface and the soil in the following stage. Tensile strength no longer plays a role once the fiber has been completely released. Therefore, fiber length also contributes significantly to soil stabilization using this material [11], ensuring that both components remain effective.

Figure 5 shows the relationship between the tensile strength and the maximum tension of OPEFB fibers. The tensile strength data appears random as a function of diameter, as shown in Figure 5(a). Even though it looks like it tends to get smaller, this is because fiber diameter is one of the parameters used in the calculation of tensile strength, where tensile strength is the tensile force divided by the cross-sectional area of the fiber being tested. In addition, the tensile strength of the fiber, which is its mechanical property, is not only influenced by its physical properties, including fiber diameter, but also by the chemical properties contained in the fiber in one bunch, even though it is taken from the same tree [19,20]. As shown in Figure 5(b), this also occurs in the relationship between fiber diameter and fiber maximum stretch.

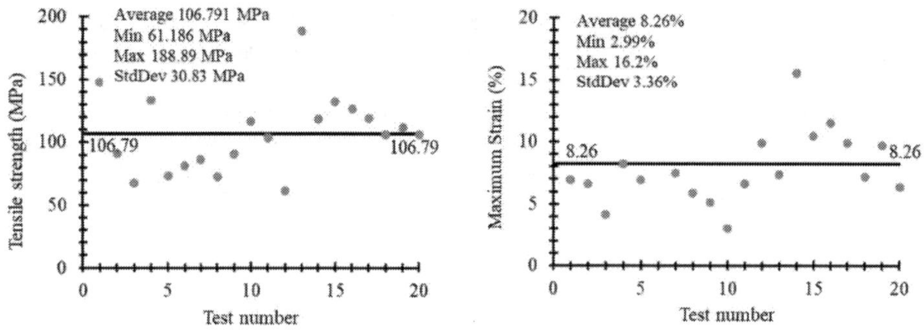

(a) (b)

Figure 4 (a) Tensile strength and (b) maximum strain of OPEFB fiber

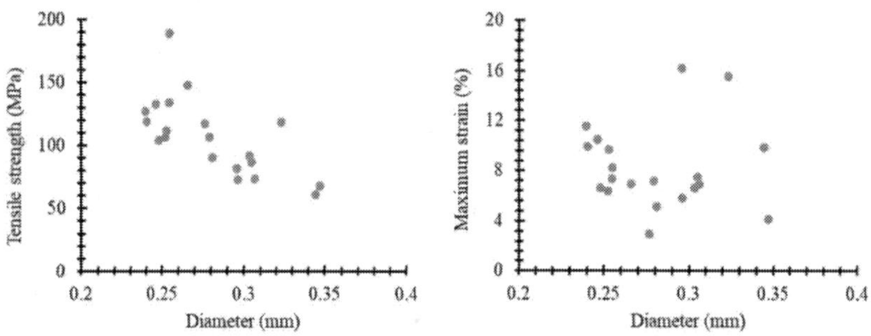

(a) (b)

Figure 5. (a) Tensile strength, and (b) maximum strain as a function of fiber diameter

5. Conclusions

The equipment used in this study was successful in measuring the tensile strength of OPEFB fiber. The tool was developed by modifying fiber clamps and was tested to ensure that it did not cause damage to the fiber at the two ends. There were also two calibrations performed to make entirely sure that the values that were measured for force and vertical deformation were accurate. The average diameter of the fiber of OPEFB used is 0.28 mm. All of the fibers tested in this study show three regions in the stress-strain relationship. These are the elastic, plastic, and rupture regions. The average tensile strength of the fiber is 106.79 MPa. The average maximum strain of the fiber is 8.26%. The tensile strength and maximum strain of OPEFB fiber are independent of the sample diameter.

Acknowledgement

The author acknowledges that this research was funded by the University of Lambung Mangkurat through the *Dosen Wajib Meneliti* Program in 2022 with the contract number of 026.18/UN8.2/PL/2022.

References

[1] Rama Rao P and Ramakrishna G 2022 Oil palm empty fruit bunch fiber: surface morphology, treatment, and suitability as reinforcement in cement composites- A state of the art review *Clean. Mater.* **6** 100144

[2] Bateni F, Ahmad F, Yahya A S and Azmi M 2011 Performance of oil palm empty fruit bunch fibres coated with acrylonitrile butadiene styrene *Constr. Build. Mater.* **25** 1824–9

[3] Rama Rao P and Ramakrishna G 2021 Experimental Investigation on Mechanical Properties of Oil Palm Empty Fruit Bunch Fiber Reinforced Cement Mortar *Mater. Today Proc.* **46** 471–7

[4] Kroehong W, Jaturapitakkul C, Pothisiri T and Chindaprasirt P 2018 Effect of Oil Palm Fiber Content on the Physical and Mechanical Properties and Microstructure of High-Calcium Fly Ash Geopolymer Paste *Arab. J. Sci. Eng.* **43** 5215–24

[5] Bakar A A, Hassan A and Mohd Yusof A F 2006 Mechanical and Water-Absorption Properties of Sisal Fiber (Agave Sisalana)-Reinforced Polyester C *Int. J. Polym. Mater.* **55** 627–41

[6] Danso H 2017 Properties of Coconut, Oil Palm and Bagasse Fibres: As Potential Building Materials *Procedia Eng.* **200** 1–9

[7] Abdullah C I, Azzahari A D, Rahman N M M A, Hassan A and Yahya R 2019 Optimizing Treatment of Oil Palm-Empty Fruit Bunch (OP-EFB) Fiber: Chemical, Thermal and Physical Properties of Alkalized Fibers *Fibers Polym.* **20** 527–37

[8] Ramli M and Dawood E T 2010 Effects of Palm Fiber on the Mechanical Properties of Lightweight Concrete Crushed Brick *Am. J. Eng. Appl. Sci.* **3** 489–93

[9] Yusoff M Z M, Salit M S and Ismail N 2009 Tensile properties of single oil palm empty fruit bunch (OPEFB) fibre *Sains Malaysiana* **38** 525–9

[10] Norul Izani M A, Paridah M T, Anwar U M K, Mohd Nor M Y and H'Ng P S 2013 Effects of fiber treatment on morphology, tensile and thermogravimetric analysis of oil palm empty fruit bunches fibers *Compos. Part B Eng.* **45** 1251–7

[11] Li C and Zornberg J G 2013 Mobilization of Reinforcement Forces in Fiber-Reinforced Soil *J. Geotech. Geoenvironmental Eng.* **139** 107–15

[12] Diambra A and Ibraim E 2014 Modelling of fibre–cohesive soil mixtures *Acta Geotech.* **9** 1029–43

[13] Arifin Y F, Arsyad M, Siswanto R, Astawa I K T F, Ridha M H and Ramadhani M R 2022 Tensile Strength and Durability of Oil Palm Empty Fruit Bunch Fiber in Soft Soil *Int. J. GEOMATE* **23** 72–81

[14] ASTM-D3379-75 1989 Standard Test Method for Tensile Strength and Young's Modulus for High-Modulus Single-Filament Materials *ASTM Int. West Conshohocken, PA, www.astm.org* **75** 1–5

[15] Omar F N, Mohammed M A P and Baharuddin A S 2014 Effect of silica bodies on the mechanical behaviour of oil palm empty fruit bunch fibres *BioResources* **9** 7041–58

[16] Ramlee N A, Jawaid M, Zainudin E S and Yamani S A K 2019 Tensile, physical and morphological properties of oil palm empty fruit bunch/sugarcane bagasse fibre reinforced phenolic hybrid composites *J. Mater. Res. Technol.* **8** 3466–74

[17] Omar F N, Mohammed M A P and Baharuddin A S 2014 Microstructure modelling of silica bodies from oil palm empty fruit bunch (OPEFB) fibres *BioResources* **9** 938–51

[18] Danso H, Martinson D B, Ali M and Williams J B 2017 Mechanisms by which the inclusion of natural fibres enhance the properties of soil blocks for construction *J. Compos. Mater.* **51** 3835–45

[19] Komuraiah A, Kumar N S and Prasad B D 2014 Chemical Composition of Natural Fibers and its Influence on their Mechanical Properties *Mech. Compos. Mater.* **50** 359–76

[20] Peponi L, Biagiotti J, Torre L, Kenny J M and Mondragon I 2008 Statistical Analysis of the Mechanical Properties of Natural Fibers and Their Composite Materials. I. Natural Fibers *Polym. Compos.* **29** 313–20

IOP Conf. Series: Earth and Environmental Science 1184 (2023) 012003 doi:10.1088/1755-1315/1184/1/012003

Stability Analysis of Toll Road Embankment Induced by Liquefaction In Klaten Regency Based On FEM Numerical Simulation

F Hasiholan[1,2], S Ismanti[1*] and A Rifa'i[1]

[1] Department of Civil and Environmental Engineering, Faculty of Engineering, Universitas Gadjah Mada, Yogyakarta 55281, Indonesia
[2] Directorate General of Highways, The Ministry of Public Works and Housing, South Jakarta 12110, Indonesia

*Corresponding author: sito.ismanti@ugm.ac.id

Abstract. The construction of toll roads in Klaten Regency supports community connectivity in increasing economic growth and accessibility to tourist areas around the southern part of Java Island. The location of this study is in Polanharjo District, Klaten Regency. The study area has geological conditions, sandy volcanic ash, and shallow groundwater levels and is accompanied by a history of earthquakes that have occurred greater than Mw. 5.0. These three things are triggers for liquefaction. This study discusses the stability analysis of toll road embankments at the STA. 10+724 on the construction of the Solo – Yogyakarta – YIA Kulon Progo Toll Road due to the potential for liquefaction reaching 14 meters. The road embankment construction is placed on the ground surface, which has the potential for liquefaction with a height of up to 5 meters. Calculate the safety factor using the numerical simulation finite element (Plaxis v.8.2 software) with static and pseudostatic analysis. Numerical calculations show that the stability analysis of toll road embankments placed on the ground with the liquefaction potential is still safe for construction with a safety factor on the static analysis (SF: 1.989 > 1.5) and pseudostatic analysis (SF: 1.912 > 1.1).

1. Introduction

The construction of toll roads will affect regional and economic development and improve the mobility and accessibility of people and goods in the area. Klaten Regency is one of the areas under construction of a toll road. The toll road will connect Klaten Regency with Surakarta City, Yogyakarta City, and Yogyakarta International Airport (YIA) Kulon Progo [1]. The length of this toll road from Interchange Kartasura – YIA Kulon Progo is about ± 96 kilometers. In section 1.1, especially STA. 10+724 (Figure 1) [2], the construction of this toll road is a soil embankment with a height of up to 5 meters. The location of the toll road embankment is the location of this study. The study location is in Polanharjo District, Klaten Regency [3].

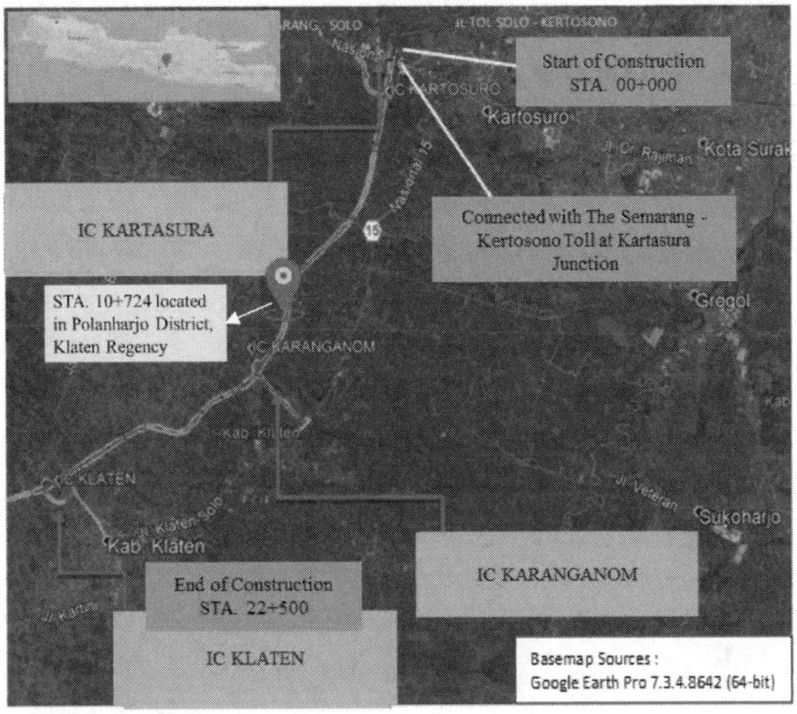

Figure 1. Location of Solo – Yogyakarta – YIA Kulon Progo Toll Road Section 1.1 construction ([4] with modification)

Based on the history of earthquakes, several earthquakes were felt to be greater than the moment magnitude (Mw) 5.0 around the Polanharjo District, Klaten Regency. The biggest earthquake occurred on May 27, 2006, in Bantul Regency, Yogyakarta Special Region Province, with Mw 6.3 at a depth of 12.5 Km [5]. The seismicity in the Yogyakarta and surrounding areas originate from the sea due to the subduction zone activity system from the collision between the Indian Ocean Plate - Australia and the Eurasian Continental Plate and on land due to active fault movements [6]. According to Green and Bommer, 2019, Earthquakes that can trigger liquefaction are at least Mw 5.0 [7].

The liquefaction phenomenon observed after the 2006 Yogyakarta earthquake is marked by sand boils along soil cracks in Paseban Village, Bayat District, Klaten Regency [6]. Since this earthquake, there have been many studies on the potential for liquefaction in this district. Still, there has yet to be a study on the potential for liquefaction in Polanharjo District, which is the location for constructing this toll road embankment [8].

Liquefaction is a geotechnical phenomenon. The soil type susceptive to liquefaction is loose sand with a high water level. During an earthquake, cyclic shear stress at the ground surface due to shear waves increases the pore water pressure of loose sand. The seismic shaking of the soil occurs so rapidly that it causes non-cohesive soils to experience undrained loads [9]. Liquefaction susceptibility depends on soil density, groundwater level, and distribution of soil particles [10]. The more significant earthquake, the greater the liquefaction potential that can happen [11].

This study examines the liquefaction potential and analyzes the stability of toll road embankments Induced by liquefaction at STA. 10+724 in Polanharjo District, Klaten Regency. The potential liquefaction analysis in this study uses a point Standard Penetration Test (SPT). Calculation of the factor of safety for the analysis of potential liquefaction using the Idriss-Boulanger (2008) method. Calculation of the stability analysis of the toll road embankment uses Numerical Simulation Finite Element Method (FEM) with

IOP Conf. Series: Earth and Environmental Science 1184 (2023) 012003 doi:10.1088/1755-1315/1184/1/012003

comparative static analysis (soil condition without seismic load) and pseudostatic analysis (soil property condition with seismic load).

2. Research Method

2.1. Condition of study area

In the map of the geological Surakarta – Girintontro, Java, which Surono and Sudarno made in 1992, Polanharjo District is above the Merapi volcanic rock formation (Qvm), which belongs to the Holocene quarter. The formation of his rock is made of volcanic breccia, lava, and tuff (Figure 2) [12]. The Holocene Quaternary began ten thousand years ago and continues into the present day [13]. Volcanic rock formation in Klaten Regency is composed of sand, gravel, boulders, and some igneous rocks that resulted from the eruption of Mount Merapi [14]. An area with geological conditions like tuff typically has a more serious potential hazard to vibration effects caused by amplification and interaction of soil vibrations [15]. Figure 3 shows the condition of the existing ground surface at the study site.

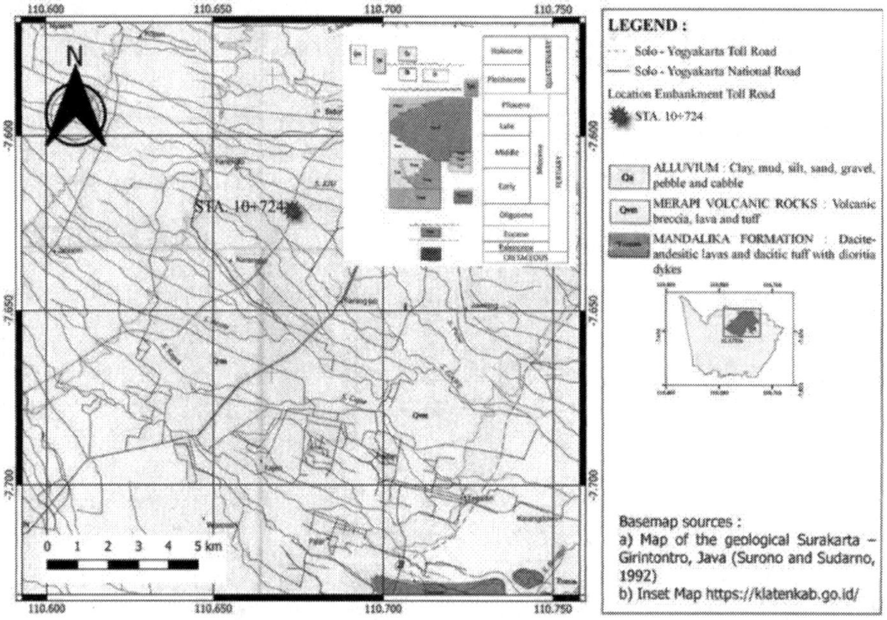

Figure 2. Map of geological of SPT points in the study area ([12] with modification)

Figure 3. Condition of the existing ground surface at STA. 10+724

This study uses SPT data from Solo – Yogyakarta – YIA Kulon Progo Toll Road construction (STA. 10+724) in December 2020. STA. 10+724 is at the latitude coordinates (- 7.621) and longitude coordinates (110.673). The groundwater level at this location is at a depth of 0.06 m. This area's soil is sandy and has loose to dense textures [16]. Table 1 shows the result of a geotechnical test in this study area.

Table 1. The result of a geotechnical test in STA. 10+724

Depth (m)	Description	N_{SPT}	Classification symbol (USCS)
0 - 2	Stiff brownish grey to dark brown sandy SILT	15	ML
2 - 4	Stiff brownish grey to dark brown sandy SILT	13	ML
4 - 6	Medium dense to dense brownish grey SAND with silt	34	SP-SM
6 - 8	Medium dense to dense brownish grey SAND with silt	19	SP-SM
8 - 10	Medium dense to dense brownish grey SAND with silt	18	SP-SM
10 - 12	Medium dense to dense brownish grey SAND with silt	11	SP-SM
12 - 14	Medium dense to dense brownish grey SAND with silt	37	SP-SM
14 - 16	Loose blackish grey poorly-graded SAND with silt	5	SP-SM
16 - 18	Medium dense blackish grey poorly-graded SAND	15	SP
18 - 20	Dense blackish grey to brown silty SAND	48	SM
20 - 22	Very dense grey silty SAND	60	SM
22 - 24	Brownish grey silty SAND	60	SM
24 - 26	ANDESITE	60	
26 - 28	Very stiff brownish grey CLAY	17	CL
28 - 30	Very dense brownish grey clayey SAND	60	SC

Based on the United States Geological Survey (USGS), Figure 4 shows historical data on earthquakes around the study area. Figure 4 shows two historical earthquakes around the location that can trigger liquefaction. The earthquake occurred on 8 October 1979 with an Mw 5.2 (source of the Benioff earthquake and a depth of 180 Km) and on 27 May 2006 with an Mw of 6.3 (source of the Opak fault and a depth of 12.5 Km).

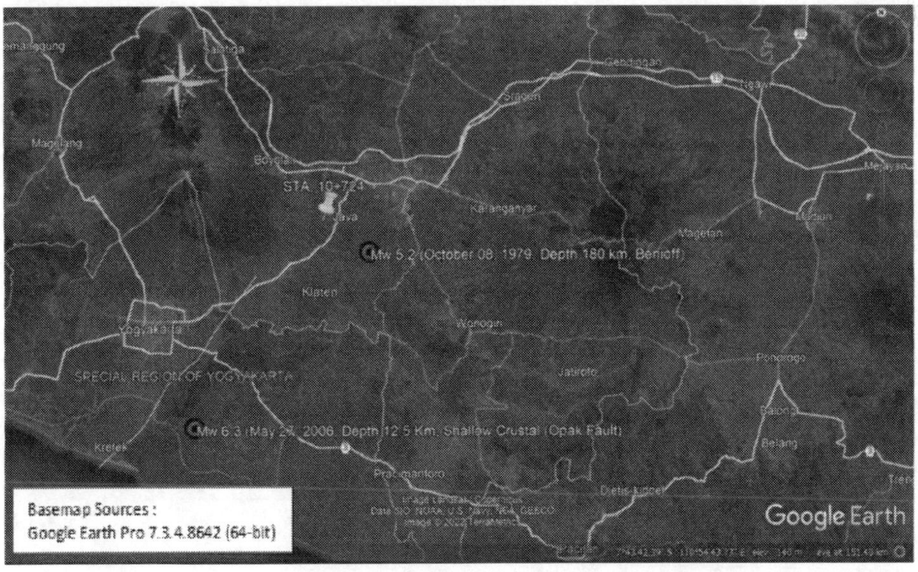

Figure 4. Earthquake history Mw 5.0 around the study area ([4] and [17] with modification)

2.2. Seismic design analysis

There are two methods of assessing seismic hazards at a certain point: Probabilistic Seismic Hazard Analysis (PSHA) and Deterministic Seismic Hazard Analysis (DSHA) [18]. The PSHA method uses the help of the Lini application developed by the Directorate General of Highways, Ministry of Public Works and Housing, by entering the latitude and longitude coordinates of the study area [19]. PSHA method for toll road construction uses a 7% probability in 75 years with a 1,000-year return period [20].

The DSHA method uses the attenuation equation to assign Peak Ground Acceleration (PGA) that causes the most significant risk. Several parameters have a considerable effect on the DSHA method: earthquake source distance, earthquake magnitude, earthquake source mechanism, and conditions of earthquake recording location [18]. Table 2 shows the attenuation equation to assign PGA in this study area.

Table 2. The attenuation to assign Peak Ground Acceleration (PGA) in this study area [18], [21], [22]

No.	Mw	Latitude	Longitude	Depth of earthquake (km)	Source of earthquake	Attenuation equation	Application
1.	5.2 (October 08, 1979)	- 7.670	110.760	180.0	*Benioff*	1. *Geomatrix slab seismicity rock, 1997* 2. *AB 2003 intraslab seismicity worldwide*	Manual
2.	6.3 (May 27, 2006)	- 7.960	110.450	12.5	*Fault*	1. *Boore-Atkinson NGA.* 2. *Campbell-Bozorgnia NGA.* 3. *Chio-Youngs NGA.*	Using NGA-West 2

Combination of the PSHA and DSHA methods to identify PGA of bedrock with the most significant risks. Based on SNI 8460:2017, a site's modified peak ground acceleration (PGA_M) is calculated by multiplying the PGA at bedrock with the site coefficient for that class of site (F_{PGA}) [20]. Table 3 shows the site coefficient for that class of site (F_{PGA}).

Table 3. The site coefficient for that class of site (F_{PGA}) [20]

Class of site	PGA ≤ 0,1	PGA = 0,2	PGA = 0,3	PGA = 0,4	PGA = 0,5
Hard rock (SA)	0.8	0.8	0.8	0.8	0.8
Rock (SB)	1.0	1.0	1.0	1.0	1.0
Very dense soil/soft rock (SC)	1.2	1.2	1.1	1.0	1.0
Stiff soil (SD)	1.6	1.4	1.2	1.1	1.0
Soft soil (SE)	2.5	1.7	1.2	0.9	0.9
Special soils requiring site-specific evaluation (SF)	SS	SS	SS	SS	SS

Note : SS = sites requiring geotechnical investigation and site-specific response analysis

2.3. Analysis of potential liquefaction

This study's liquefaction potential analysis uses the simplified procedure method developed by Idriss - Boulanger in 2008. The equation for calculating the liquefaction potential safety factor is shown in Table 4.

Table 4. The equation for calculating the liquefaction safety factor [23],[24]

Parameter	Equation
Cyclic Stress Ratio (CSR) method by Seed and Idriss	$CSR = 0.65 \text{ x } (PGA_M) \text{ x } \left(\frac{\sigma_{vc}}{\sigma'_{vc}}\right) \text{ x } r_d$ (1) $r_d = \exp[\alpha(z) + \beta(z) . M]$ (2) $\alpha(z) = -1.012 - 1.126 \sin\left(\frac{z}{11.73} + 5.133\right)$ (3) $\beta(z) = 0.106 + 0.118 \sin\left(\frac{z}{11.28} + 5.142\right)$ (4) where PGA_M is site modified of PGA, σ_{vc} is total stress at a depth of z meters, σ'_{vc} is effective stress at a depth of z meters, r_d is shear stress reduction factor, z is the depth of each borehole below ground level in meters, and M is earthquake magnitude.
$CRR_{M=7.5,\sigma'_{vc}=1}$ (Cyclic Resistance Ratio for the earthquakes with M_w 7.5 and the effective overburden stress = 1)	$CRR_{M=7.5,\sigma'_{vc}=1} = \exp\left(\frac{(N_1)_{60CS}}{14.1} + \left(\frac{(N_1)_{60CS}}{126}\right)^2 - \left(\frac{(N_1)_{60CS}}{23.6}\right)^3 + \left(\frac{(N_1)_{60CS}}{25.4}\right)^4 - 2.8\right)$ (5) $(N_1)_{60CS} = (N_1)_{60} + \Delta(N_1)_{60}$ (6) $\Delta(N_1)_{60} = \exp\left(1.63 + \frac{9.7}{FC + 0.01} - \left(\frac{15.7}{FC + 0.01}\right)^2\right)$ (7) $(N_1)_{60} = C_N.N_{60} = C_N.C_E. C_B.C_R.C_S . N_M$ (8) where $(N_1)_{60CS}$ is the emendation of value N-SPT for fines content, FC is fines content, C_N is the factor of overburden stress, C_E is the factor of energy ratio emendation, C_B is borehole diameter emendation, C_R is rod length emendation, C_S is sample tube emendation, and N_M is the worth of N-SPT in field.
$CRR_{M,\sigma'_{vc}}$ (Cyclic Resistance Ratio for the earthquakes with other M_w and the effective overburden stress)	$CRR_{M,\sigma'_{vc}} = CRR_{M=7.5,\sigma'_{vc}=1} . MSF . K_\sigma$ (9) $K_\sigma = 1 - C_\sigma \ln\left(\frac{\sigma'_{vc}}{P_a}\right)$ (10) $C_\sigma = \frac{1}{18.9 - 2.55\sqrt{(N_1)_{60}}}$ (11) $MSF = 6.9 \exp\left(\frac{-M_w}{4}\right) - 0.058$ (12) where MSF is factor of scale magnitude, K_σ is factor of the overburden emendation that restricted to maximum worth of 1.1, and C_σ is factor of the overburden correction which corrected N-SPT and the tool with maximum value of 0.3.
Safety Factor of Liquefaction (FS_{Liq})	$FS_{Liq} = \frac{CRR_{M,\sigma'_{vc}}}{CSR}$ (13) If the value of $FS_{Liq} < 1$, the soil layer is potential liquefaction but $FS_{Liq} \geq 1$ means that the soil layer is not potential liquefaction.

2.4. Stability analysis of toll road embankment

Earthquakes can cause secondary effects on toll road embankments in the form of slope movements or slope landslides. Slope stability disturbance due to earthquakes can damage the toll road embankment failure [25]. Therefore, it is necessary to analyze the stability of the toll road embankment, which aims to determine the safety factor from landslides on the toll road embankment. The toll road embankment is said to be stable or unstable, judging by the magnitude of the safety factor [26].

Slope stability analysis can be carried out with the stages of the static and seismic methods. In the analysis of static slope stability, the slope is considered unstable if the acting shear stress exceeds the shear resistance in the potential landslide area. The method for analyzing slope stability under static conditions is generally carried out using limit equilibrium analysis [25].

One of the seismic slope analysis methods is usually the pseudostatic limit equilibrium method. Pseudostatic limit equilibrium analysis is done by applying a static force which represents the inertial force caused by the earthquake. In the stability analysis of the toll road embankment, the static force is expressed in the seismic coefficient multiplied by the mass weight that will make the landslide. In the pseudostatic analysis method, the safety factor against slope failure due to earthquakes is carried out through an analysis similar to the study of static equilibrium [25]. In the analysis using the pseudostatic method, the effect of the earthquake is described by horizontal and/or vertical acceleration. This analysis shows the seismic effect with pseudostatic acceleration that produces inertial forces, F_h and F_v, acting on the center of mass of collapse. The horizontal seismic coefficient (kh) is determined at 0.5 of the horizontal peak ground acceleration with site class determination and factor amplification (PGA_M). Table 5 shows the safety factor for the stability of the toll road embankment.

Table 5. Safety factor for the stability of the toll road embankment [20]

Method	Safety factor	Requirement
Static analysis	1.5	Without seismic load
Pseudostatic analysis	1.1	With horizontal seismic coefficient (0.5 of the horizontal peak ground acceleration with site class determination and factor amplification (PGA_M)

2.5. Finite Element Method (FEM) Numerical Simulation

For more complex slope stability calculations, modeling static, pseudostatic, and dynamic conditions in the loading system total can use the Finite Element Method (FEM) numerical simulation. In this study, The FEM numerical simulation uses Plaxis v.8.2 software [20].

Figure 5 shows the geometric model of the toll road embankment at STA. 10+724. Table 6 shows the dead load and live load in this study. Table 7 shows the input parameters for each soil layer and toll road embankment in the Plaxis software [20] [27]. Table 8 shows this study's stages of static and pseudostatic analysis.

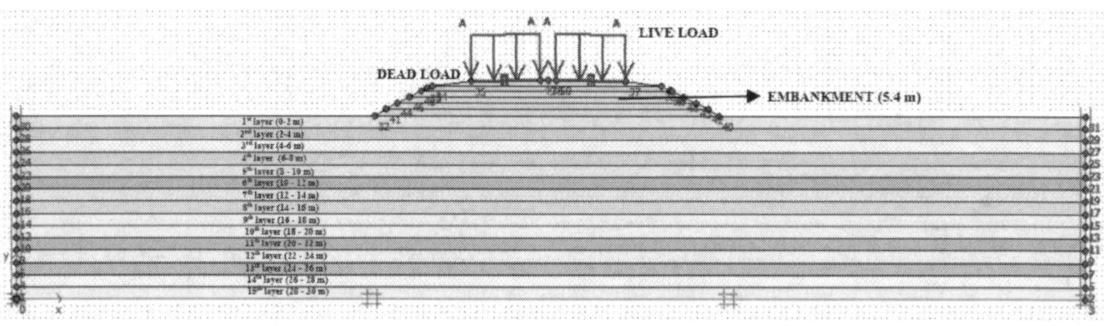

Figure 5. Geometric modeling of toll road embankment at STA. 10+724

IOP Conf. Series: Earth and Environmental Science 1184 (2023) 012003 doi:10.1088/1755-1315/1184/1/012003

Table 6. Dead load and live load in this study [20] [27]

Load	Parameter	Name	Value	Unit
Dead	Type of behaviour	Type	Elastic	-
	Normal stiffness	EA	1.220E+06	kN/m
	Flexural rigidity	EI	6,850.00	kNm^2/m
	Equivalent thickness	d	0.260	M
	Weight	w	0.471	kN/m/m
	Poisson's ratio	v	0.150	-
Live	Traffic load	Class I (toll road)	15	kPa

Table 7. Soil parameter data at STA. 10+724 [27]

Layer	Depth (m)	Material Mode	Material Type	γ_{unsat} (kN/m^3)	γ_{sat} (kN/m^3)	k_x (m/day)	k_y (m/day)	E_{REF} (kN/m^2)	v (nu)	c_{REF} (kN/m^2)	φ (°)	Ψ (°)
Layer 1	0 – 2	Mohr-Coulomb	UnDrained	17.3	14.495	8.64E-04	8.64E-04	1.75E+04	0.30	70	0	0
Layer 2	2 – 4	Mohr-Coulomb	UnDrained	12.043	17.17	8.64E-04	8.64E-04	1.75E+04	0.30	70	0	0
Layer 3	4 – 6	Mohr-Coulomb	Drained	13.147	18.55	8.64E-02	8.64E-02	5.10E+04	0.30	1	37	0
Layer 4	6 – 8	Mohr-Coulomb	Drained	13.199	17.56	8.64E-02	8.64E-02	2.40E+04	0.30	1	37	0
Layer 5	8 – 10	Mohr-Coulomb	Drained	13.199	17.5	8.64E-02	8.64E-02	2.40E+04	0.30	1	37	0
Layer 6	10 – 12	Mohr-Coulomb	Drained	13.199	17.04	8.64E-02	8.64E-02	2.40E+04	0.30	1	37	0
Layer 7	12 – 14	Mohr-Coulomb	Drained	13.199	18.75	8.64E-02	8.64E-02	5.50E+04	0.30	1	37	0
Layer 8	14 – 16	Mohr-Coulomb	Drained	12.522	16.64	8.64E-01	8.64E-01	7.50E+03	0.30	1	37	0
Layer 9	16 – 18	Mohr-Coulomb	Drained	13.108	17.3	8.64E-02	8.64E-02	2.25E+04	0.30	1	37	0
Layer 10	18 – 20	Mohr-Coulomb	Drained	13.377	19.47	8.64E-02	8.64E-02	7.20E+04	0.30	1	37	0
Layer 11	20 – 22	Mohr-Coulomb	Drained	12.616	20.27	8.64E-02	8.64E-02	1.41E+05	0.30	1	38	0
Layer 12	22 – 24	Mohr-Coulomb	Drained	12.616	20.27	8.64E-02	8.64E-02	1.41E+05	0.30	1	38	0
Layer 13	24 – 26	Mohr-Coulomb	Drained	12.616	20.27	8.64E-02	8.64E-02	1.41E+05	0.30	1	38	0
Layer 14	26 – 28	Mohr-Coulomb	UnDrained	12.616	19.13	8.64E-04	8.64E-04	2.13E+04	0.30	85	0	0
Layer 15	28 – 30	Mohr-Coulomb	Drained	12.616	20.27	8.64E-02	8.64E-02	1.08E+05	0.30	1	38	0
Toll Road Embankment	0 – 5.4	Mohr-Coulomb	Drained	18	18	1.00E+04	1.00E+00	1.10E+04	0.30	5	30	0

Table 8. Work stage analysis in this study

No.	Work Stage	Static Analysis	Pseudostatic Analysis	Description
1.	General settings with 0.5 PGA_M	-	v	0.5 PGA_M in x-acceleration
2.	Modelling geometry	v	v	From Detail Engineering Design toll road embankment
3.	Generate mesh	v	v	-
4.	Input initial condition	v	v	Groundwater level
5.	Calculation with stages construction of embankment (Plastic analysis with stage construction)	v	v	From layer to layer embankment until the structure pavement. In this study : 1. Embankment 1 (0 – 1 m) 2. Embankment 2 (1 – 2 m) 3. Embankment 3 (2 – 3 m) 4. Embankment 4 (3 – 4 m) 5. Embankment 5 (4 – 5 m) 6. Embankment 6 (5 – 5.4 m) 7. Dead load + live load (Table 6)
6.	Seismic load (Plastic analysis with total multipliers)	-	v	Define 0.5 PGA_M in Σ-Maccel
7.	Safety factor analysis	v	v	Phi/c reduction with incremental multipliers

3. Yield and Discussion

3.1. Seismic design analysis
Based on Figure 4, Table 2 and Table 3, value of modified PGA at site (PGA_M) as shown in Table 9.

Table 9. Modified Peak Ground Acceleration at site (PGA_M)

SPT Points	$\overline{N_{SPT}}$	Site Class	PGA Max (g)	F_{PGA}	PGA_M (g)
STA. 10+724	18.47	SD	0.3150	1.1850	0.3733

Table 9 shows the average NSPT at STA. 10+724 is 18.47, included in the SD site class (stiff soil) [20]. PGA Max is the PGA of bedrock with the most significant risks (combination of the PSHA and DSHA methods). Like previous studies, the bigger the earthquake, the greater the potential for liquefaction [11]. The worth of modified PGA at site (PGA_M) is 0.3733g. This PGA_M value is the basis for calculating the pseudostatic stability analysis method to describe the seismic conditions in the study area.

3.2. Analysis of potential liquefaction
Based on Table 4 in Equation (1) – Equation (13), the result of analysis of potential liquefaction as shown in Table 10.

Table 10. Resume of analysis potential liquefaction at STA. 10+724

SPT Points	Groundwater Level (m)	$\overline{N_{SPT}}$	PGA_M (g)	Dominant Soil	Potential Liquefaction at Depth (m)
STA. 10+724	- 0.06	18.47	0.3733	silty SAND	0 – 4, 6 - 12, and 14-18

The potential for liquefaction to happen in each soil layer is represented in Figure 6.

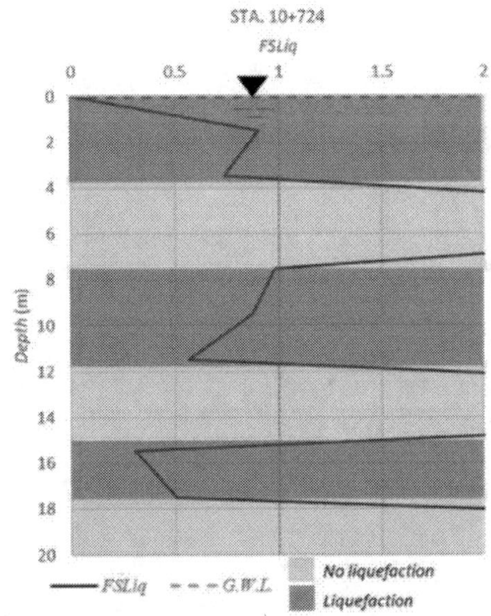

Figure 6. Potential liquefaction in each soil layer at STA. 10+724

Table 10 and Figure 6 describe the potential for liquefaction to occur at STA. 10+724, which reaches 14 meters. It can happen because the groundwater level in this location is very shallow, and the average N_{SPT} in the site class is medium soil (SD), with sand as the dominant soil type.

3.3. Stability analysis of toll road embankment

For the description of the analysis of the stages of work in Table 8, it can be seen in Figure 7. In the pseudostatic analysis, the soil layer with the potential for liquefaction, which was initially drained, turned into undrained. This happens because the excess pore pressure of the soil layer with the liquefaction potential will be considered in the analysis. The seismic load calculated in this study is 0.5 PGA_M, which is 0.1867g.

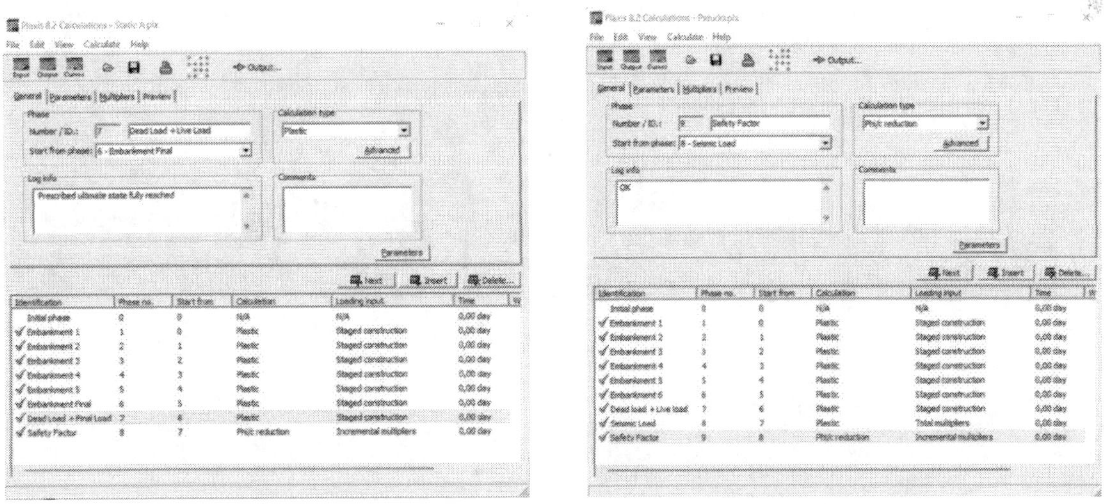

(a) Static Analysis (b) Pseudostatic Analysis

Figure 7. Work stage analysis in this study

The results of total displacement in the toll road embankment can be seen in Figure 8 and Figure 9.

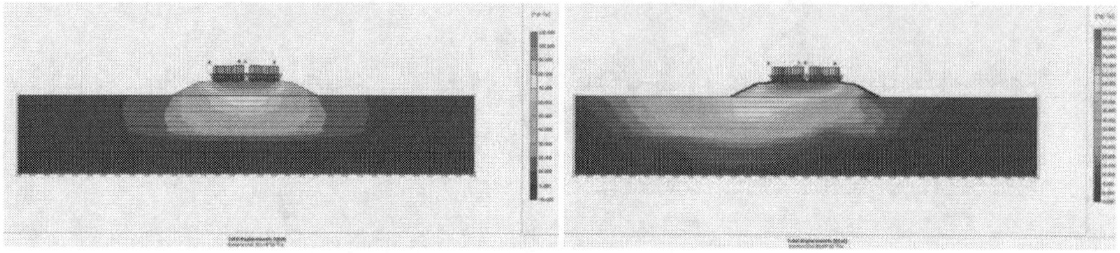

(a) Static Analysis (b) Pseudostatic Analysis

Figure 8. The results of total displacement in the toll road embankment

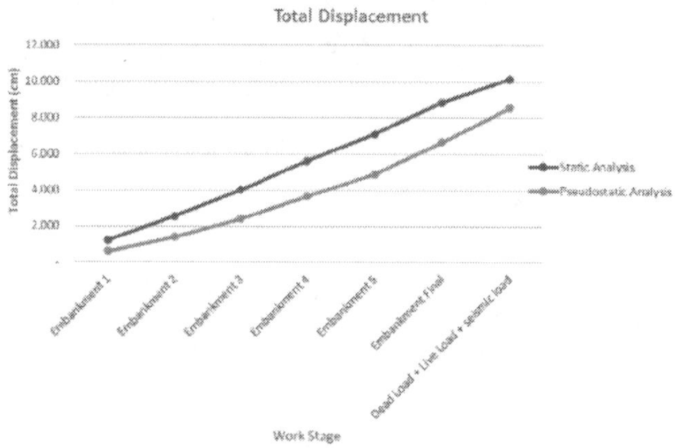

Figure 9. Total displacement result graph

Figure 8 and Figure 9 show the total displacement that occurred in this study. Both analyzes show the most significant total displacement at the end of the embankment work. For static analysis, the total displacement is 10.148 cm towards the bottom of the toll road embankment. As for the pseudostatic analysis, the total displacement that occurs is 8.50 cm to the bottom and left of the toll road embankment slope, and this is due to a seismic load of 0.1867g.

Figure 10 and Figure 11 show the results of excess pore pressures in the toll road embankment.

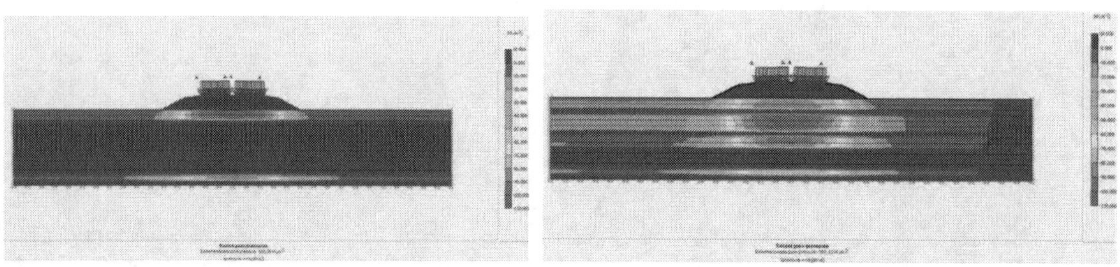

(a) Static Analysis (b) Pseudostatic Analysis

Figure 10. The results of excess pore pressures in the toll road embankment

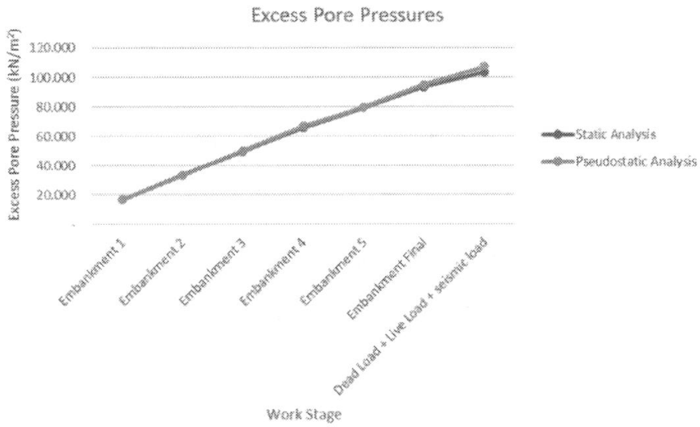

Figure 11. Excess pore pressures result graph

Figure 10 and Figure 11 show the excess pore pressure in this study. Both analyzes show the most significant excess pore pressures at the tip of the embankment work. For pseudostatic analysis, excess pore pressure also occurs in soil layers with the potential for liquefaction and an increase of 3.83 kN/m^2 from static analysis. This happens due to a seismic load of 0.1867g.

Figure 12 shows the description of safety factors in the toll road embankment.

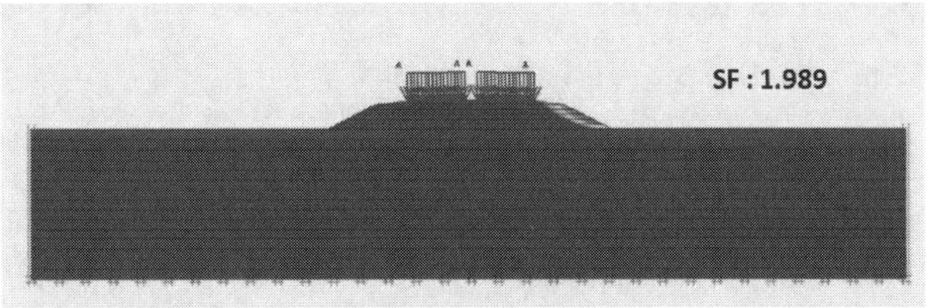

(a) Static Analysis

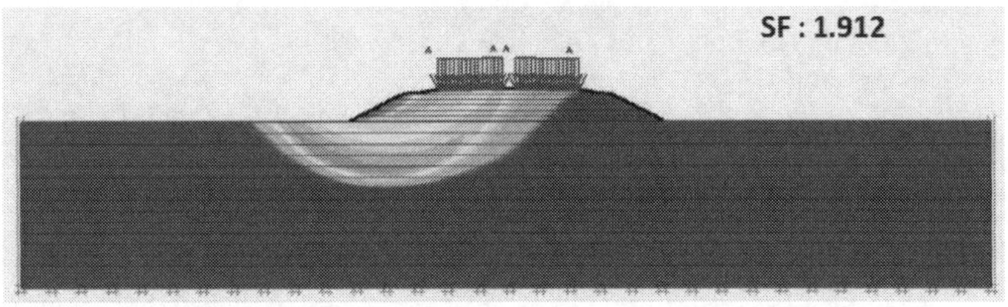

(b) Pseudostatic Analysis

Figure 12. Safety factors in the toll road embankment

Figure 12 shows the safety factor that occurs in this study. The safety factor on the static analysis (SF: 1.989 > 1.5) and pseudostatic analysis (SF: 1.912 > 1.1) indicates the construction of the toll road embankment in STA. 10+724 is safe against these two conditions and is ready for construction. The direction of the slope stability that will occur during an earthquake is to the left and below the embankment, as shown in Figure 12 (b).

4. Conclusions

Based on the condition of the study area and liquefaction safety factor analysis, STA. 10+724, the location of the toll road embankment in this study, indicates the potential for liquefaction reaching 14 meters (at depths 0 – 4, 6 - 12, and 14-18). For static analysis, the total displacement is 10.148 cm towards the bottom of the toll road embankment. As for the pseudostatic analysis, the total displacement that occurs is 8.50 cm to the bottom and left of the toll road embankment slope, and this is due to a seismic load of 0.1867g.

Excess pore pressure has increased in the soil layer with a liquefaction potential of 3.83 kN/m^2 from static analysis. This happens due to a seismic load of 0.1867g. The safety factor on the static analysis (SF: 1.989 > 1.5) and pseudostatic analysis (SF: 1.912 > 1.1) indicates the construction of the toll road embankment in STA. 10+724 is safe against these two conditions and is ready for construction.

5. Acknowledgment

The authors are grateful for the support of Ministry of Public Works and Housing and PT Jogjasolo Marga Makmur as operator of Solo – Yogyakarta – YIA Kulon Progo Toll Road Construction Project.

6. References

[1] BPJT 2022 Badan Pengatur Jalan Tol Kementerian Pekerjaan Umum dan Perumahan Rakyat
[2] Google Earth Pro 2022 Location of Solo – Yogyakarta – YIA Kulon Progo Toll Road Section 1.1 construction
[3] PT. Jogjasolo Marga Makmur and PT. Adhi Karya (Persero) Tbk. 2021 *Laporan Pendahuluan Rencana Teknik Akhir (RTA) Jalan Tol Solo - Yogyakarta - NYIA Kulon Progo Seksi 1.1 (STA 0+000 - 22+500)* (Surakarta)
[4] Google Earth Pro 2022 Google Earth Pro 7.3.4.8642 (64-bit)
[5] Setiyono U, Gunawan I, Priyobudi, Yatimantoro T, Imananta R T, Ramdhan M, Hidayanti, Anggraini S, Rahayu R H, Hawati P, Yogaswara D S, Julius A M, Apriani M, Harvan M, Simangunsong G and Kriswinarso T 2019 *Katalog Gempabumi Signifikan dan Merusak 1821 - 2018* (Jakarta: Pusat Gempabumi dan Tsunami Kedeputian Bidang Geofosika Badan Meteorologi Klimatologi dan Geofisika)
[6] Supartoyo 2006 Gempabumi Yogyakarta Tanggal 26 mei 2006 *Buletin Berkala Merapi* **Volume 3, No. 2** 36–55
[7] Green R A and Bommer J J 2019 What is the Smallest Earthquake Magnitude that Needs to be Considered in Assessing Liquefaction Hazard? *Earthquake Spectra* **35** 1441–64
[8] PT. Jogjasolo Marga Makmur 2021 *Overview Pembangunan Jalan Tol Solo - Yogyakarta - Kulon Progo* (Surakarta)
[9] Day R W 2002 *Geotechnical Earthquake Engineering Handbook* (New York: McGraw-Hill)
[10] Sauri S, Rifa'i A and Hardiyatmo H C 2021 Liquefaction Vulnerability Analysis using N-SPT Value and Grain Size Analysis on Gumbasa Irrigation Canal in the Post-Disaster Petobo Area, Sulawesi *IOP Conf. Ser.: Earth Environ. Sci.* **930** 012081
[11] Fauzan, Rifa'i A and Ismanti S 2021 Study of Liquefaction Potential at Sabo dam Construction on Poi and Bangga River, Sigi Regency, Central Sulawesi *IOP Conf. Ser.: Earth Environ. Sci.* **930** 012083
[12] Surono B T and Sudarno I 1992 Peta Geologi Lembar Surakarta dan Girintoro, Jawa
[13] William M A J, Dunkerley D L, De Decker P, Kershaw A P and Stokes T J 1993 Quaternary Environments *Edward Arnold, A division of hodder & Stoughton, London New York Melbourne Auckland* 329pp
[14] Rahmahnia A 2017 *Mikrozonasi Kegempaan Berdasarkan Efek Lokal, Indeks Kerentanan Seismik, dan Percepatan Gerakan Tanah di Kabupaten Klaten, Jawa Tengah* (Surabaya: Institut Teknologi Sepuluh Nopember)
[15] Nakamura Y 2000 Clear identification of fundamental idea of Nakamura's technique and its applications *Proceedings of the XII World Conference Earthquake Engineering* **Paper no 2656**

[16] PT Jogjasolo Marga Makmur, PT Adhi Karya (Persero) Tbk. and PT Perentjana Djaja 2021 *Rencana Teknik Akhir Jalan Tol Solo - Yogyakarta - NYIA Kulon Progo Seksi 1.1 (STA. 0+000 - 22+300) - Laporan Studi Geologi dan Sesar Aktif Rencana Centerline*
[17] USGS 2022 Latest Earthquakes
[18] Tim Pusat Studi Gempa Nasional 2017 *Peta Sumber dan Bahaya Gempa Indonesia Tahun 2017* (Bandung: Pusat Penelitian dan Pengembangan Perumahan dan Permukiman Badan Penelitian dan Pengembangan Kementerian Pekerjaan Umum dan Perumahan Rakyat)
[19] Bina Marga PUPR 2022 Aplikasi LINI versi Beta untuk Peta Sumber dan Bahaya Gempa Indonesia Untuk Jembatan
[20] SNI 8460:2017 *Persyaratan Perancangan Geoteknik* (Jakarta: Badan Standardisasi Nasional)
[21] Mase L Z 2019 Performance of NGA Models in Predicting Ground Motion Parameters of The Strong Earthquake *Journal of the Civil Engineering Forum* **5** 227–42
[22] PEER Center 2014 PEER Ground Motion Database - Pacific Earthquake Engineering Research Center
[23] Idriss I M and Boulanger R W 2008 *Soil Liquefaction During Earthquake* (Earthquake Engineering Research Institute)
[24] Seed H B and Idriss I M 1970 *A Simplified Procedure For Evaluating Soil Liquefaction Potential* (California: College of Engineering University of California)
[25] Hardiyatmo H C 2022 *Rekayasa Gempa Untuk Analisis Struktur & Geoteknik* (Yogyakarta: Gadjah Mada University Press)
[26] Hardiyatmo H C 2018 *Mekanika Tanah 2* (Yogyakarta: Gadjah Mada University Press)
[27] PT Jogjasolo Marga Makmur, PT Adhi Karya (Persero) Tbk. and PT Perentjana Djaja 2022 *Rencana Teknik Akhir (RTA) Jalan Tol Solo - Yogyakarta - NYIA Kulon Progo Seksi 1 Paket 1.1 (STA. 0+000 - 22+300) Report Analisis Stabilitas Timbunan Pada Mainroad dan Simpang Susun Kartasura, Simpang Susun Karanganom, Simpang Susun Klaten Seksi 1 Paket 1.1 STA. 0+000 - STA. 22+300* (Solo - Yogykarta)

IOP Conf. Series: Earth and Environmental Science 1184 (2023) 012004 doi:10.1088/1755-1315/1184/1/012004

Gelam Wood (*Melaleuca cajuputi* Powell) as a Substructure Construction Material

Muhammad Afief Ma'ruf[1,3*], Yudi Firmanul Arifin[2], Mufidah Asy'ari[1], Rusdiansyah[3]

[1]Doctoral Program of Agricultural Science, Postgraduate Program, University of Lambung Mangkurat, Banjarbaru, Indonesia, [2]Faculty of Forestry, University of Lambung Mangkurat, Banjarbaru, Indonesia. [3]Faculty of Engineering, University of Lambung Mangkurat, Banjarbaru, Indonesia

*)afief.maruf@ulm.ac.id

Abstract. *Gelam* wood (*Melaleuca cajuputi* Powell) which is widely used for substructure construction materials in South Kalimantan is classified into strong class II and durable wood class III. This study aims to provide more detailed information regarding what forms of use of *gelam* wood as substructure construction material and why *gelam* wood was chosen for it. This information will later be able to support the preparation of future utilization plans for the use of *gelam* wood considering that currently *gelam* wood is starting to decrease in availability. The method to identified the use of *gelam* wood as a substructure construction material is field survey method combine with other references. *Gelam* wood with a large diameter of 10-12 cm and a length of 4 m is generally used as a construction foundation for 1-storey building with an ultimate bearing capacity of 729,25 kg for single pile foundation. In terms of cost, the foundation with *gelam* wood material is cheaper than the mini-pile concrete foundation for the construction of the same building. The large size of *gelam* wood is also used as sheet pile or retaining wall with an average flexural strength of 100,13 MPa. The small size of *gelam* wood is generally used as wooden scaffolding where the cost of *gelam* wood scaffolding is 26.85% cheaper than iron scaffolding in the rental price, and 15 times the cost of the iron scaffold in the purchase price.

1. Background

Gelam wood, which belongs to the *Melaleuca* genus of the *Myrtaceae* family, is a plant that grows naturally in the peatlands of South Kalimantan, Central Kalimantan, and other areas. In this case, the species commonly marketed as gelam wood on Kalimantan Island is *Melaleuca cajuputi* Powell [1]. *Gelam* wood as shown in Figure 1 and 2 is capable of growing up to 40 m in height and up to 35 cm in diameter. *Gelam* wood has distinctive characteristics, one of which is able to adapt in different ecosystem environments and adapt to environmental changes that occur, so it is very suitable to be used as construction material for reclamation or regional development. Besides being used in the form of logs (small diameter logs), *gelam* can also be used in the form of sawn wood.

Gelam wood (*Melaleuca cajuputi* Powell) which is widely used for construction materials in South Kalimantan is classified into strong class II and durable wood class III [2]. The chemical composition of gelam wood itself consists of pentosan, cellulose, lignin, and a small amount of subtractive material, with an average density of 0.85. The strength of *gelam* wood itself will increase if it is used in a swamp area that is always submerged, or if it is not exposed to direct sunlight. Until now, the use of *gelam* wood is still very widely carried out in South Kalimantan, especially Banjarmasin City and its surroundings, because the typical type of soil in the area is tidal swamp land which is very suitable for the use of *gelam* wood as foundation construction, especially the foundation of residential buildings. From the results of interviews conducted in related research, information was obtained that for the South Kalimantan area, small *gelam* wood with a diameter of 2 cm to 3 cm is generally used for construction scaffolding, cattle pens, or plant marker poles. Meanwhile, gelam wood with a diameter of up to 10 cm and larger is used as pillars or building foundations for the construction of houses on stilts that require pegs to stilt pillars to resist soil erosion, and as construction material, namely for wooden beams and boards, as well as for making wooden crates in packaging. in freight forwarding, while the remaining wood is used for firewood [3].

The large number of uses of *gelam* wood is due to the large population of *gelam* wood which in the past was abundant and quite easy to find in the peat swamp land of South Kalimantan and other areas. In South Kalimantan itself, the largest producer of *gelam* wood is Barito Kuala District, but it should be noted that the area of the *gelam* forest tends to shrink. This depreciation was mainly caused by several factors such as land conversion into settlements and tidal rice fields. This happens because the product from *gelam* forest is relatively only in the form of *gelam* wood, so it is considered less productive, when compared to products obtained from other forest types such as meranti from mountain forests [4]

Figure 1. Gelam Forest in Swamplands

Figure 2. *Gelam* wood (*Melaleuca Cajuputi* Powell)

This study tries to provide more detailed information about the use of *gelam* wood as a substructure construction material in South Kalimantan and its surroundings. This is necessary to provide more detailed information regarding what forms of use of *gelam* wood as construction material and why *gelam* wood was chosen for it. This information will later be able to support the preparation of future utilization plans for the use of *gelam* wood considering that currently *gelam* wood is starting to decrease in availability.

2. Method

The method to identified the use of *gelam* wood as a substructure construction material is field survey method combine with other references. The area of field survey is Banjarmasin City and Banjar District which *gelam* wood is widely used. The field survey aims to described what kind of form that *gelam* wood is utilize in terms of log diameter and length of *gelam* log. The survey results then compared with other research on the physical and mechanical characteristics of *gelam* wood.

3. Results

In general, the Banjarmasin area and its surroundings have a very soft soil structure, because some of the land is inundated by water or influenced by the tides of river water. with a hard soil depth of Banjarmasin City which is at a depth of 28 – 42.4 m [5]. In this case, the use of a suitable foundation for simple buildings is a type of floating foundation with foundation materials that can adapt to the conditions of physical and environmental characteristics of the soft soil. In this case, *gelam* wood is one of the most suitable foundation materials because it is able to adapt to the surrounding conditions [6].

The field survey conducted to obtain what kind of form that *gelam* wood is utilize in terms of log diameter and length of *gelam* log. The result of the field survey is shown in Table 1.

Table 1. Log Diameter and Length of gelam log used

	Type of Building	Location	Log Diameter (cm)	Length (m)
1	2-storey house residence	A. Yani Road Km. 11,8, Banjar District	10-12	6
2	1-storey restaurant	A. Yani Road Km. 7,8, Banjar District	10-12	7
3	1-storey house residence	Sunga Bakung Road, Banjar District	10-12	4
4	1-storey house	Sunga Bakung Road,	10-12	4

	residence	Banjar District		
5	2-storey warehouse	M.T. Haryono Road, Banjarmasin City	10-15	9
6	1-storey house residence	AMD Road, Banjarmasin City	10-12	4
7	1-storey house residence	Sungai Andai Road, Banjarmasin City	10-12	4
8	1-storey house residence	Peradapan Karya Road, Banjarmasin City	10-12	4
9	residential fence	Tarakan Road, Banjarmasin City	10-12	3
10	scaffolding	Perdagangan Road, Banjarmasin City	6-8	3

From the survey results shows in Table 1, for 1-storey house residence the 4 m length gelam wood with diameter 10-12 cm is widely used. For 2-storey building used a 6-9 m gelam wood with 10-15 cm diameter, and for other use such as fence is using a 3 m length gelam wood.

The result from Table 1 shows that *gelam* wood is still widely used, especially for 1-storey house residence, as illustrated in Figure 3. In the city of Banjarmasin and its surroundings, generally the type of piled-*gelam* foundation is used for a 1 (one) floor house made of wood. Recent developments, for concrete buildings using reinforced concrete slab foundations on a pile foundation of cerucuk gelam wood as well. The addition of the *gelam* wood foundation is intended to improve the carrying capacity of the swamp soil [7].

Figure 3. *Gelam* wood foundation for residential houses

Research conducted on residential buildings of the Red Valerian type in the Citra Garden Residence in Banjar District tried to compare the use of a *gelam* wood foundation with a concrete mini pile foundation. The results showed that the bearing capacity of the *gelam* wood foundation was 1368.27 kg with a maximum load received by the pile of 1260.55 kg for the size of the *gelam* wood with a diameter of 10 cm and a length of 6 m. For the same length with a size of 20 cm x 20 cm, the concrete mini pile provides a bearing capacity of 6942.58 kg with a maximum load received by the pile of 6280.03 kg. In this case, the cost needed to make a foundation with *gelam* wood is Rp. 100,079,380,00, while the cost needed to make a mini pile foundation is Rp. 102,837,362.00 [8]. From the results of this study, it can be seen that the cost required for a foundation with *gelam* wood material is still smaller than that of a mini pile concrete foundation for the construction of the same building.

IOP Conf. Series: Earth and Environmental Science 1184 (2023) 012004 doi:10.1088/1755-1315/1184/1/012004

From Table 1 we can also see that 2-storey building is using gelam wood as foundation material. Research conducted on 2 different 3-storey shophouses in Banjarmasin City which experienced structural failure showed that the *gelam* wood foundation used in structural calculations was able to carry the load of the 3-storey shophouse [9] [10]. Another research based on the results of the ANSYS software modeling also shows that *gelam* wood foundation itself is in principle able to withstand the load of high-rise buildings, as long as it is planned properly [11].

When talking about *gelam* wood as a construction material, the bearing capacity of the pile is an important factor in the use of *gelam* wood as a construction material, especially for pile foundations. A laboratory-scale study that combines *gelam* wood pile with woven bamboo in peat swamp soil has given very satisfactory results as an alternative to soil reinforcement. The research was conducted by comparing the bearing capacity of the *gelam* wood pile without bamboo woven reinforcement and with woven bamboo. The results of laboratory tests showed that the maximum soil bearing capacity for the *gelam* wood pile without woven bamboo reinforcement was 39.33 kPa with the variation of the *gelam* wood pile slanted on the left and right sides of the foundation. Reinforcement with *gelam* wood pile combined with 3 layers of woven bamboo itself was able to increase the bearing capacity of the soil up to almost 10 times the condition without woven bamboo reinforcement. This shows that the use of *gelam* wood pile, either with woven bamboo or not, is able to increase the bearing capacity of the foundation on peat soil [12].

In addition to laboratory-scale research, previous studies have also conducted field loading tests to see the ultimate strength of the single pile bearing capacity of the *gelam* wood foundation. The study was conducted in a location which is a soft soil that has a highly water content and is dominated by peat soil using a sample of *gelam* wood foundation piles with a diameter of 12 cm and 14 cm and a length of 3.5 m. The applied load uses a dead load of press brick material by giving the load gradually until the pile settlement reaches 10% of the diameter of the pile. The results showed that the ultimate bearing capacity of *gelam* wood ranged from 567.75 kg to 729.25 kg for 12 cm and 14 cm diameter and 3.5 m length. The results of this study indicate that the *gelam* wood foundation is the right material for soft soil because it is able to transmit the load through the attachment of the pile blanket to the soil [13]

The survey result also shows that *gelam* wood used as foundation for fence, which is to withstand not an axial load, but rather lateral load. The ability of *gelam* wood to withstand lateral forces or horizontal ground movements is closely related to the flexural strength of *gelam* wood itself. Research on the flexural strength of wood using a method based on SNI 03-3959-1995 [14] and ASTM D-198 [15] which compares the flexural strength of *gelam* wood and *sengon* wood shows that the average flexural strength of *gelam* wood is 100.13 MPa and is higher than the average flexural strength. The average of *sengon* wood is 82.62 MPa. In addition, the results of the loading modeling on a simple house structure with the help of the SAP 2000 software show that the flexural stress capacity possessed by *gelam* wood and *sengon* wood is still greater than the flexural stress that must be resisted due to the combination of loading in simple building modeling. This indicates that *gelam* and sengon wood have the potential to be used in simple construction such as simple roof trusses [16].

Besides being used to carry vertical or axial loads such as in building foundations, *gelam* wood is also used as soil reinforcement in resisting soil slides due to lateral forces or horizontal movements of the soil layer, as illustrated in figure 4. Research conducted by regarding the prevention of landslides on road embankments using gelam wood sheet pile walls stated that the effect of strengthening gelam wood sheet pile walls increased soil stability and reduced the occurrence of landslides [17]. The results of this study are in agreement with other studies that use wooden chimneys to increase the shear strength of soft soils where the use of wooden crevices can increase the shear strength of the soil. In this case, the results of the study indicate that the longer the wood recess is used, the strength (shear resistance) of the soil increases [18].

Figure 4. Gelam wood as lateral reinforcement

The survey result also shows that gelam wood is also widely used as a supporting material for building construction, namely as wooden scaffolding, as illustrated in Figure 5. Related research compares the budget for the use of scaffolding made of *gelam* wood with scaffolding made of iron. This research was motivated by the availability of *gelam* wood in Banjarmasin which is increasingly scarce, so that alternative materials are needed that are more durable, easy to obtain and can be used repeatedly as scaffolds. One of them in this case is the iron scaffolding. The location for data collection is the construction of the Central Statistics Agency (BPS) South Kalimantan Province on Trikora Road Banjarbaru and a 3-floor shophouse on Benua Anyar Road, Banjarmasin. From the results of the study, it was found that the cost of work for *gelam* wood scaffolding was 26.85% cheaper than iron scaffolding in the rental price. and 15 times the cost of iron scaffolding in the purchase price [19]. However, it should be noted that iron scaffolding can be used over and over again, while gelam wood scaffolding can only be used during one construction work period. In addition to scaffolding, in the city of Banjarmasin itself, *gelam* wood is also used for road pavement foundations, pond foundations, and many other types of construction.

Figure 5. Use of gelam wood as a scaffolding

4. Conclusion

Based on these previous studies, conclusion of this study can be given as follows:

1. *Gelam* wood with a large diameter of 10-12 cm and a length of 4 m is generally used as a 1-storey house residence foundation with an ultimate bearing capacity of 729.25 kg for 1 pile foundation.
2. In terms of cost, the foundation with *gelam* wood material is still smaller than the mini pile concrete foundation for the construction of the same building.
3. For 2-storey house building, the gelam wood used is a 6-9 m gelam wood with 10-15 cm diameter
4. Large size gelam wood is also used as fence or retaining wall with an average flexural strength of 100.13 MPa of gelam wood.
5. Small-sized *gelam* wood is generally used as wooden scaffolding where the work cost of galam wood scaffolding is 26.85% cheaper than iron scaffolding in the rental price. and 15 times the cost of the iron scaffold in the purchase price.

References

[1] Banjarbaru Environmental and Forestry Research and Development Center. 2020. *Komoditas Paludikultur Di Kalimantan*. IPB Press. C.01/06.2020
[2] Ministry of Forestry, 2003. *Budidaya kayu Putih* (*Melaleuca cajuputi*) (silviculture of *kayu putih*). Departemen Kehutanan, Jakarta. 29p.
[3] Arifin, Y. F., Hamidah, S., Arifin, Y. F 2014. The Analysis of Management and Timber Trade System of *Gelam* (*Melaleuca cajuputi*) From Peat Swamp Forest in South Kalimantan. Journal of Wetlands Environmental Management 2, 79 – 83.
[4] Raja, O. 2010. *Hutan Galam atau Kayu Galam* [Online]. Available: http://onanraja.blogspot.com/2010/09/hutan-galam-atau-kayu-galam.html [Accessed 27 Januari 2021 2021].
[5] Ma'ruf, M. A., Rusliansyah, Fitriati, U., & Rachman, A. A. 2019. Digital Mapping of Hard Soil Depth in Banjarmasin City. International Journal of Engineering and Technology, 11, 316–320.
[6] Masitah, M., A.R. Shamsul Bahri, Jamilah M.S. And Salwani Ismail 2014. Histological observation of Gelam (Melaleuca cajuputi Powell) in different ecosystems of Terengganu. Journal of Biology, Agriculture and Healthcare, 4.
[7] Nurfansyah, M. I. S., Prima Widya Wastuti, Ahmad Qurrata Aini, & Nursyarif 2020. *Perkembangan Pondasi Tiang Kayu Pada Rumah Tinggal Di Banjarmasin*. Info-Teknik 21, 199-214.
[8] Rifky, A. 2014. *Studi Perbandingan Pondasi Kayu Galam Dan Mini Pile Pada Bangunan Perumahan Tipe Red Valerian Komplek Citra Garden Banjarmasin*. Jurnal Rekayasa Sipil 2.
[9] Marzuki, A., Alpiannor. *Evaluasi Kegagalan Pondasi Pada Gedung Bertingkat (Studi Kasus: Proyek Pembangunan Ruko 3 Lantai – Banua Anyar Banjarmasin*). Seminar Nasional Geoteknik 2016 PS S1 Teknik Sipil Unlam, 01 Oktober 2016 2016 Banjarmasin. Teknik Sipil Unlam.
[10] Mulyana. 2021. *Studi Daya Dukung Pondasi Kayu Galam Pada Bangunan Ruko Yamaha Di Jalan Ahmad Yani Km.2,5 Kota Banjarmasin*. Diploma, Universitas Islam Kalimantan MAB.
[11] Anggraini, E., Irwandy Muzaidi 2020. *Pemodelan Perilaku Keruntuhan Balok Tinggi Akibat Penurunan Pondasi Di Atas Tanah Lunak Kota Banjarmasin Menggunakan Metode Elemen Hingga 3D*. Jurnal Konstruksia, 11.
[12] Dewi, R., Yulia Hastuti, Yulindasari Sutejo, Maulid M. Iqbal, Muhammad Arfan 2020. Peat Soil Improvement Method Using Woven Bamboo and Cerucuk. International Journal on Advance Science Engineering Information Technology, 10.
[13] Khaliq, A. 2014. *Penentuan Beban Batas Tiang Galam Dengan Loading Test*. Jurnal Poros Teknik, 6.
[14] National Standardization Body. 1995. SNI 03-3959-1995. *Metode Pengujian Kuat Lentur Kayu Di Laboratorium*

IOP Conf. Series: Earth and Environmental Science 1184 (2023) 012004 doi:10.1088/1755-1315/1184/1/012004

[15] American Society For Testing And Materials. 2016. Standard Test Methods of Static Tests of Lumber in Structural Sizes. ASTM D 198 - 15.
[16] Basyaruddin, A. S., Jatmoko Awali 2019. *Potensi Pemanfaatan Kayu Gelam Dan Kayu Sengon Dalam Dunia Konstruksi Berdasarkan Uji Kuat Lentur*. Jurnal Rekayasa Sipil, 13.
[17] Gazali, A. *Analisis Penanggulangan Longsoran Pada Tepi Timbunan Jalan Menggunakan Perkuatan Dinding Turap Galam Di Daerah Rawa Kabupaten Barito Kuala*. Seminar Nasional Tahunan VI Program Studi Magister Teknik Sipil ULM, 26 Oktober 2019 2019 Banjarmasin. Program Studi Magister Teknik Sipil ULM.
[18] Rusdiansyah. *Asumsi Sistem Cerucuk Sebagai Alternatif Solusi Dalam Penanganan Kelongsoran Lereng Jalan Diatas Tanah Lunak*. Seminar Nasional Geoteknik 2016 PS S1 Teknik Sipil Unlam, 1 Oktober 2016 2016 Banjarmasin. Teknik Sipil Unlam.
[19] Rafik, A., Rinova Firman Cahyani 2018. *Analisis Perbandingan Biaya Penggunaan Perancah Kayu Galam Dan Perancah Besi* (Scaffolding). Jurnal Gradasi Teknik Sipil, 2.

Liquefaction Mitigation of Ramp On Kretek 2 Bridge

C T S A Galag[1, 2], F Faris[1] and A D Adi[1]

[1] Department of Civil and Environment Engineering, Faculty of Engineering, Universitas Gadjah Mada, Yogyakarta 55281, Indonesia
[2] Directorate General of Highways, The Ministry of Public Works and Housing, South Jakarta 12110, Indonesia

fikri.faris@ugm.ac.id

Abstract. The Kretek 2 Bridge was built over the Opak River where there is an active fault. According to bore log data and embankment site analysis, there is potential for liquefaction at a depth of 0 - 3 meters. Due to space limitations, ramp of bridge must be able to stand upright. An alternative treatment that can be implemented in this condition is to reinforce the embankment at the bridge ramp with the Mechanically Stabilized Earth Wall system. Analysis of liquefaction potential using the method of Idriss and Boulanger. Analysis of the application of the planned MSE wall was carried out based on the external, internal, and global stability in dynamic conditions in the subgrade after replacing the subgrade in the detected layer of liquefaction potential. The method for calculating the lateral pressure caused by an earthquake uses the pseudostatic method. The MSE wall with soil replacement meets the requirements for external, internal, and global stability under dynamic conditions to be an alternative solution to mitigate liquefaction potential and increase soil bearing capacity and limitation area problems. Reinforcement of friction tie using granular embankment gets good results in the calculation of safety factor breaking, pull out, overturning, and sliding.

1. Introduction

The Kretek 2 Bridge was built across the Opak River where around the Opak River, there is an active fault (Opak fault) which is one of the sources of earthquakes in Indonesia. From a careful examination of the soil layers, indications of the Opak Fault line were found in the eastern part of the trench. Earthquake fault fractures are seen cutting the lower part of the body of the sand dune bed (see **Figure 1**). So the indications of the Opak Fault path, which are interpreted from the results of the geoelectric survey, can be proven visually with this trench test. With the shift of the structure to the west to avoid faults, it has implications for the addition of the bridge ramp embankment height behind Abutment 2. The embankment height due to this shift is approximately 8 m. The subgrade must withstand the forces caused by the embankment load.

IOP Conf. Series: Earth and Environmental Science 1184 (2023) 012005 doi:10.1088/1755-1315/1184/1/012005

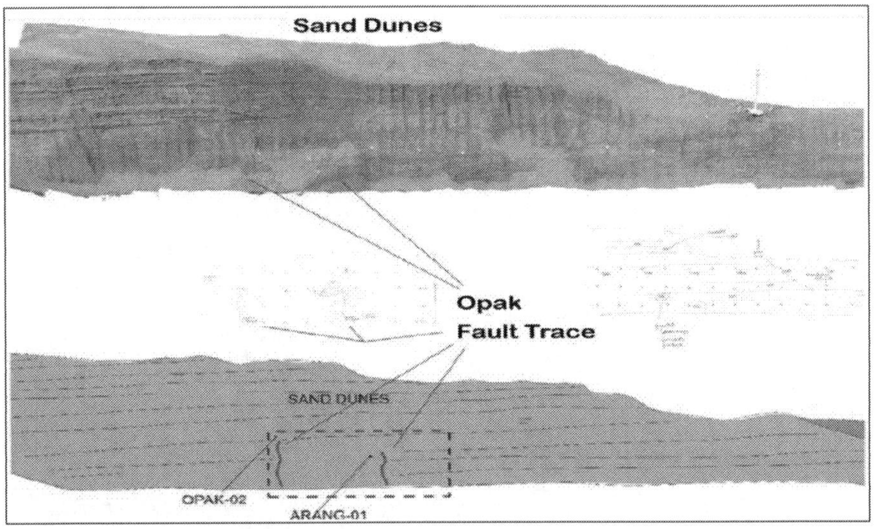

Figure 1. Interpretation of Trench Walls Based on Observations, Descriptions. Field Sketches and 3D Orthophoto [1]

According to bore log data and analysis of the embankment location, at borehole point number 8, there is a potential for liquefaction at a depth of 0 - 3 meters from the ground surface [2]. Liquefaction is a phenomenon of loss of shear strength of soil in a saturated layer of water due to an earthquake so that the soil collapses and behaves like a liquid. Liquefaction generally occurs in non-cohesive (granular) soils saturated with water and receiving cyclic loads in the form of earthquakes [3]. Based on the results of the analysis of soil grain gradation to determine the potential for soil liquefaction [4], it is clear that the soil composition in BH-8 is in an area that has the potential for liquefaction (**Figure 2**).

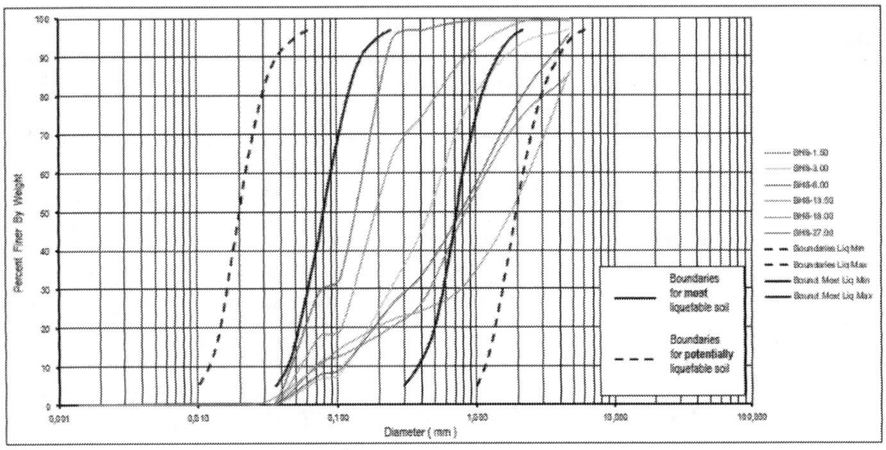

Figure 2. BH-8 grain gradation analysis chart [2]

Analysis of liquefaction potential was also carried out using a semi-empirical approach based on SPT data to evaluate the Cyclic Resistance Ratio (CRR) on the soil as well as the Cyclic Stress Ratio (CSR) [5]. The results of the calculation of the liquefaction potential for each layer at the drill point using the simplified procedure by Idriss and Boulanger there are FS values less than 1, as shown in **Figure 3**.

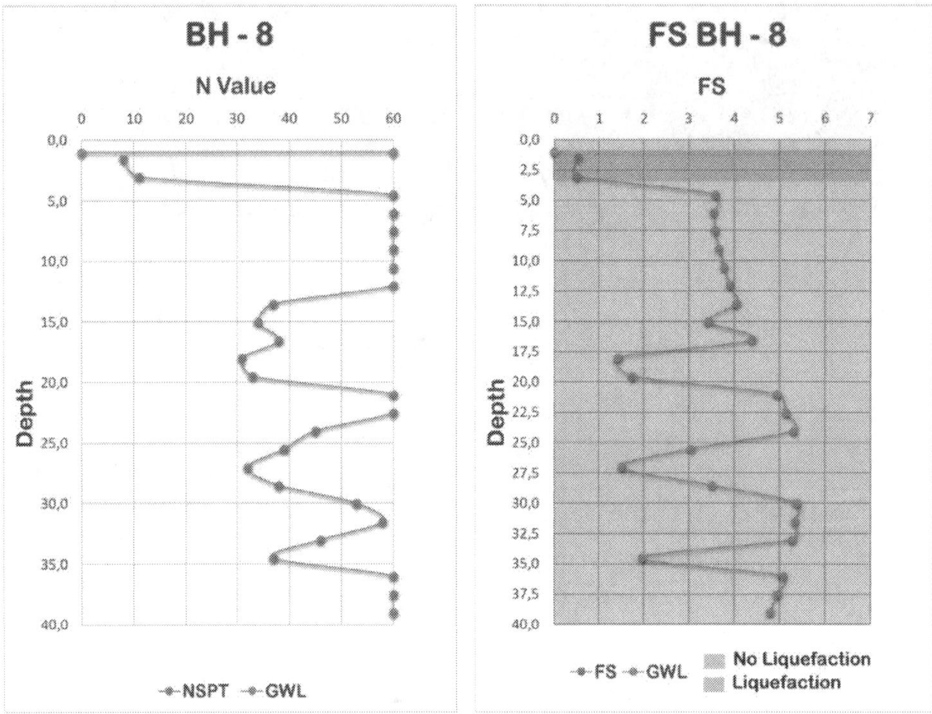

Figure 3. Graph of FS value on BH-8 [2]

Based on the two analyses of liquefaction potential above, it can be concluded that the ramp location of the East side of the bridge has the liquefaction potential. In addition, if the embankment ramp uses ordinary soil without reinforcement, an embankment slope of 1V:1.5H is required. However, due to limited land with a Corrugated Concrete Sheet Pile (CCSP) building and a Water Police Unit (Polair) fence on the north side of the bridge, this option is not possible to implement. Due to limited land, the bridge ramp must be able to stand upright. An alternative treatment that can be done with this condition is the reinforcement of the bridge ramp embankment using the MSE (Mechanically Stabilized Earth) Wall system.

2. Research Method

2.1. Soil Classification

According to Das [6], the soil classification system is a means of arranging soil types that are distinct but similar into groups and sub-groups depending on their usage. The classification system provides a simple language for describing the basic characteristics of highly variable soils without going into detail [7]. The majority of soil classification systems created for practices in the field of engineering are in reference to the simple soil index characteristics including grain size distribution and plasticity. Although there are currently various soil classification systems, none of these systems provides a clear explanation of all possible uses. This is due to the highly variable soil properties.

Table 1. Cohesion value based on soil classification [8]

USCS	Description	Cohesion Value (kPa)
GW	Well graded gravel, sandy gravel, with little or no fines	0
GP	Poorly graded gravel, sandy gravel, with little or no fines	0
GM	Silty gravels, silty sandy gravels	0
GC	Clayey gravels, clayey sandy gravels	0
SW	Well graded sands, gravelly sands, with little or no fines	0
SP	Poorly graded sands, gravelly sands, with little or no fines	0
SM	Silty sands	0
SC	Clayey sands	0
ML	Inorganic silts, silty or clayey fine sands, with slight plasticity	0
CL	Inorganic clays, silty clays, sandy clays of low plasticity	10 – 30
OL	Organic silts and organic silty clays of low plasticity	5 – 15
MH	Inorganic silts of high plasticity	0 – 10
CH	Inorganic clays of high plasticity	15 – 35
OH	Organic clays of high plasticity	5 – 15
Pt	Peat and other highly organic soils	0 – 10

2.2. MSE WALL

One type of retaining wall is a mechanically stabilized retaining wall or what is referred to as a Mechanically Stabilized Earth Wall (MSE WALL). This MSE consists of reinforcement within the soil embankment which helps to resist lateral earth pressure [9]. When compared to ordinary retaining walls, MSE walls usually have several advantages [10]. MSE walls are more flexible than ordinary retaining walls, such as concrete cantilever walls or gravity-type retaining walls [11]. Therefore, MSE walls are more suitable for areas with poor foundation soil and seismically active areas [12].

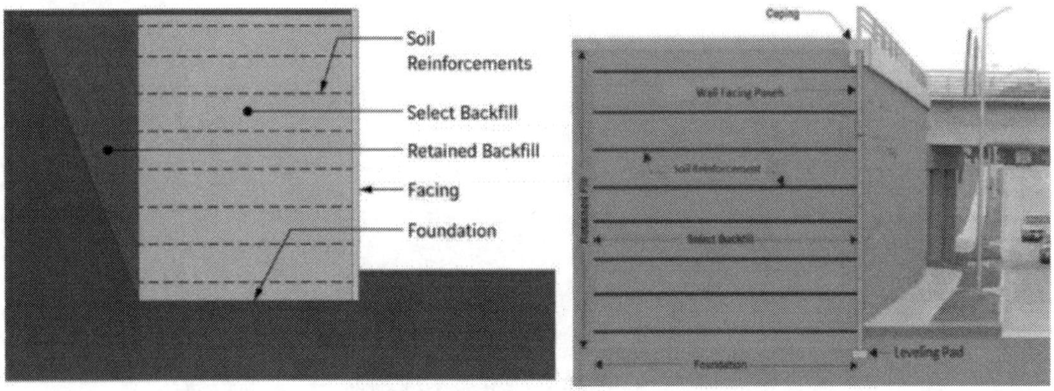

Figure 4. Components of MSE WALL [9]

MSE walls use several reinforcing materials, including mild steel, which is galvanized or coated with epoxy and geosynthetic [12]. MSE Wall is a soil construction made with artificial reinforcement, it is a type of retaining wall that combines facing the wall (a wall where the panels face and supports are so bonded to exert general action under load) and lateral reinforcement [13]. In constructing this MSE wall, the core reinforcement, Friction Tie, is made of high-tensile polyester material wrapped in textured high-density polyethylene (HDPE), with this concept the friction tie utilizes the overburden pressure of the embankment as a force that keeps the wall upright [14]. MSE Wall is a composite structure based on the interaction of soil friction and friction tie as core reinforcement [14].

The recommended embankment soil for MSE walls is granular soil because granular soil has the ability to transmit stress, resistance, and good drainage compared to clay soil [15]. The granular material requirements on the MSE wall detailed in **Table 2**.

Table 2. MSE Wall Select Granular Reinforced Fill Requirements [12]

Gradation: (AASHTO T-27)	U.S. Sieve Size	Percent Passing(a)
	4 in (102 mm) (a,b)	100
	No. 4 ((0.425mm)	0 – 60
	No. 200 (0,075 mm)	0 – 15
Plasticity Index, PI (AASHTO T-90)	PI ≤ 6	
Soundness: (AASHTO T-104)	The materials shall be substantially free of shale or other soft, poor durability particles. The material shall have a magnesium sulfate soundness loss of less than 30 percent after four cycles (or a sodium sulfate value less than 15 percent after five cycles).	

[a] To apply default F* values, Cu, should be greater than or equal to 4.
[b] As a result of recent research on construction survivability of geosynthetics and epoxy coated reinforcements, it is recommended that the maximum particle size for these materials be reduced to ¾-in. (19 mm) for geosynthetics, and epoxy and PVC coated steel reinforcements unless construction damage assessment tests are or have been performed on the reinforcement combination with the specific or similarly graded large size granular fill. Prequalification tests on reinforcements using standard agency fill materials should be considered.

2.3. Land Failure

Based on Hardiyatmo [15], the results of observations on a full-scale reinforced soil wall structure show that the locations of the maximum tensile forces that occur in the reinforcement are located almost close to the landslide area (slide area) when the structure is loaded until it collapses as shown. Shown in **Figure 5**.

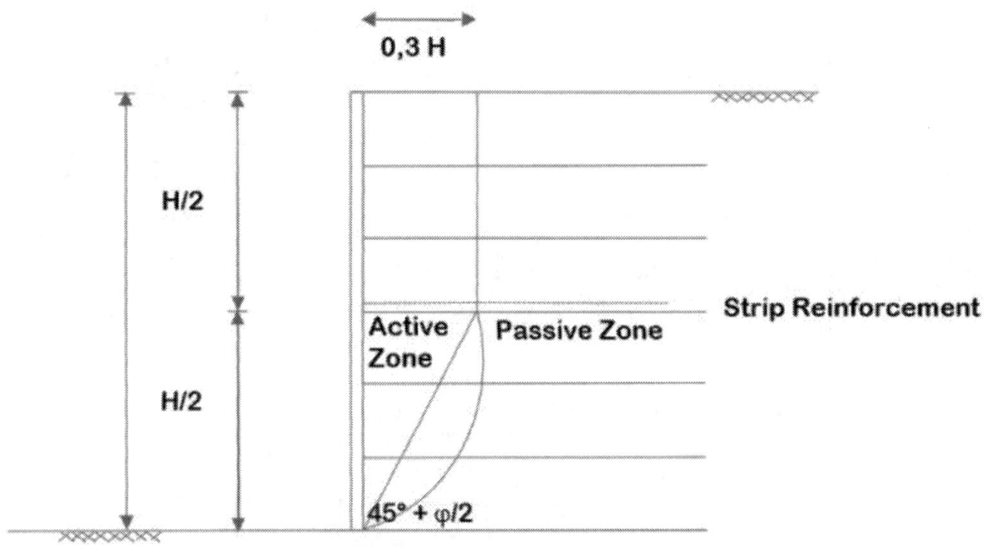

Figure 5. The location of the landslide field on a metal-striped soil wall [15]

The surface of the landslide area for vertical walls with soil reinforced with easily stretchable reinforcement (such as geotextiles) is generally considered to coincide with the Rankine landslide area, namely the failure that occurs in a area at an angle (45° + φ/2) to the horizontal area therefore for soils reinforced with geotextiles, the active soil coefficient (Ka) is used in calculating the lateral soil pressure [16].

2.4. Loading criteria

For pseudostatic analysis of excavated slopes, as well as embankments, a more specific earthquake load is suggested according to geological conditions and seismicity area, and slope importance. The minimum safety factor required for analysis using the pseudostatic model is greater than 1.1 (SF>1.1) using the seismic coefficient obtained from the peak acceleration on the surface (PGA) [11].
Requires the amount of traffic load used in the stability analysis described in **Table 3**.

Table 3. Traffic loads for stability analysis (DPU, 2001) and outside loads [11]

Highway Class	Traffic Load (kPa)	Outside Loads (kPa)
I	15	10
II	12	10
III	12	10

3. RESEARCH METHODS

3.1. Research sites

This research was carried out in the Kretek 2 Bridge area in the Parangtritis heap work segment which uses MSE Wall as embankment protection. Kretek Bridge 2 itself is located in District of Kretek, Regency of Bantul, Yogyakarta Special Region Province. The location of the Kretek Bridge is right above the opak active fault so it is very vulnerable to earthquake events as can be seen in **Figure 6**.

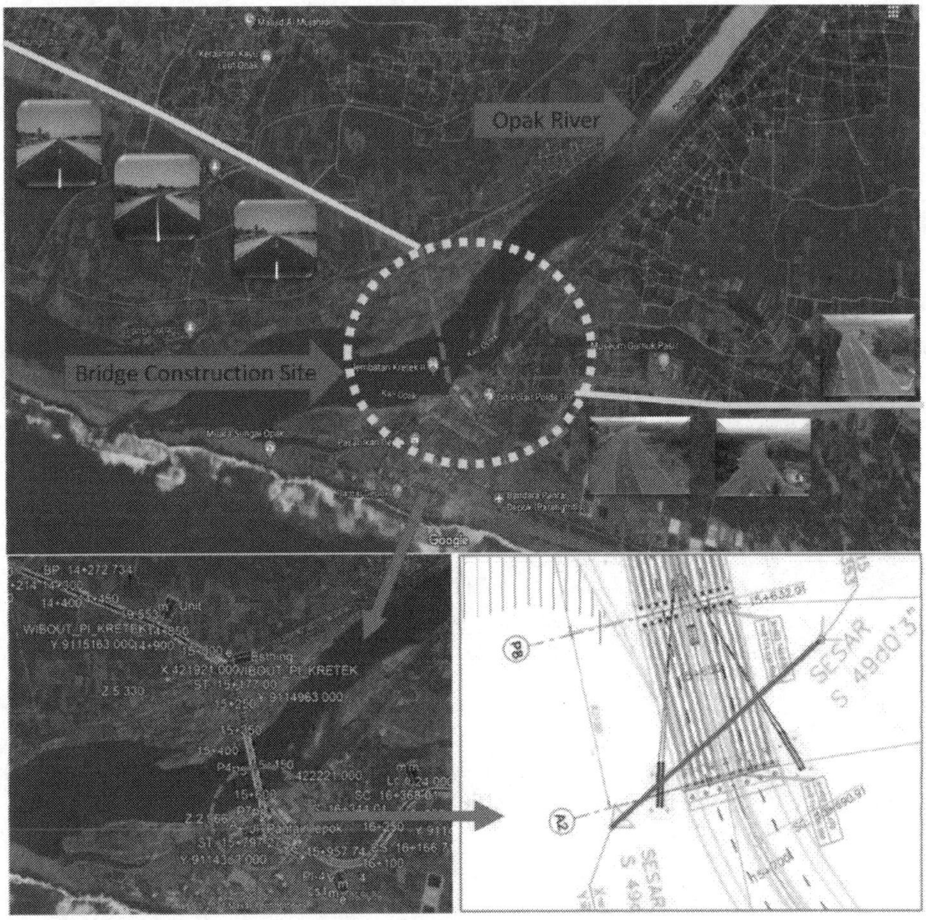

Figure 6. The location of the Kretek 2 Bridge is right above the Opak Fault [17]

3.2. Research data

This research was conducted using data obtained from PPK 1.4 as the owner of the Kretek Bridge 2 project. Drilling data obtained as many as 14 drill points. In the research conducted on the embankment work segment in the Parangtritis direction, the data used are drilling data for BH-8.

3.3. Active Earth Pressure

Based on a study done by Hardiyatmo, active earth pressure is lateral earth pressure that can reduce the stability of the retaining wall [18]. Active earth pressure coefficient (Ka) for flat ground is calculated using Equation 1.

$$K_a = \frac{(1 - \sin\varphi)}{(1 + \sin\varphi)} = tan^2\left(45° - \frac{\varphi}{2}\right) \tag{1}$$

Equation 2 calculates the coefficient of active earth pressure (Ka) for sloping ground.

$$K_a = \cos\beta \frac{\cos\beta - \sqrt{cos^2\beta - cos^2\varphi}}{\cos\beta + \sqrt{cos^2\beta - cos^2\varphi}} \tag{2}$$

Where, φ = ground shear angle (°), β = angle of inclination of the ground surface (°). For determine active earth pressure (Pa) using Equation 3..

$$P_a = \frac{1}{2} K_a.\gamma.H^2 \tag{3}$$

Equation 4 used for due to cohesive soil.

$$P_a = \frac{1}{2} K_a.\gamma.H^2 - 2C\sqrt{K_a}.H \tag{4}$$

Due to ground water level is using Equation 5.

$$P_a = \frac{1}{2} K_a.\gamma'.H^2 + \frac{1}{2}\gamma_w.H^2 \tag{5}$$

Due to evenly distributed load is obtained using(6.

$$P_a = \frac{1}{2} K_a.\gamma.H^2 + K_a.q.H \tag{6}$$

Where, γ = weight soil volume (kN/m^3), γ' = weight of the volume of soil submerged in water (kN/m^3), γ_w = specific gravity of water (9,81 kN/m^3), H = soil depth (m), c = soil cohesion (kN/m^2), and q = distributed load (kN/m^2).

3.4. Stability analysis against external forces

As in the stability study of classical retaining structures, the review is always based on stability against external forces, namely stability against shear, overturning and bearing capacity hazards. In this case, it is assumed that the entire construction is a unitary mass of soil that works together.

IOP Conf. Series: Earth and Environmental Science 1184 (2023) 012005 doi:10.1088/1755-1315/1184/1/012005

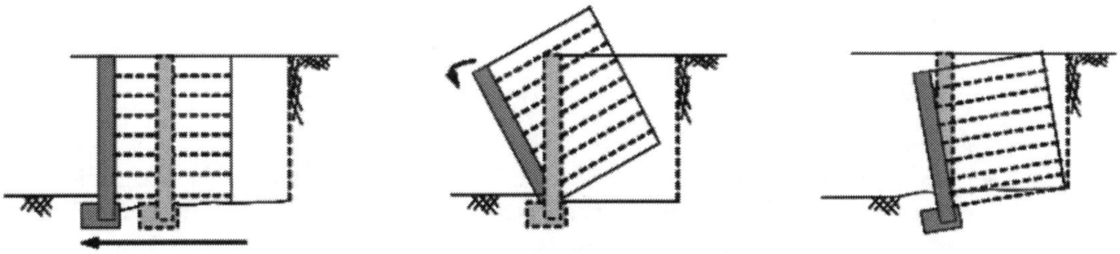

Figure 7. Potential external failure mechanisms for MSE wall[12]

3.4.1. Sliding stability

Sliding stability is defined as the ratio of the resisting force to the pushing force. By comparing the resisting and pushing forces, a factor of safety for shear stability will be obtained. If the analysis is reversed, where we determine the desired safety factor, then the minimum anchorage length will be obtained according to a certain safety factor. Shear stability (sliding) can be calculated by this equation below.

$$SF_{Sliding} = \frac{\sum P_R}{\sum P_D} \geq 1{,}5 \quad (7)$$

$$P_R = (V_1 + V_2 + F_T \sin\beta).\mu \quad (8)$$

$$P_D = F_h = F_T \cos\beta \quad (9)$$

With, P_R = horizontal resistance forces, P_D = horizontal driving forces where μ is the smallest value of (tan ϕ_f, tan ϕ_r or tan ρ).

3.4.2. Overturning stability

Overturning stability is the stability that is reviewed based on the overturned soil condition. The moment that causes overturning is the active thrust of the soil and the load on the center of the roll. While the force that resists overturning is the soil's own load and the passive force of the front of the soil to the center of the roll.

$$SF_{Overturning} = \frac{M_{RO}}{M_O} \geq 2 \quad (10)$$

Where, M_{RO} = working resistance moments, M_O = overturning moments.

3.4.3. Bearing capacity

The subgrade used to support the embankment load has a limited strength called the soil's bearing capacity. The large bearing capacity of the soil can be employed to determine its allowable load for planning an embankment structure or to find out the safe number of a structure in the analysis of a design activity. According to Terzaghi [19], the equation for ultimate bearing capacity and net ultimate capacity can be written as follows:

$$q_u = c.N_c + D_f.\gamma.N_q + 0{,}5.\gamma.B.N_\gamma \quad (11)$$

and

$$q_{un} = c.N_c + D_f.\gamma.(N_q - 1) + 0{,}5.\gamma.B.N_\gamma \quad (12)$$

IOP Conf. Series: Earth and Environmental Science 1184 (2023) 012005 doi:10.1088/1755-1315/1184/1/012005

With, q_u = capacity of ultimate bearing (kN/m^2), q_{un} = capacity of net ultimate bearing (kN/m^2), c = cohesion (kN/m^2), D_f = foundation depth (m), γ = weight volume of soil (kN/m^3), B = foundation width (m), N_c, N_q, N_γ = bearing capacity factor of soil.

3.5. Stability analysis of internal forces

The design process for reviewing internal stability consists of calculating the maximum tensile force that occurs, determining the critical slip surface location and the resistance rendered by the reinforcement material, namely the tensile and pull out capacities [10].

Tensile is the tensile force exerted on the reinforcement (reinforcing bar) as a result of the vertical pressure acting. The working vertical stress must not exceed the tensile strength of the reinforcement so that it does not break which can cause structural failure.

$$SF_{breaking} = \frac{T_{allow}\, strip}{Full\ Force\ Strip} \tag{13}$$

Where, T_{allow} strip = allowable tensile strength of strip.

$$T_{allow} = T_{ult}\left(\frac{1}{RF_{CR}RF_{ID}RF_{D}}\right) \tag{14}$$

With, RF_{CR} = creep reduction factor (1,3), RF_{ID} = installation damage reduction factor (1,2), RF_D = installation durability reduction factor (1,1), $FF_{strip} = total\ load\ x\ S_v\ x\ average\ width\ panel$, and S_v = vertical spacing (m).

The tensile force that occurs in the reinforcement material becomes greater than the pull out resistance needed to extract the reinforcement material from the soil mass. This failure is referred to as a pull out failure.

$$SF_{pullout} = \frac{Full\ Force\ Resistance}{Full\ Force\ Strip} \tag{15}$$

With,

$$FF_{resistance} = 2(\gamma . H) x \tan\delta\ x\ W_{strip}\ x\ \left(L_{design} - L_r\right) x\ S_h \tag{16}$$

Where, W_{strip} = strip thickness (m), S_h = horizontal spacing (m).

3.6. The pseudostatic method of Mononobe Okabe

The Coulomb active earth pressure coefficient equation can be modified to calculate lateral pressure caused by an earthquake. This modification is known as the pseudostatic method of Mononobe Okabe [20].

IOP Conf. Series: Earth and Environmental Science 1184 (2023) 012005 doi:10.1088/1755-1315/1184/1/012005

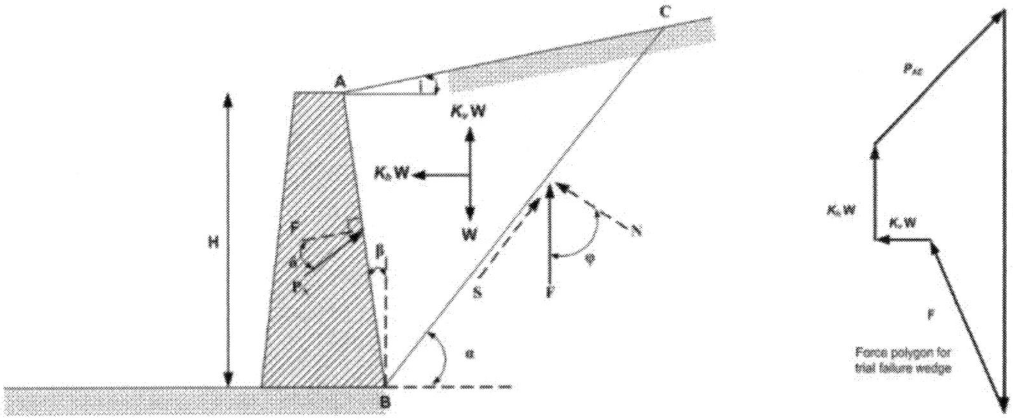

Figure 8. Earthquake forces for external stability analysis [3]

Where the active earth pressure coefficient equation becomes:

$$K_{AE} = \frac{cos^2(\varphi' - \theta - \beta)}{\cos\theta\, cos^2\beta cos(\delta+\beta+\theta)\left[1+\sqrt{\frac{sin(\delta+\varphi')sin(\varphi'-\theta-i)}{cos(\delta+\beta+\theta)cos(\beta-i)}}\right]^2} \quad (17)$$

$$\theta = \tan^{-1}\left(\frac{k_h}{1-k_v}\right) \quad (18)$$

$$P_{AE} = \frac{1}{2}\gamma H^2(1-k_v)K_{AE} \quad (19)$$

Mononabe-Okabe modified approach by considering soil cohesion. In this method, it is assumed that the coefficient of vertical seismic (Kv) = 0, and the coefficient of horizontal seismic (Kh) = PGA which has been adjusted to the site conditions [11].

4. RESULT AND DISCUSSIONS

Area around east side of bridge ramp location is the location that will be used as the MSE WALL construction. Based on the location in the field, which identified the potential for liquefaction at a depth of 0–3 meters below the ground surface, soil replacement is done to a depth of 3 meters to increase the carrying capacity of the soil by using the selected embankment material. The plan for the embankment material also uses selected soil material. The calculation of the loads is carried out in stages in accordance with the implementation procedures in the field which are carried out layer by layer with a height of 7.5 m.

Table 4. Granular Soil Data

Granular Soil Data	
ϕ	29 °
c (cohesion)	4 kN/m2
ϒ (unit weight)	16.00 kN/m3
KAE	0.59
Δ (delta)	19.33
Traffic Load	20 kN/m2
Dead Load	7 kN/m2

IOP Conf. Series: Earth and Environmental Science 1184 (2023) 012005 doi:10.1088/1755-1315/1184/1/012005

Table 5. Properties of MSE Wall

Properties of MSE Wall	
Height Panel (h)	1.44 m
Wall Height (H)	7.56 m
Height of Backfill (H')	0.36 m
Panel Length (l)	1.5 m
Thickness Minimum (t)	0.14 m

Table 6. Friction Tie Specification

Friction Tie Specification	
Type	Polyester strap structure high tenacity polyester coated (high density polyethelene) HDPE resin
Width Dimension (W)	± 9cm
Friction Tie Thickness (I)	5 mm
Friction Tie Tensile Strength	50 kN (Ultimate Tensile Strength)

Table 7. Properties of Friction Tie

Properties of Friction Tie		
Strip	50	kN
T alow strip	29.14	kN
W	0.09	m
L	0.005	m
Sh	0.7	m
Sv	0.72	m
Wall Height (H)	7.56	m
Panel Width (l)	1.5	m
Spacing Load	0.75	m

The findings of the calculation for internal stability are displayed in **Table 8**.

Table 8. Internal Stabiliy Calculation

Wall Height (m)	L design (m)	Pull Force (kN)	Pull Resistance (kN)	Safety Pullout	CEK OK	Safety Breaking	CEK OK
7,56	7	7,75	9,71	1,25	OK	26,33	OK
6,84	6	14,77	17,55	1,19	OK	11,84	OK
6,12	5,5	21,80	25,79	1,18	OK	7,35	OK
5,40	5	28,81	33,41	1,16	OK	5,06	OK
4,68	5	35,81	49,71	1,39	OK	4,07	OK
3,96	5	42,79	69,51	1,62	OK	3,40	OK
3,24	5	49,77	92,83	1,87	OK	2,93	OK
2,52	4,5	56,74	104,14	1,84	OK	2,31	OK
1,80	5	63,72	149,99	2,35	OK	2,29	OK
1,08	5,5	70,69	203,48	2,88	OK	2,27	OK
0,36	6	77,67	264,62	3,41	OK	2,25	OK

Based on **Table 8**, SF value at each review point can be depicted in graphic form as depicted in **Figure 9** and **Figure 10**.

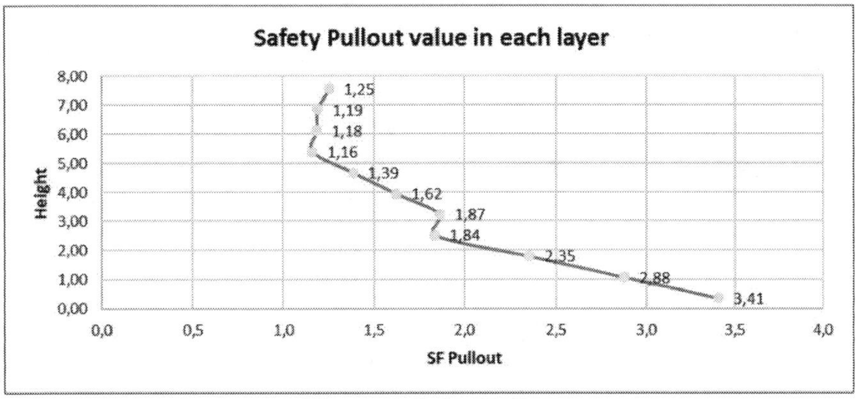

Figure 9. Safety Pullout

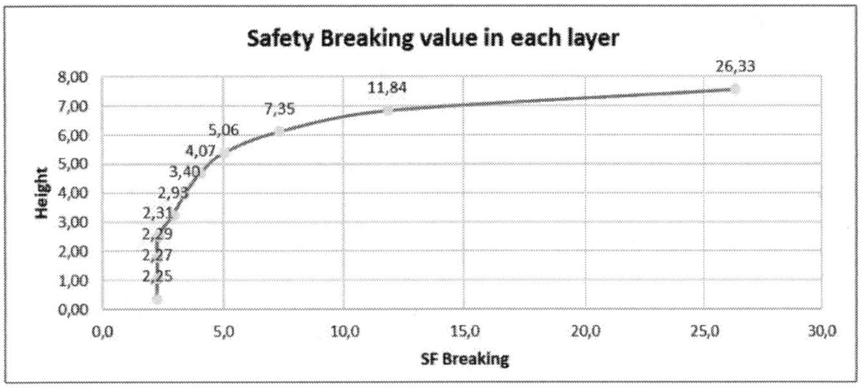

Figure 10. Safety Breaking

The calculation result of external stability for overturning and sliding can be seen in **Table 9**.

Table 9. External Stability Calculation for Overturning and Sliding

Wall Height (m)	Ldesign (m)	Total Pressure Load	Total Force Position	SF Overturning	CEK OK	SF Sliding	CEK OK
7,56	7	7,47	0,0432	437,1211	OK	5,3953	OK
6,84	6	14,24	0,03888	56,1537	OK	7,2775	OK
6,12	5,5	21,02	1,0800	19,1807	OK	7,5328	OK
5,40	5	27,79	2,1168	8,5657	OK	7,2528	OK
4,68	5	34,54	3,4992	5,3605	OK	7,5029	OK
3,96	5	41,27	5,2272	3,6702	OK	7,6739	OK
3,24	5	48,00	7,3008	2,6702	OK	7,7978	OK
2,52	4,5	54,73	9,7200	1,6441	OK	7,1025	OK
1,80	5	61,46	12,4848	1,5949	OK	7,9648	OK
1,08	5,5	68,18	15,5952	1,5563	OK	8,8257	OK
0,36	6	74,91	19,0512	1,5252	OK	9,6854	OK

Based on **Table 9**, SF value for overturning and sliding can be depicted in graphic form as depicted in **Figure 11** and **Figure 12**.

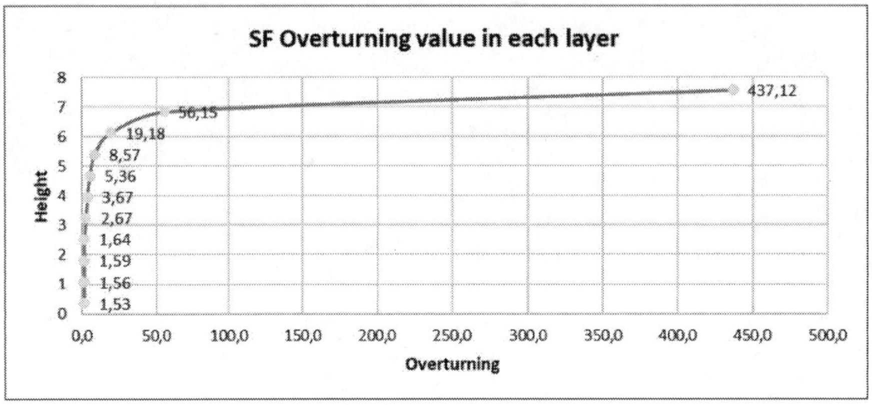

Figure 11. SF Overturning

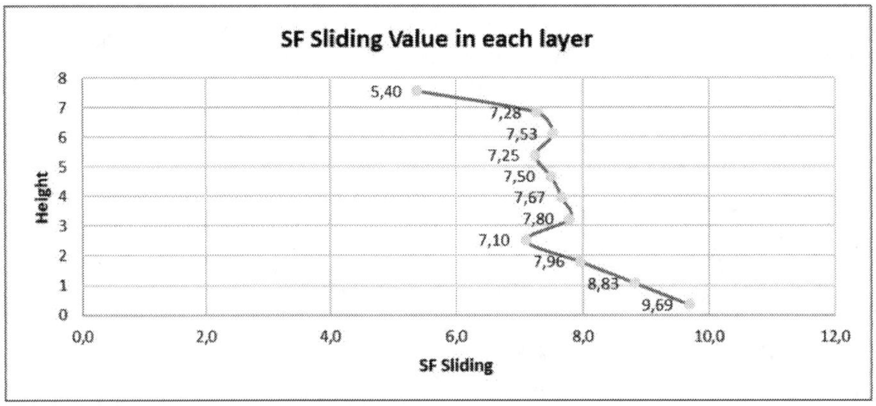

Figure 12. SF Sliding

The calculation result of external stability for bearing capacity can be seen in **Table 10**.

Table 10. Bearing Capacity Calculation

Height (m)	Embankment Load (kPa)	Live Load (kPa)	Dead Load (kPa)	Total Load (kPa)	Total Load <Qall CEK
1	16,00	20	7	43,00	OK
1,5	24,00	20	7	51,00	OK
2	32,00	20	7	59,00	OK
2,5	40,00	20	7	67,00	OK
3	48,00	20	7	75,00	OK
3,5	56,00	20	7	83,00	OK
4	64,00	20	7	91,00	OK
4,5	72,00	20	7	99,00	OK
5	80,00	20	7	107,00	OK
5,5	88,00	20	7	115,00	OK
6	96,00	20	7	123,00	OK
6,5	104,00	20	7	131,00	OK
7	112,00	20	7	139,00	OK
7,5	120,00	20	7	147,00	OK

Based on **Table 10**, SF value for bearing capacity can be depicted in graphic form as shown in **Figure 13**.

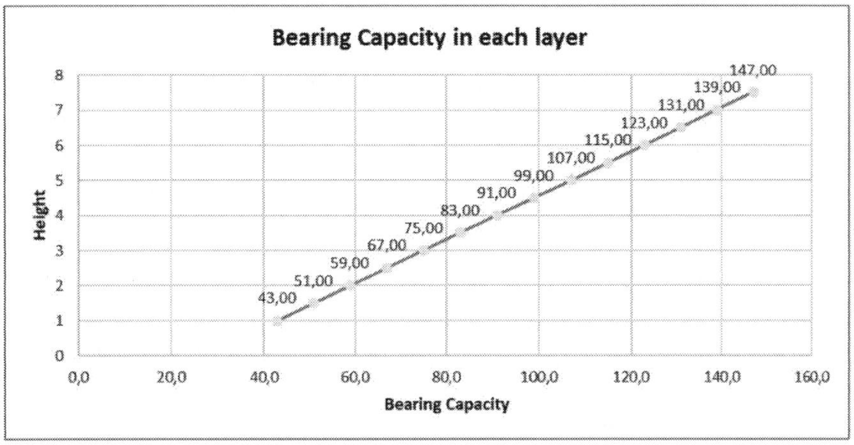

Figure 13. Bearing Capacity

5. Conclusion

The plan for this retaining wall structure will be at the location where the subgrade replacement has been carried out.

The bearing capacity of the subgrade was able to meet the carrying capacity of the embankment soil with a height of 7.5 meters.

The analysis that has been carried out is in the form of internal stability and external stability with the value of the safety number for static conditions > 1.5 and the value with an earthquake > 1.1 in accordance with SNI 8460-2017.

Reinforcement of friction tie using granular embankment gets good results in the calculation of safety factor breaking, pull out, overturning and sliding.

Acknowledgment

The authors would like to specify their true appreciation to the Ministry of Public Works and Housing and Gadjah Mada University for funding this research.

References

[1] Natawidjaja D, 2021 Uji Paritan Sesar Aktif Opak di Jembatan Kretek II , Yogyakarta p. 1–36.
[2] Galag C T S A Faris F and Adi A D, 2022 A Study Of Liquefaction Potential On Retaining Wall And Embankment In Parangtritis Village in 5th International Conference on Earthquake and Disaster Mitigation.
[3] Kramer S L, 1996 Geotechnical Earthquake Engineering .
[4] Tsuchida H, 1970 Prediction and Countermeasure against Liquefaction in Sand Deposit Abstr. Semin. Port Harb. Res. Institute, Minist. Transp. Yokosuka, Japan.
[5] Idriss I M and Boulanger R W, 2008 Soil Liquefaction During Earthquakes 160, 4057 .
[6] Das B, 2006 Principles of Geotechnical Engineering 7th .
[7] Das B, 2008 Fundamentals of Geotechnical Engineering Third Edition Third Edit United States.
[8] Sito Ismanti K B S, 2012 Analisis Perilaku Timbunan Dengan Perkuatan Geosintetik Menggunakan Software Plaxis Univ. Gadjah Mada.
[9] Muteb H H and Falah M W, 2021 Mechanically Stabilized Earth MSE Walls Applications: Review IOP Conf. Ser. Earth Environ. Sci. 856, 1.
[10] AASHTO, 2007 AASHTO LRFD Bridge Design Specifications: SI Unit 4th Edition 2007 .
[11] Badan Standarisasi Nasional Indonesia, 2017 Persyaratan Perancangan Geoteknik SNI 8460-2017 8460 .

[12] Federal High Way Administration (FHWA), 2009 Design and Construction of Mechanically Stabilized Earth Walls and Reinforced Soil Slopes–Volume I I, November .
[13] Badan Standardisasi Nasional Indonesia, 2016 Perencanaan Jembatan Terhadap Beban Gempa SNI 2833:2016 p. 1–70.
[14] Standard B, 2010 BSI Standards Publication Code of Practice for Strengthened / Reinforced Committee ISBN 978 0 580 53842 1 .
[15] Hardiyatmo H C, 2017 Geosintetik Untuk Perkerasan Jalan Raya Kedua Yogyakarta: Gadjah Mada University Press.
[16] Hardiyatmo H C, 2020 Analisis dan Perancangan Fondasi I Keempat Yogyakarta: Gadjah Mada University Press.
[17] Karya W, 2021, Laporan Evaluasi Data Teknis Acuan Pelaksanaan.
[18] Hardiyatmo H C, 2018 Mekanika Tanah II Keenam Yogyakarta: Gadjah Mada University Press.
[19] Terzaghi K Peck B R and Mesri G, 1996 Soil Mechanics in Engineering Practice.pdf .
[20] Yazdani M Azad A Farshi A and Talatahari S, 2013 Extended "' Mononobe-Okabe '" Method for Seismic Design of Retaining Walls 2013.

IOP Conf. Series: Earth and Environmental Science 1184 (2023) 012006 doi:10.1088/1755-1315/1184/1/012006

Influence of Groundwater Table Fluctuation on Liquefaction Potential Analysis Using Cyclic Stress Approach

W Dwiyantoro[1,2], T F Fathani[1] and A D Adi[1]

[1] Department of Civil and Environmental Engineering, Universitas Gadjah Mada, Daerah Istimewa Yogyakarta 55284, Indonesia.
[2] Directorate General of Human Settlements, Ministry of Public Works and Housing, Jakarta 12110, Indonesia.

Email: wahyu.dwiyantoro@mail.ugm.ac.id

Abstract. In addition to earthquakes, groundwater table (GWT) also contributes to uncertainty in liquefaction analysis, so it needs to be considered. This research was conducted by analyzing the liquefaction potential with a cyclic stress approach using standard penetration test data and cone penetration test data with varied GWT in a case study of building design in Sleman, Yogyakarta Special Region. The liquefaction safety factor (FS_L) was calculated to obtain the liquefaction potential index (LPI) and liquefaction severity index (LSI). Based on the survey results, the GWT ranges from -6 m to -8 m. Referring to the FS_L, there are several soil layers that have liquefaction potential at all GWT variations. The shallower the GWT, the greater the LPI and LSI values at the same point will be. Changes in GWT from -8 m to -6 m cause changes in the LPI and LSI classifications at several points. Based on LPI classification, the location belongs to high to very high category at -8 m and -7 m GWT and very high category at -6 m GWT. Based on LSI classification, the location belongs to low category at -8 m GWT, and low to moderate category at -7 m and -6 m GWT.

1. Introduction

Building design needs to include an analysis of liquefaction potential, especially if it is built in a location that is indicated to have liquefaction potential based on historical records and geological conditions. In Indonesia, the initial indication refers to the Liquefaction Vulnerability Map from the Geological Agency of the Ministry of Energy and Mineral Resources (2019) [1]. Liquefaction is one of the earthquakes induced phenomena that can take place in saturated cohesionless soils [2]. Besides the earthquake, the groundwater table (GWT) also contributes to liquefaction potential analysis. The fluctuation of GWT level provides an uncertainty in liquefaction potential analysis. Therefore, the effect of GWT fluctuation needs to be taken into account in the liquefaction potential analysis. In this study, liquefaction potential analysis was carried out at a building design site in Sleman, Yogyakarta Special Region, whose location included moderate vulnerability according to the Liquefaction Vulnerability Map (figure 1), by using GWT fluctuations data based on actual conditions. Soil investigation data used are standard penetration test (SPT) and cone penetration test (CPT) that is shown in figure 1. The effect of GWT fluctuations on the liquefaction potential analysis was measured using the liquefaction safety factor (FS_L), liquefaction potential index (LPI), and liquefaction severity index (LSI) parameters.

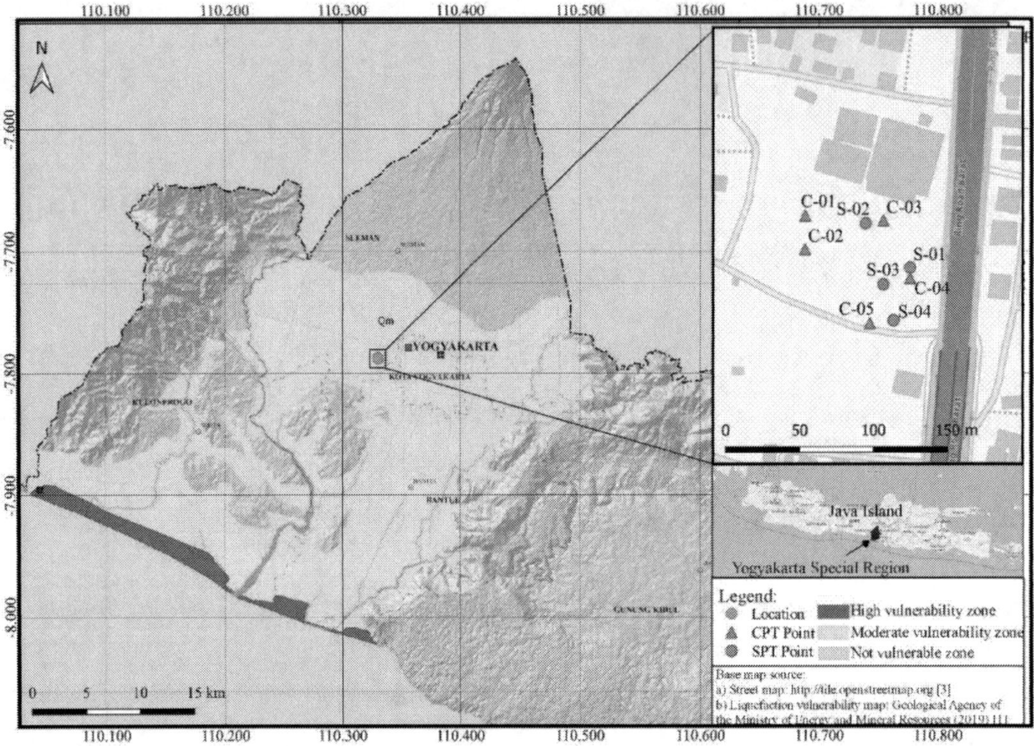

Figure 1. Research location and soil test distribution in the Yogyakarta Special Region [3,11].

2. Method

Liquefaction potential analysis used a cyclic stress approach with a simplified method of Idriss and Boulanger (2008) [4] on 4 SPT data and 5 CPT data. Each of the soil investigation data is far from each other and none of them coincides. Groundwater table was obtained from measurements and historical records at several points around the site. The surface peak ground acceleration (PGA) which is used as input for analysis is calculated from the bedrock PGA multiplied by the amplification factor according to the site class based on the Indonesian seismic code, SNI 1726:2019 [5]. The earthquake source for determining the moment magnitude is used source which has the most dominant influence on the location based on the results of disaggregation. The output of the analysis based on the cyclic stress approach in the form of cyclic resistance ratio (CRR) and cyclic stress ratio (CSR) is then compared to produce the value of the safety factor against liquefaction (FS_L). The CSR value is obtained from the earthquake loading characterization which is calculated on each soil layer, so it does not depend on the SPT and CPT values. The CRR value shows the available cyclic resistance in each soil layer which can be calculated from the SPT-based and CPT-based methods. Formulas of CRR at M_w 7.5 ($CRR_{7.5}$) for the SPT-based and CPT-based of Idriss and Boulanger (2008) [4] methods are shown in equation 1 and equation 2, respectively.

$$\mathrm{CRR}_{7.5} = \exp\left(\frac{(N_1)_{60cs}}{14.1} + \left(\frac{(N_1)_{60cs}}{126}\right)^2 - \left(\frac{(N_1)_{60cs}}{23.6}\right)^3 + \left(\frac{(N_1)_{60cs}}{25.4}\right)^4 - 2.8 \right) \tag{1}$$

$$\mathrm{CRR}_{7.5} = \exp\left(\frac{q_{cNcs}}{540} + \left(\frac{q_{cNcs}}{67}\right)^2 - \left(\frac{q_{cNcs}}{80}\right)^3 + \left(\frac{q_{cNcs}}{114}\right)^4 - 3 \right) \tag{2}$$

The SPT-based uses the equivalent clean sand corrected standard penetration resistance parameter, $(N_1)_{60cs}$ based on the corrected standard penetration dan fines content. The CPT-based uses the equivalent clean sand value of the corrected tip resistance parameter, q_{cNcs} based on the corrected CPT tip resistance dan fines content. The FS_L values are further processed to produce LPI and LSI values. Each LPI and LSI value are classified according to their category. LPI analysis and its classification were using the method Sonmez (2003) [6]. LSI analysis and its classification were using the method Sonmez and Gokceoglu (2005) [7]. Each data was analysed using the GWT which varied according to the measurement results and historical records obtained. Ground water table affects soil pore water pressure. The changes in GWT depth will affect the pore water pressure value at a point in a soil layer. At soil layers that are initially saturated, the shallower the groundwater table, the higher the pore water pressure. Under static conditions, the pore water pressure affects the effective soil stress value. The effective soil stress value in static conditions affects the value of CSR and CRR. The results of the analysis on each variation of the depth of the GWT are then compared to each other on each of the data and parameters reviewed.

3. Result and Discussion

3.1. Site class, peak ground acceleration and moment magnitude determination

Site class determination according to Indonesian seismic code, SNI 1726:2019 can use N-SPT data, shear wave velocity, and undrained shear strength averaged to a depth of 30 m [5]. The N-SPT and CPT data used in this study can be seen in figure 2 and figure 3 respectively. Neither N-SPT nor CPT data reaches a depth of 30 m, so they cannot be used for site class determination. Therefore, secondary data is used in the form of shear wave velocity in 30 m depth average (v_{s30}) from Muzli *et al.* (2016) [8]. The value of v_{s30} at the research site is in the range of 230 m/s - 290 m/s based on Muzli *et al.* (2016) [8]. According to the code [5], this site belongs to site class D.

The bedrock PGA value based on PGA map of Maximum Considered Earthquake Geometric Mean at Bedrock from the code [5] is 0.48 g. The amplification factor value for site class D at PGA 0.48 g is 1.116. The value of surface PGA (PGA_M) is calculated as 1.116 × 0.48 g = 0.535 g. The moment magnitude value is obtained from the largest value from the most dominant earthquake source in the research location. Based on the results of disaggregation carried out by Sunardi (2015) [9], the Opak fault is the dominant earthquake source for the Yogyakarta location. The moment magnitude value is determined from the largest value according to the Indonesian Earthquake Hazard Map and Source book [10], which is M_w 6.6. The PGA_M value and moment magnitude are then used to input CSR analysis.

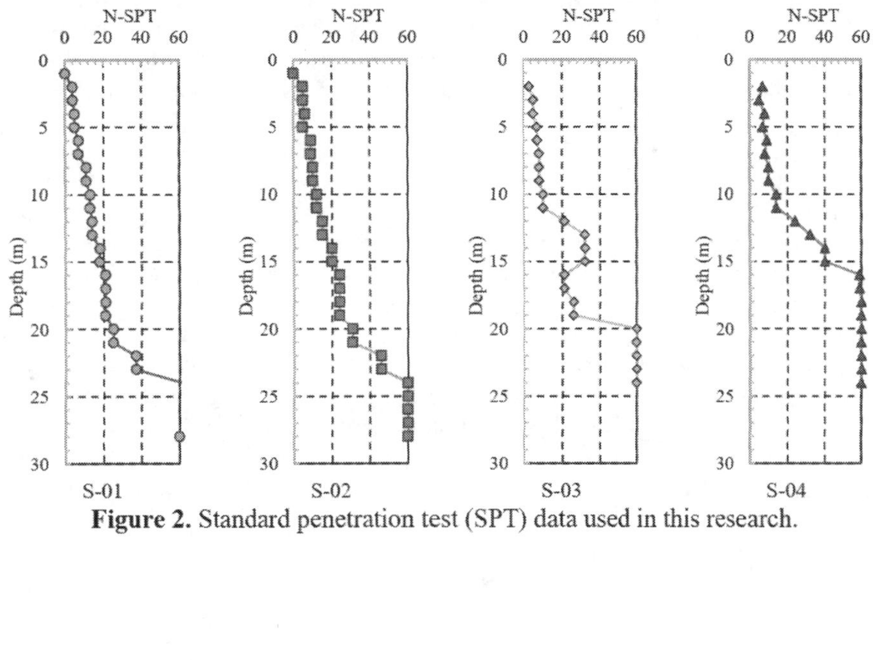

Figure 2. Standard penetration test (SPT) data used in this research.

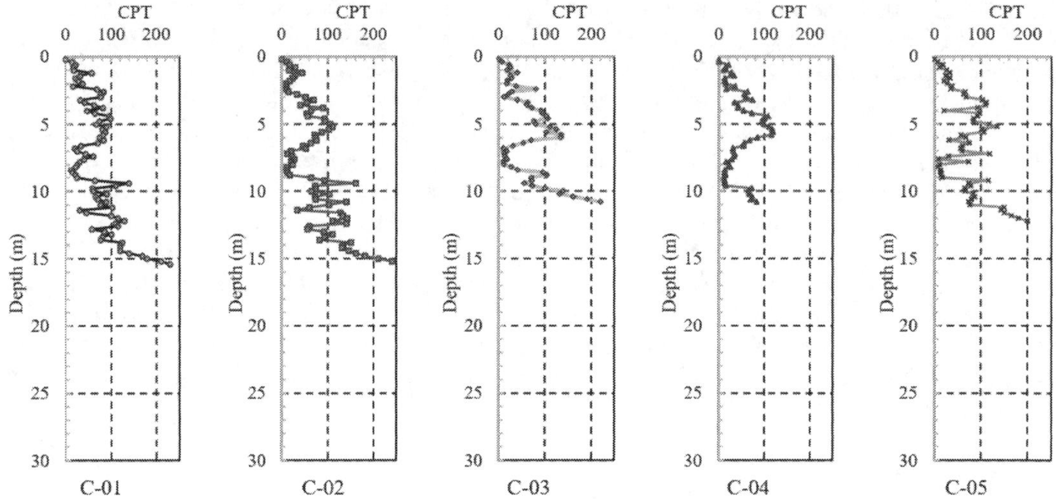

Figure 3. Cone penetration test (CPT) data used in this research.

3.2. Groundwater table fluctuation data

Groundwater table data were obtained from direct measurements at several wells around the soil investigation point. The value of fluctuations is obtained from residents' information based on each well's historical records. Based on direct measurements at the soil investigation location, the depth of the GWT is -7 m. Based on historical records, the depth of the water level around the soil investigation location was recorded to have raised 1 m and lowered 1 m from the existing condition. Thus, the GWT depths used for further analysis are -6 m, -7 m, and -8 m.

3.3. Liquefaction safety factor

The liquefaction safety factor is used to assess the liquefaction potential at each soil layer. Liquefaction occurs when the safety factor value is less than 1.0 (Kramer, 1996) [2]. The liquefaction potential analysis was only carried out on saturated granular soils, so the safety factor against liquefaction in unsaturated granular soils and cohesive soils was uncalculated and not available (n.a.). The analysis was carried out on each SPT and CPT data with the depth of the varied GWT according to GWT fluctuation conditions. One of the simplified method limitations on the cyclic stress approach is the calculation of the CSR value is only limited to a depth of 20 m. This is because the value of the stress reduction coefficient in CSR analysis only applies to a depth of 20 m [4]. Therefore, the analysis is limited to a depth of 20 m. The liquefaction safety factor values from CPT-based and SPT-based method are presented in table 1 and table 2 respectively. All soil layers are sand, except for the 1 m clay layer at surface. The CPT-based method safety factor analysis is carried out every 0.2 m depth interval. In table 1, the FS_L is only shown per 1 m depth interval for a depth of less than 10 m. For the analysis of LPI and LSI on the CPT-based method using data with a depth interval of 0.2 m.

Based on the table 1 and table 2, it is known that the upper limit of the liquefied layer depends on the GWT depth, because the layer at a depth of more than 6 m has soil properties that is susceptible to liquefy. So, the deeper the GWT, the deeper of the upper boundary of the liquefied layer. The deeper the GWT, the higher the safety factor at a certain depth. This is due to the lower initial pore water pressure conditions in the deeper GWT. The research by [11] showed that, in addition to the initial effective vertical stress, the initial pore water pressure also affects the value of excess pore water pressure. The smaller the initial pore water pressure, the smaller the excess pore water pressure value. At a certain depth, the smaller the excess pore water pressure, the smaller the ratio between the pore water pressure and the initial effective vertical stress. So, the possibility of liquefaction at certain depth is also getting smaller.

The results of the CPT-based analysis in table 1 did not reach a depth of 20 m. The CPT test data did not reach 20 m due to the test was stopped at each final depth in each data because the cone resistance value had reached 250 kg/cm^2. The safety factor tendency and characteristics in CPT-based result are the same as SPT-based result.

Table 1. Liquefaction safety factor (FS_L) from CPT-based method with varied GWT.

Depth (m)	C-01			C-02			C-03			C-04			C-05		
	-6 m	-7 m	-8 m	-6 m	-7 m	-8 m	-6 m	-7 m	-8 m	-6 m	-7 m	-8 m	-6 m	-7 m	-8 m
1.0	n.a.	n.a.	n.a.	n.a.	n.a.	n.a.	n.a.	n.a.	n.a.	n.a.	n.a.	n.a.	n.a.	n.a.	n.a.
2.0	n.a.	n.a.	n.a.	n.a.	n.a.	n.a.	n.a.	n.a.	n.a.	n.a.	n.a.	n.a.	n.a.	n.a.	n.a.
3.0	n.a.	n.a.	n.a.	n.a.	n.a.	n.a.	n.a.	n.a.	n.a.	n.a.	n.a.	n.a.	n.a.	n.a.	n.a.
4.0	n.a.	n.a.	n.a.	n.a.	n.a.	n.a.	n.a.	n.a.	n.a.	n.a.	n.a.	n.a.	n.a.	n.a.	n.a.
5.0	n.a.	n.a.	n.a.	n.a.	n.a.	n.a.	n.a.	n.a.	n.a.	n.a.	n.a.	n.a.	n.a.	n.a.	n.a.
6.0	0.57	n.a.	n.a.	0.5	n.a.	n.a.	0.48	n.a.	n.a.	0.46	n.a.	n.a.	0.27	n.a.	n.a.
7.0	0.28	0.31	n.a.	0.2	0.22	n.a.	0.19	0.22	n.a.	0.25	0.27	n.a.	0.76	0.81	n.a.
8.0	0.20	0.22	0.24	0.2	0.20	0.22	0.21	0.23	0.25	0.20	0.22	0.24	0.17	0.19	0.21
9.0	0.31	0.33	0.35	0.4	0.46	0.49	0.34	0.36	0.39	0.17	0.19	0.20	0.58	0.61	0.64
10.0	0.40	0.42	0.44	0.5	0.48	0.50	0.63	0.65	0.68	0.30	0.31	0.33	0.36	0.38	0.40
10.2	0.29	0.30	0.32	0.3	0.33	0.35	0.98	0.99	1.01	0.32	0.33	0.35	0.37	0.39	0.41
10.4	0.30	0.32	0.34	0.3	0.33	0.35	2.00	1.98	1.93	0.30	0.32	0.34	0.34	0.36	0.38
10.6	0.38	0.39	0.41	0.7	0.70	0.72	2.00	2.00	2.00	0.35	0.37	0.38	0.31	0.32	0.34
10.8	0.32	0.33	0.35	0.4	0.45	0.47	2.00	2.00	2.00	0.34	0.35	0.37	0.32	0.33	0.35
11.0	0.42	0.44	0.46	0.3	0.27	0.29	-	-	-	0.30	0.32	0.33	0.75	0.77	0.79
11.2	0.18	0.20	0.21	0.2	0.20	0.21	-	-	-	1.06	1.06	1.07	0.67	0.69	0.71
11.4	0.21	0.23	0.24	0.5	0.56	0.59	-	-	-	0.64	0.66	0.68	0.72	0.74	0.76
11.6	0.40	0.42	0.44	0.6	0.59	0.61	-	-	-	0.57	0.59	0.61	0.90	0.91	0.93
11.8	0.47	0.49	0.51	0.6	0.64	0.66	-	-	-	0.52	0.54	0.56	1.20	1.20	1.20
12.0	0.55	0.57	0.59	0.4	0.45	0.47	-	-	-	0.32	0.34	0.35	2.00	1.99	1.93
12.2	0.43	0.45	0.47	0.6	0.62	0.64	-	-	-	0.43	0.45	0.47	2.00	2.00	2.00
12.4	0.45	0.47	0.49	0.2	0.26	0.27	-	-	-	0.73	0.75	0.76	-	-	-
12.6	0.23	0.25	0.26	0.2	0.24	0.25	-	-	-	1.13	1.12	1.12	-	-	-
12.8	0.32	0.33	0.35	0.3	0.35	0.37	-	-	-	1.91	1.83	1.77	-	-	-
13.0	0.38	0.39	0.40	0.4	0.43	0.45	-	-	-	2.00	2.00	2.00	-	-	-
13.2	0.30	0.31	0.32	0.3	0.35	0.37	-	-	-	-	-	-	-	-	-
13.4	0.28	0.29	0.31	0.3	0.31	0.33	-	-	-	-	-	-	-	-	-
13.6	0.48	0.49	0.51	0.6	0.64	0.66	-	-	-	-	-	-	-	-	-
13.8	0.45	0.46	0.48	0.5	0.53	0.54	-	-	-	-	-	-	-	-	-
14.0	0.45	0.46	0.47	0.5	0.51	0.52	-	-	-	-	-	-	-	-	-
14.2	0.44	0.46	0.47	0.6	0.59	0.61	-	-	-	-	-	-	-	-	-
14.4	0.54	0.56	0.57	0.7	0.70	0.71	-	-	-	-	-	-	-	-	-
14.6	0.79	0.80	0.81	0.9	0.93	0.93	-	-	-	-	-	-	-	-	-
14.8	0.92	0.93	0.93	1.9	1.84	1.78	-	-	-	-	-	-	-	-	-
15.0	1.94	1.85	1.79	2.0	2.00	2.00	-	-	-	-	-	-	-	-	-
15.2	2.00	2.00	2.00	2.0	2.00	2.00	-	-	-	-	-	-	-	-	-
15.4	2.00	2.00	2.00	-	-	-	-	-	-	-	-	-	-	-	-

IOP Conf. Series: Earth and Environmental Science 1184 (2023) 012006 doi:10.1088/1755-1315/1184/1/012006

Table 2. Liquefaction safety factor (FS_L) from SPT-based method with varied GWT.

Depth (m)	S-01			S-02			S-03			S-04		
	-6 m	-7 m	-8 m	-6 m	-7 m	-8 m	-6 m	-7 m	-8 m	-6 m	-7 m	-8 m
1	n.a.	n.a.	n.a.	n.a.	n.a.	n.a.	n.a.	n.a.	n.a.	n.a.	n.a.	n.a.
2	n.a.	n.a.	n.a.	n.a.	n.a.	n.a.	n.a.	n.a.	n.a.	n.a.	n.a.	n.a.
3	n.a.	n.a.	n.a.	n.a.	n.a.	n.a.	n.a.	n.a.	n.a.	n.a.	n.a.	n.a.
4	n.a.	n.a.	n.a.	n.a.	n.a.	n.a.	n.a.	n.a.	n.a.	n.a.	n.a.	n.a.
5	n.a.	n.a.	n.a.	n.a.	n.a.	n.a.	n.a.	n.a.	n.a.	n.a.	n.a.	n.a.
6	0.48	n.a.	n.a.	0.56	n.a.	n.a.	0.48	n.a.	n.a.	0.56	n.a.	n.a.
7	0.42	0.46	n.a.	0.49	0.53	n.a.	0.45	0.50	n.a.	0.45	0.50	n.a.
8	0.51	0.55	0.59	0.47	0.51	0.55	0.41	0.44	0.48	0.47	0.51	0.55
9	0.49	0.52	0.55	0.45	0.48	0.52	0.39	0.42	0.45	0.45	0.48	0.52
10	0.52	0.55	0.58	0.49	0.52	0.55	0.42	0.45	0.48	0.55	0.59	0.62
11	0.49	0.52	0.55	0.46	0.49	0.51	0.40	0.43	0.45	0.52	0.55	0.58
12	0.50	0.53	0.55	0.54	0.56	0.59	0.83	0.86	0.88	1.15	1.17	1.19
13	0.48	0.50	0.53	0.51	0.54	0.56	2.00	2.00	2.00	2.00	2.00	2.00
14	0.60	0.62	0.64	0.69	0.71	0.74	2.00	2.00	2.00	2.00	2.00	2.00
15	0.58	0.60	0.62	0.67	0.69	0.71	2.00	2.00	2.00	2.00	2.00	2.00
16	0.69	0.71	0.73	0.90	0.91	0.92	0.69	0.70	0.72	2.00	2.00	2.00
17	0.67	0.69	0.71	0.86	0.87	0.88	0.67	0.68	0.70	2.00	2.00	2.00
18	0.66	0.67	0.69	0.82	0.84	0.85	0.98	0.99	1.00	2.00	2.00	2.00
19	0.64	0.66	0.67	0.80	0.81	0.82	0.94	0.95	0.96	2.00	2.00	2.00
20	0.84	0.85	0.86	1.76	1.73	1.70	2.00	2.00	2.00	2.00	2.00	2.00

3.4. Liquefaction potential index

The LPI value is used to indicate the potential of liquefaction at a site [12] by considering liquified layer from the surface to a depth of 20 m. LPI is calculated based on the value of the liquefaction safety factor from the results of the previous analysis with a depth-dependent function. The LPI value is calculated up to a depth of 20 m, but the CPT data does not reach a depth of 20 m. The value of the safety factor at the lowest layer in each CPT data based on table 1 reaches 2.0. For a safety factor value greater than 1.2, it does not contribute to the addition of the LPI value. Thus, so that the LPI value in the CPT-based method can be calculated, then for each layer below the lowest CPT-test layer depth it is assumed that liquefaction does not occur or has a safety factor equal to 2. So, this assumption will give no contribution to the addition of the LPI value. This assumption is also used for LSI calculation from the CPT-based liquefaction safety factor. The CPT-based method LPI value is not used for primary use, but it is used as an additional consideration. This assumption will still be valid if the soil layer that is not CPT-tested above 20 m is denser than the layer above it. In other words, the soil layer has better parameters to against liquefaction than the layer above it. The LPI values from SPT-based and CPT-based method and their classification can be seen in table 3 and table 4 respectively.

Based on table 3 and table 4, it is known that from both SPT-based and CPT-based method results, the lower the GWT, the lower the LPI value. The difference in LPI value for each 1 m GWT difference is about 4 to 5. It contributed to the accumulated safety factor difference that takes into account in LPI calculation. At several points, a 1-m-change in the GWT, in a change in the LPI classification. This is because the LPI value in the previous category is at the upper or lower limit of the range. From the SPT-based method, the location belongs to the high to very high category at -8 m and -7 m GWT and very high at -6 m GWT.

Table 3. Liquefaction potential index (LPI) using FS_L from SPT-based method.

GWT (m)	LPI	S-01	S-02	S-03	S-04
-6	value	25.53	23.54	21.52	17.18
	category	very high	very high	very high	very high
-7	value	20.59	19.19	16.85	13.17
	category	very high	very high	very high	high
-8	value	15.98	15.10	12.77	9.17
	category	very high	very high	high	high

Table 4. Liquefaction potential index (LPI) using FS_L from CPT-based method.

GWT (m)	LPI	C-01	C-02	C-03	C-04	C-05
-6	value	28.18	26.87	17.03	24.39	20.40
	category	very high	very high	very high	very high	very high
-7	value	23.13	21.93	11.89	19.40	15.62
	category	very high	very high	high	very high	very high
-8	value	18.12	16.53	6.61	14.24	11.34
	category	very high	very high	high	high	high

3.5. Liquefaction severity index

The liquefaction severity index (LSI) is used to evaluate the liquefaction susceptibility [7]. LSI values from SPT-based and CPT-based method and their classification can be seen in table 5 and table 6 respectively.

Table 5. Liquefaction severity index (LSI) using FS_L from SPT-based method.

GWT (m)	LSI	S-01	S-02	S-03	S-04
-6	value	49.19	47.48	39.82	33.85
	category	moderate	moderate	moderate	low
-7	value	41.85	40.39	32.68	26.96
	category	moderate	moderate	low	low
-8	value	34.89	33.68	26.09	20.32
	category	low	low	low	low

Table 6. Liquefaction severity index (LSI) using FS_L from CPT-based method.

GWT (m)	LSI	C-01	C-02	C-03	C-04	C-05
-6	value	41.85	40.63	25.05	34.74	31.11
	category	moderate	moderate	low	low	low
-7	value	35.09	33.80	18.21	27.96	24.24
	category	moderate	low	low	low	low
-8	value	28.67	27.31	11.77	21.58	18.25
	category	low	low	very low	low	low

Same as LPI, the LSI value is also calculated until 20 m depth, but the CPT data doesn't reach 20 m depth. The LSI value using FS_L from CPT-based method is also use assumption as well as LPI calculation. The CPT-based LSI value is used only for additional consideration, not for primary use.

According to table 5 and table 6, the lower the GWT, the lower the LSI value. The difference in LSI value for each 1 m GWT difference is about 6 to 7. At several points, a 1 m change in the GWT, in a change in the LSI classification. From the SPT-based method, the location belongs to the low category at -8 m GWT and low to moderate at -7 m and -6 m GWT. According to [13], liquefaction susceptibility can also be measured from the GWT depth. The shallower the groundwater table, the greater the susceptibility of liquefaction. In this study also tends to be the same. The factor of safety value decreases at each layer when the GWT elevation rises. When the factor of safety value decreases, the LPI and LSI values increase. The greater the LPI and LSI values, the greater the liquefaction potential and liquefaction susceptibility.

4. Conclusion

Based on the analysis results, it can be inferred that there is liquefaction potential at various depths starting from the GWT depth. The deeper the GWT is, the deeper the upper boundary of the liquefied layer will be. The deeper the GWT is, the higher the liquefaction safety factor at a certain depth will be. The LPI value changes for each applied groundwater table depth review. The lower the GWT is, the lower the LPI value will be. According to the SPT-based method, the location belongs to the high to very high LPI category at -8 m and -7 m GWT and very high LPI category at -6 m GWT. Similar to the LPI, the LSI value also changes every time the water level review is applied. The lower the GWT is, the lower the LSI value will be. According to the SPT-based method, the location belongs to the low LSI category at -8 m GWT and low to moderate LSI category at -7 m and -6 m GWT. Both the LPI and LSI value is calculated until 20 m depth, with the CPT data not reaching 20 m depth, so the CPT-based method is only for additional consideration, not for primary use.

5. References

[1] Ministry of Energy and Mineral Resources 2019 *Atlas Zona Kerentanan Likuefaksi Nasional* (Geological Agency, Bandung, Indonesia)

[2] Kramer S L 1996 *Geotechnical Earthquake Engineering* (Prentice-Hall, Inc., United States)

[3] OpenStreetMap, accessed on November 2022, http://tile.openstreetmap.org

[4] Idriss I M and Boulanger R W 2008 *SPT-Based Liquefaction Triggering Procedures* (Earthquake Engineering Research Institute, USA)

[5] Badan Standardisasi Nasional 2019 *SNI* 1726:2019 *Tata cara perencanaan ketahanan gempa untuk struktur bangunan gedung dan nongedung* (Badan Standardisasi Nasional, Jakarta, Indonesia)

[6] Sonmez H 2003 Modification of the liquefaction potential index and liquefaction susceptibility mapping for a liquefaction-prone area (Inegol, Turkey) *Environmental Geology* **44**(7), 862-871

[7] Sonmez H and Gokceoglu C 2005 A Liquefaction Severity Index Suggested to Engineering Practice *Environmental Geology*. **48**, 81-89

[8] Muzli M, Mahesworo R P, Madijono R, Siswoyo, Pramono S, Dewi K R, Budiarta, Sativa O, Sulistyo B, Swatikarani R, Oktavia N, Moehajirin, Efendi N, Wijaya T A, Subadyo B, Mujianto and Suwarto 2016 Mapping of v_{s30} using MASW Method for Yogyakarta Region *Jurnal Meteorologi dan Geofisika*. **17**, 25-32

[9] Sunardi B 2015 Percepatan Tanah Sintetis Kota Yogyakarta Berdasarkan Deagregasi Bahaya Gempa *Jurnal Lingkungan dan Bencana Geologi*. **6**(3), 211-228

[10] National Center of Earthquake Studies 2017 *Earthquake Source and Hazard Maps* (Research and Development Agency of the Ministry of Housing and Public Works, Bandung, Indonesia)

[11] Saha P, Horikoshi K, and Takahashi A 2020 Performance of sheet pile to mitigate liquefaction-induced lateral spreading of loose soil layer under the embankment *Soil Dynamics and Earthquake Engineering*. **139**, 1-14

[12] Iwasaki T, Tokida K, and Tatsuoka F 1981 Soil Liquefaction Potential Evaluation with Use of the Simplified Procedure *International Conferences on Recent Advances in Geotechnical Earthquake Engineering and Soil Dynamics*. **12**, 209-214

[13] Youd Y L and Hoose S N Liquefaction Susceptibility and Geologic Setting *Proceedings 6th World Conference on Earthquake Engineering, New Delhi, India*. **6**, 37-42

Acknowledgement
The authors would like to express sincerely gratitude for Ministry of Public Works and Housing and *Balai Prasarana Permukiman Wilayah Daerah Istimewa Yogyakarta* which have support the data and all parts of this article.

Axial and lateral bearing capacity assessment of bored piles on medium-dense sand and liquefiable potential based on numerical simulation

A Zakariya[1,2], A Rifa'i[1,*], S Ismanti[1], M. S. Hidayat[2]

[1]Department of Civil and Environmental Engineering, Universitas Gadjah Mada, Sleman, Indonesia, 55281
[2]Directorat General of Highways, Ministry of Public Works and Housing, Jakarta, Indonesia, 12110

*Corresponding email: ahmad.rifai@ugm.ac.id

Abstract. Liquefaction on bridges can cause structural failure due to loss of soil strength. The Showa Bridge that collapsed in 1964 Niigata earthquake was caused by buckling in piles as the soil layer liquefied and friction losses. In addition, there was a case of Palu IV bridge collapse in Indonesia during the recent 2018 Palu earthquake, which indicated have been affected by earthquakes and liquefaction. This study examines the reliability of the bored pile foundation of the Kretek 2 Bridge in Bantul Regency, Yogyakarta Special Region Province, for axial and lateral bearing capacity. This research method uses the soil structure interaction and 3D numerical simulation by Midas Civil and Midas GTS NX pile modeling. The pile driving analysis and lateral static loading test were also conducted, which can be compared to MIDAS results. For the foundation modeling result, there is an increase in axial and moment forces but a decrease in shear forces during liquefaction. While the lateral displacement increases significantly in the liquefaction state, foundation performance is still stable. Therefore, the foundation of the kretek 2 bridge is sufficient to support axial and lateral load both in a static state and a liquefaction state.

Keywords: liquefaction, bored pile foundation, soil structure interaction, 3D numerical simulation, pile modeling.

1. Introduction

One of the earthquake events in which the bridge foundation lost its load-bearing capacity due to liquefaction namely Niigata earthquake in 1964 and Palu earthquake in 2018, which led to the Showa Bridge collapse in Japan and the Palu IV Bridge collapse in Indonesia, respectively. According to Dash et al (2010), due to lateral loading and the addition of free piles, Showa Bridge collapses when the upper soil layer is liquefied, generating significant moment forces that lead to buckling failure [1]. Meanwhile, Imran et al (2019) concluded that Palu IV bridge failed in the bridge-supporting bearing system due to the near-fault effect [2]. After Palu earthquake, Geological Agency, ESDM (2019) published an Atlas of Liquefaction Vulnerability Zones for all provinces in Indonesia [3]. This atlas combines soil type, groundwater depth, and proximity to the dominant earthquake source. Based on the map, an adequate foundation calculation must be carried out for areas with medium to high liquefaction potential, considering the soil layer with liquefaction potential. Therefore, like the collapse of Palu IV bridge, which caused road network disconnection and severely affected the city's economy, it is essential to assess the bridges built in the medium to high liquefaction hazard area.

2. Bridge research location and specification

2.1. Research location

This study examines the reliability of the Kretek 2 Bridge bored pile against the potential for high liquefaction and strong earthquakes due to its proximity to an opaque fault as a shallow earthquake. The Kretek 2 Bridge at Opak River Estuary, Bantul Regency, Yogyakarta Special Region, Indonesia, has high liquefaction susceptibility due to cohesionless soil types, shallow water tables, and strong seismic sources. In a previous study, Zakariya et al (2022) showed that the thickness at point BH-6 has a liquefied layer at depths of 0-1.5 m and 4.5-6.0 m [4]. The thickness of liquefiable layer is crucial as it will become a reference in later soil and foundation modeling. As a consideration, the liquefaction layer at a depth of 0–6.0 m is assumed for the worst case. According to Muntafi et al (2020), in a case study of 2006 Yogyakarta earthquake, the closer a building is to the epicenter, the higher damage value [5]. In previous research, Sari (2017) designed the Kretek 2 bridge [6]. Yet, the design differs from the built bridge, and there was also no analysis of the lateral capacity of the pile with liquefied soil layer.

2.2. Bridge structure specification

The Kretek 2 Bridge is a standard span reinforced concrete bridge using PCI girders as the main span structure with a length of 40 m per span (total 8 spans) and a total bridge length of 342 m. This study models the structure of Pier number 7, which has a bored pile foundation depth of 34 m long by 1.0 m diameter with a total of 24 piles in a pile cap (see Figure 1). This bridge is located in a very close earthquake source <1 km (Zakariya et al, 2022), so it has enormous potential for lateral forces.

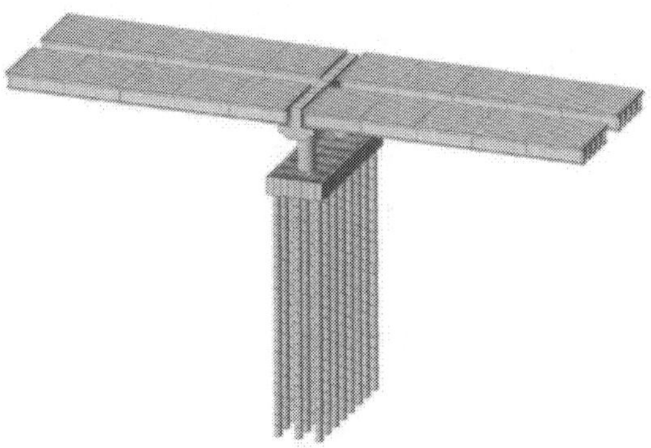

Figure 1. Kretek 2 Bridge model

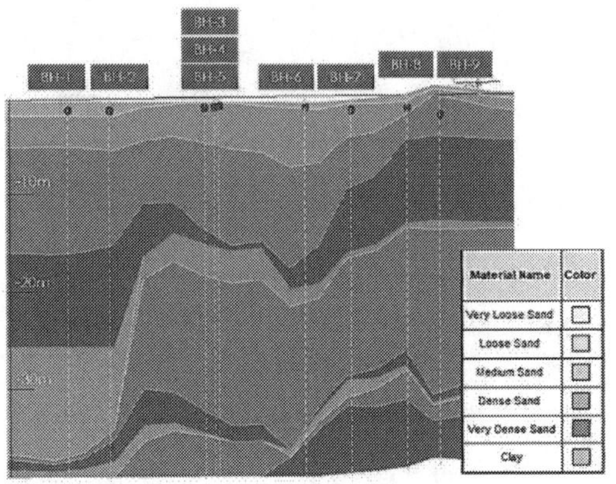

Figure 2. Soil Stratigraphy [4]

2.3. Pile bearing capacity

2.3.1. Axial bearing

The axial bearing capacity of a pile can be determined by estimating the skin resistance and the end bearing of the pile, in which an interface between pile face and soil occurred. According to Hardiyatmo (2018), the pile bearing capacity can be calculated using the following formula:

$$Q_u = Q_b + Q_s - W_P \tag{1}$$

where Q_u is ultimate bearing capacity, Q_b is end bearing capacity, Q_s is skin friction capacity, W_P is pile weight [7]. When estimating values of Q_b and Q_s, it is determined by the type of soil, shape of the pile, kind of pile construction, and depth of the pile. The approaches commonly used for sandy soils in Indonesia are those of Meyerhof (1976), Reese and Wright (1977), Poulos and Davis (1980), and O'neil and Reese (1989). Some of these researchers conducted an empirical relationship between the value of N-SPT with the ultimate resistivity and surface friction [8][9][10][11]. This study uses N-SPT soil survey data and soil properties as a reference for bearing capacity calculations.

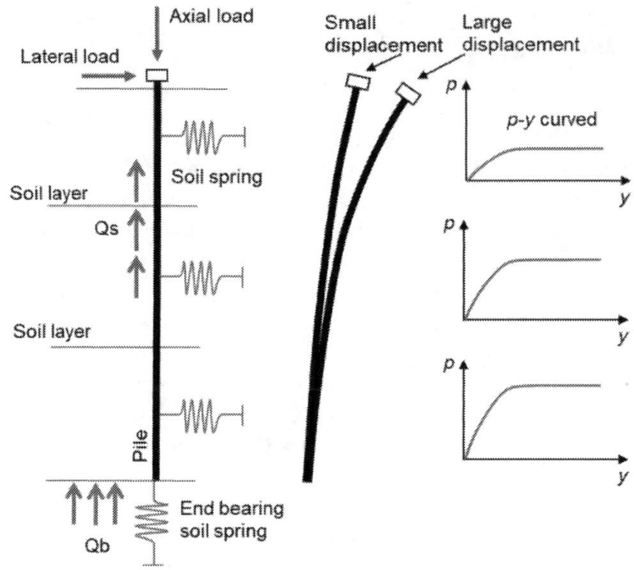

Figure 3. Effect of the shape of p–y curve on the mobilised strength and stiffness of a pile foundation in a liquefied layer (Kardogan et al, 2019) with modification [12]

2.3.2. Lateral bearing

Some researchers, Hansen (1961) and Broms (1964), have developed an approach to calculate the ultimate lateral soil to get the maximum soil moment and deflection that occur when the pile is loaded [13][14]. The disadvantage of this method is that the calculation of different soil layers becomes very complicated. Currently, the p-y method is a method that is widely used using nonlinear analysis. In this method, the lateral load versus deflection was related in the graphic (see Figure 3). This relation ratio is influenced by the type of soil, the shape of the pile, the coefficient of friction between the pile and the soil, and the pile installed method. Evan and Duncan (1982) compiled a graph of the results of p-y method analysis using a computer program. This method can be used in the analysis provided that the pile stiffness is constant throughout the pile (the pile does not change), the soil properties are constant at all depths, and the pile is long enough to be considered perfectly wedge to the pile base to be considered [15].

The lateral load-bearing capacity of foundation piles, especially bridges, is limited by the displacement value of the piles under load. Based on Indonesian Standard Code (SNI) 8460:2017, the displacement value of 80 cm diameter piles is no more than 25.4 mm, while for pile diameters > 80 cm, the pile displacement value is 4% of the pile diameter (diameter of 1 m = 40.0 mm) [16]. Meanwhile, according to Xiao et al (2016), if the horizontal displacement is less than 25.4 mm, the damage that occurs is always tolerable, while if the horizontal displacement is more than 54 mm, the damage that occurs endangers the structure, it is recommended that the maximum value of 38 .1 mm on the building without significant damage [17]. Therefore, a lateral displacement limit of 25.4 mm was used for this study to ensure satisfactory foundation conditions during earthquakes and liquefaction.

According to Rostami et al (2020), under liquefaction conditions, the strength of seismic vibrations with lateral loads increases, and at the same time bearing capacity losses occur [18]. The pile condition will be unbearable, and buckling occurs along while the increase in axial and lateral loads. When the pile can no longer restrain the load, the substructure or superstructure collapses. Research by Huded and Dash (2021) also stated that the longer pile free stand, the lower pile's ability to resist buckling [19].

3. Bridge Modelling in MIDAS

This study uses MIDAS to bridge modeling, which is commonly used in Indonesia. MIDAS Civil applies to model the bridge's superstructure so that the internal forces and displacement value on the

IOP Conf. Series: Earth and Environmental Science 1184 (2023) 012007 doi:10.1088/1755-1315/1184/1/012007

foundation piles are obtained using the soil spring approach. Meanwhile, MIDAS GTS NX will be used to model the soil using 3D numerical simulation and soil structure interactions that occur. The objective of MIDAS GTS used is to generate internal forces on the foundation piles as well as the value of the bearing capacity of the soil in response to the occurring loads on foundation piles. Other researchers Elgridly et al (2021) used MIDAS GTS NX in their research with finite element output with reliability comparing field measurements [20].

3.1. Model input

In MIDAS Civil, the superstructure and substructure are modeled to obtain the internal forces and the bridge behavior, while for MIDAS GTS NX with added function to modeled soil layer in a 3D numerical simulation as shown in Figures 4 and 5 below:

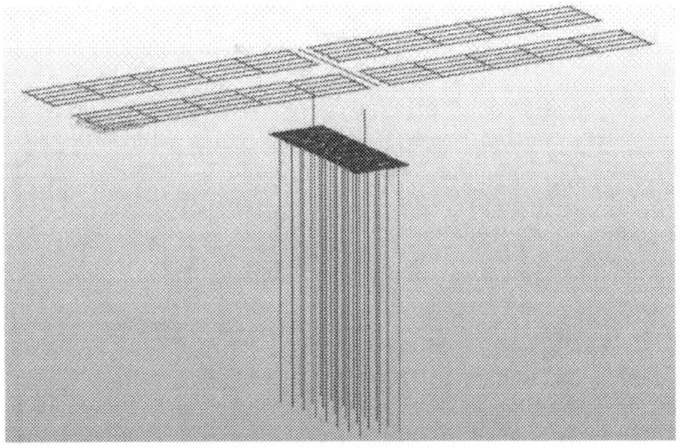

Figure 4. Modelling in MIDAS Civil

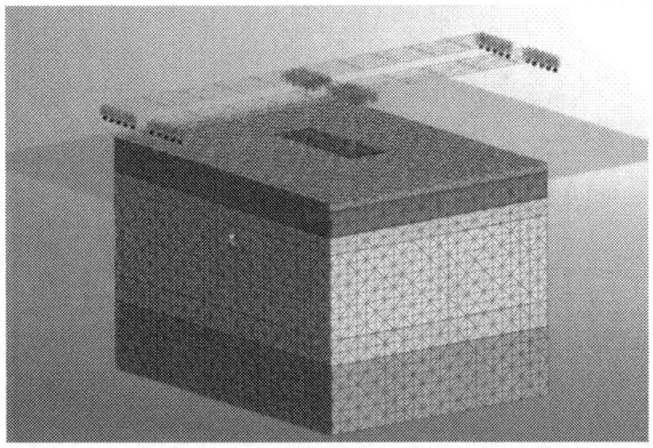

Figure 5. Modelling in MIDAS GTS NX

3.2. Soil structure interaction in MIDAS Civil

For the foundation at MIDAS Civil, the soil-structure interaction approach was used to model the pile as a pile spring. The soil around the pile is considered a spring. Lateral springs are modeled as symmetrical nonlinear elastic springs, and vertical springs are modeled as linear elastic springs. The ground spring stiffness value is automatically calculated and modeled in the point spring support. Getting the value of soil spring required the value of subgrade reaction modulus. In this study based on Bowles (1997) (see Table 1), additional soil properties were taken into account in the form of correction for N-SPT (N_{60}), bulk soil density (γ_{bulk}), saturated soil density (γ_{sat}), angle of friction (ϕ), earth pressure coefficient (K_0), subgrade reaction modulus coefficient (K_h), dan young's modulus (E), which can be seen in Table 2. Under liquefaction conditions, K_h value is considered 0 or freestanding.

Table 1. Range of subgrade reaction modulus [21]

Soil Type	K_h (kN/m^3)
Loose sand	4800 – 16000
Medium dense sand	9600 – 80000
Dense sand	64000 – 128000
Clayey medium dense sand	32000 – 80000
Silty medium dense sand	24000 – 48000
Clayey soil:	
$q_a \leq 200$ kPa	12000 – 24000
$200 < q_a \leq 800$ kPa	24000 – 48000
$q_a > 800$ kPa	> 48000

Table 2. Soil properties input

Depth (m)	$N_{60\,average}$	γ_{bulk} (kN/m³)	γ_{sat} (kN/m³)	ϕ (°)	K_0	K_h (kN/m³)	E (kN/m²)
0.0-1.5	8.40	16.840	19.050	34	0.440	10720.00	19800.00
1.5-9.0	23.28	18.328	20.538	38	0.383	56462.50	49550.00
9.0-24.0	40.02	20.002	22.212	41	0.344	96026.67	83033.33
24.0-28.5	28.00	18.800	21.010	39	0.370	69833.33	59000.00
28.5-34.5	53.33	21.333	23.543	42	0.336	117333.33	109666.67

3.3. Soil structure interaction and 3D numerical simulation in MIDAS GTS NX

MIDAS GTS NX takes into account the pile spring and the reaction of the soil layer by modeling them in the form of a 3D numerical simulation so that the soil reaction to the load applied to the pile is known. The load on the MIDAS GTS NX uses a load imported from the MIDAS Civil model. In MIDAS GTS NX, it can use the interface wizard interaction to model the stiffness of the pile interface using the following formula:

$$K_n = \frac{E_{oed,i}}{(L \times tv)} \tag{2}$$

$$K_t = \frac{G_i}{(L \times tv)} \tag{3}$$

$$E_{oed,i} = \frac{2 \times Gi \times (1-vi)}{(1-2 \times vi)} \tag{4}$$

$$G_i = R \times G_{soil} \tag{5}$$

$$G_{soil} = \frac{E}{2 \times (1+v_{soil})} \tag{6}$$

where K_n is interface normal stiffness, K_t is interface shear stiffness, $E_{oed,i}$ is interface drained soil young's modulus, L is length of pile interaction, tv is virtual thickness (generally has value between 0.01-0.1), G_i is interface shear modulus, vi is interface poisson's ratio = 0.45, G_{soil} is soil shear modulus, R is strength reduction factor, E is young's modulus, and v_{soil} is soil poisson's ratio. For general, strength reduction factor for structural members and contiguous ground properties are as follow; Sand/Steel = 0.6-0.7, Clay/Steel = 0.5, Sand/Concrete = 1.0-0.8, and Clay/Concrete = 1.0-0.7.

According to Jeongsik (2018), for sandy soil with liquefied state, 3-dimensional numerical analysis by finite element analysis was applied to the model pile inserted into the soft clay soil [22]. Another researcher, Lombardi et al (2016), stated that modeling the liquefaction conditions on the p-y curve using soft clays but replacing the undrained shear strength with the residual strength of liquefied soil [23]. In this study, the cohesion value used during the liquefaction state is 25-30 kPa. The summary of soil properties input in static and liquefaction states is shown in Table 3 and Table 4 below:

Table 3. Soil properties input in static state

Depth (m)	γ_{bulk} (kN/m³)	γ_{sat} (kN/m³)	ϕ (°)	E (kN/m²)	μ	G_{soil} (kN/m²)	E_{oed} (kN/m²)	K_n (kN/m²)	K_t (kN/m³)
0.0-1.5	16.84	19.05	34	19800.00	0.35	7342.04	64609.91	430732.72	48946.90
1.5-9.0	18.33	20.54	38	49550.00	0.36	18173.15	159923.71	1066158.09	121154.33
9.0-24.0	20.00	22.21	41	83033.33	0.38	30084.18	264740.76	1764938.43	200561.18
24.0-28.5	18.80	21.01	39	59000.00	0.37	21564.33	189766.08	1265107.21	143762.18
28.5-34.5	21.33	23.54	42	109666.67	0.39	39354.07	346315.79	2308771.93	262360.45

Table 4. Soil properties input in liquefaction state

Depth (m)	γ_{bulk} (kN/m³)	γ_{sat} (kN/m³)	ϕ (°)	E (kN/m²)	μ	G_{soil} (kN/m²)	E_{oed} (kN/m²)	K_n (kN/m²)	K_t (kN/m³)
0.0-6.0	15.00	18.00	0	500	0.45	172.41	1517.24	10114.94	1149.43
6.0-9.0	18.73	20.94	39	57541.67	0.39	39354.07	185174.24	1234494.91	140283.51
9.0-24.0	20.00	22.21	41	83033.33	0.38	30084.18	264740.76	1764938.43	200561.18
24.0-28.5	18.80	21.01	39	59000.00	0.37	21564.33	189766.08	1265107.21	143762.18
28.5-34.5	21.33	23.54	42	109666.67	0.39	39354.07	346315.79	2308771.93	262360.45

4. Bridge Loads

In the regulation of SNI 1725:2016, Bridge loading divides the load into permanent and transient loads [24]. For calculating the permanent load is generally based on the volume of structure's shape multiplied by the material's unit weight. The unit weight and mass density used as a reference are as follows Table 5 below.

4.1. Load combinations

The combination of loads in the bridge uses an ultimate combination to determine the ultimate limit, a service combination to determine the service limit, an extreme combination to determine the extreme limit of the condition or ship collision, flooding or other hydraulic loads, and a fatigue combination to determine the limit fatigue from a load of repetitions. In this study, a combination of ultimate and extreme conditions can be used to determine the maximum axial and lateral values that occur in the foundation, with details in Table 6.

Table 5. Weight and mass density material [24]

Material	Weight (kN/m³)	Mass density (kg/m³)
Compacted sand, silt or clay	17.2	1755
Rolled gravel, macadam or ballast	18.8-22.7	1920-2315
Asphalt concrete	22.0	2245
Lightweight concrete	12.25-19.6	1250-2000
Concrete f'_C<35 MPa	22.0-25.0	2320
Concrete 35 < f'_C<105 MPa	22 + 0.022 f'_C	2240 + 2.29 f'_C
Steel	78.5	7850

Table 6. Extreme and Ultimate Load Combination

Combination	Description
Extreme 1	1.3 MS + 2.0 MA + 1.0 PR + 0.5 TT + 0.5 TD + 0.5 TB + 1.0 EQ_X + 0.5 EQ_Y
Extreme 2	1.3 MS + 2.0 MA + 1.0 PR + 0.5 TT + 0.5 TD + 0.5 TB + 1.0 EQ_X – 0.5 EQ_Y
Extreme 3	1.3 MS + 2.0 MA + 1.0 PR + 0.5 TT + 0.5 TD + 0.5 TB + 1.0 EQ_Y + 0.5 EQ_X
Extreme 4	1.3 MS + 2.0 MA + 1.0 PR + 0.5 TT + 0.5 TD + 0.5 TB + 1.0 EQ_Y – 0.5 EQ_X
Ultimate 1	1.3 MS + 2.0 MA + 1.0 PR + 1.8 TT + 1.8 TD + 1.8 TB + 1.2 EU_N + 1.2 ET
Ultimate 2	1.3 MS + 2.0 MA + 1.0 PR + 1.4 TT + 1.4 TD + 1.4 TB + 1.2 EW_S + 1.2 EU_N + 1.2 ET
Ultimate 3	1.3 MS + 2.0 MA + 1.0 PR + 1.4 EW_S + 1.2 EU_N + 1.2 ET

For explanation; M_S, M_A, and P_R are permanent loads for dead load, additional dead load, and prestressed load, respectively, whereas T_T, T_D, T_B, E_Q, EU_N, ET, and EW_s are transient loads for trucks, lane loads, brake loads, pedestrian loads, earthquake loads, uniform temperature loads, gradient temperature loads, and wind on structure loads, respectively.

4.2. Earthquake loads

The main factor consideration for lateral load for this bridge is earthquake loads. Kretek 2 Bridge is located very close to opaque fault (<10 km) that requires site-specific response analysis, and it is not allowed to use the response spectrum contained in SNI 2833:2016 concerning Bridge Designing Against Earthquake Loads [25]. The earthquake load used from a research by Asrurifak (2021) [26], the site-specific response spectra (SRSS) that calculated based on the seven ground motions of the following significant earthquakes; 1) 1979 Imperial Valley USA, 2) 1987 Superstition Hills USA, 3) 1989 Loma Prieta USA, 4) 1992 Erzincan Turkiye, 5) 1995 Kobe Japan, 6) 2000 Tottori Japan, 7) 2010 Darfield New Zealand, which has been spectra matching and scaling of the amplification into a response spectrum at ground level (see Figure 6). The result is that the SRSS value is higher than all response spectra using SNI 2833:2016, which uses a 7% probability of exceedance in 75 years or equivalent to 1000 years period (see Figure 7).

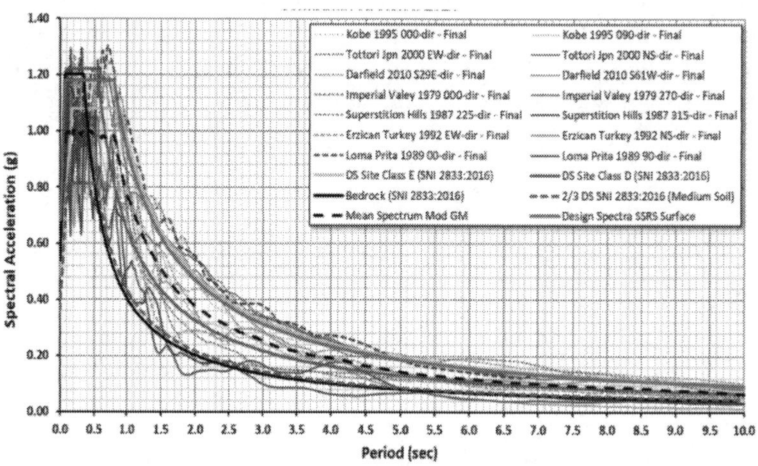

Figure 6. Response spectrum design output from spectra matching and scaling based on seven ground motions

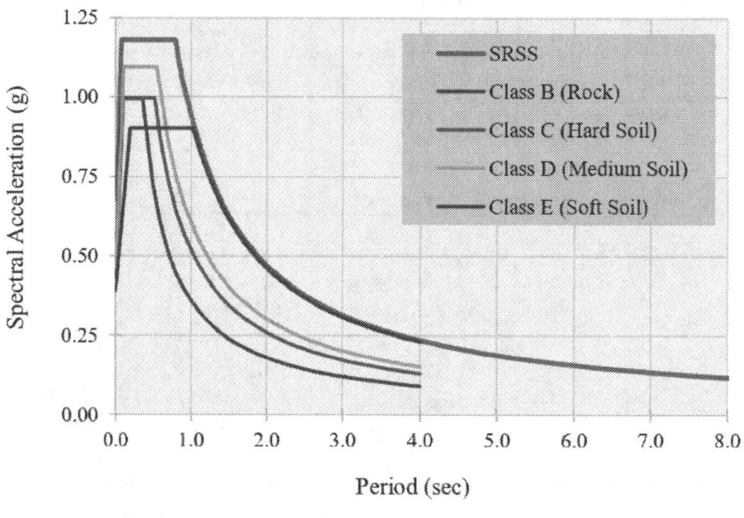

Figure 7. Response spectrum design (SRSS and SNI 2833-2016)

5. Result and Discussion

The modeling of the Kretek 2 Bridge was performed with 2 conditions, namely static state and liquefaction state. The model is performed with 7 load combinations in MIDAS Civil and GTS NX. After receiving the output from MIDAS Civil in the form of internal forces on the foundation, the value of these internal forces is reused as an ultimate load on the Midas GTS NX. This ensures that the difference in output between the two is known even though the input load is not different.

5.1. Forces output

After running the model, the forces exerted on the pile in axial, shear-y, shear-x, moment-y, and moment-x can be partially shown in Figures 8 – 10. When modeling with MIDAS GTS NX, the Solid Strain E-equivalent options can be used (see Figure 11), which shows a large solid strain for the liquefaction layer at a depth of 0–6.0 m compared to the soil layer. This shows that the liquefied soil layer can be modeled reasonably well with MIDAS GTS NX. All MIDAS Civil output forces, static and liquefied states, were maximum under extreme conditions, either Extreme 1, 2, 3, or 4 (see Tables 7 and 8). The results shows that the dominant loading in this model is seismic loads.

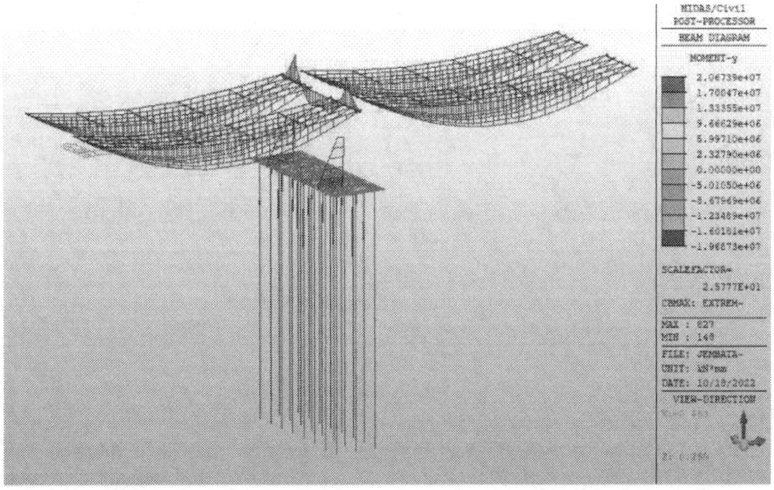

Figure 8. MIDAS Civil Output for Beam Diagram moment-y

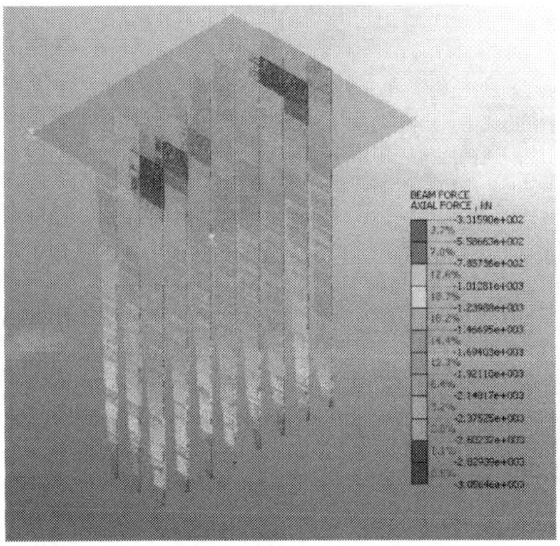

Figure 9. MIDAS GTS NX output for beam diagram axial

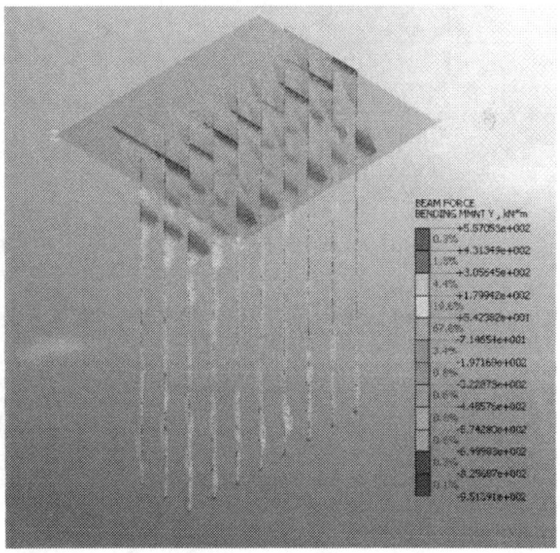

Figure 10. MIDAS GTS NX output for beam diagram moment-y

IOP Conf. Series: Earth and Environmental Science 1184 (2023) 012007 doi:10.1088/1755-1315/1184/1/012007

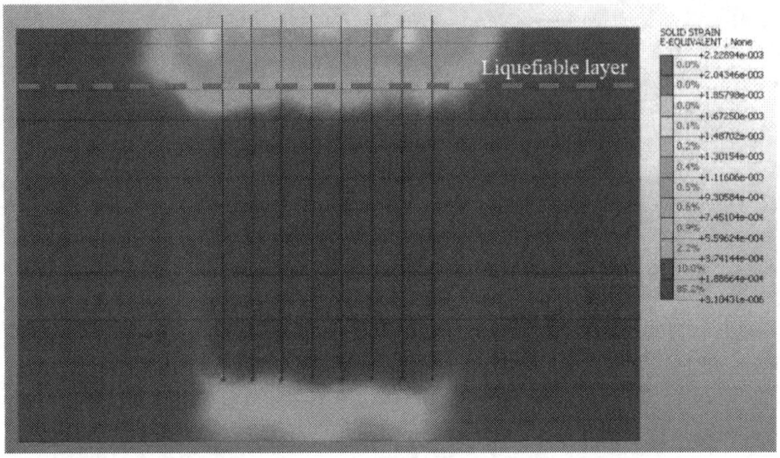

Figure 11. MIDAS GTS NX output for solid strain e-equivalent

Table 7. MIDAS Civil output in static state

Output	Load Combination						
	Extreme 1	Extreme 2	Extreme 3	Extreme 4	Ultimate 1	Ultimate 2	Ultimate 3
Forces							
Axial (kN)	2829.50	3356.80	2566.50	2694.10	3071.40	3091.40	3187.80
Shear y (kN)	125.80	123.20	229.50	229.20	27.70	28.10	56.20
Shear z (kN)	212.70	210.10	118.90	109.80	43.50	35.80	14.30
Moment y (kN.m)	601.70	590.30	351.70	313.70	148.90	128.00	62.80
Moment z (kN.m)	396.10	388.20	702.90	701.90	108.10	106.60	192.50
Displacement							
Lateral (xy) (mm)	4.25	4.19	4.16	4.13	0.57	0.49	0.73
Vertical (z) (mm)	1.85	2.16	1.68	1.77	1.99	2.00	2.06
Resultant (mm)	4.50	4.50	4.40	4.49	2.00	2.04	2.17

Table 8. MIDAS Civil output with in liquefaction state

Output	Load Combination						
	Extreme 1	Extreme 2	Extreme 3	Extreme 4	Ultimate 1	Ultimate 2	Ultimate 3
Forces							
Axial (kN)	2831.00	3395.00	2547.50	2873.80	3094.10	3116.50	3216.80
Shear y (kN)	118.40	115.00	224.50	223.50	17.60	20.30	47.30
Shear z (kN)	197.50	196.20	106.50	100.00	36.90	29.70	7.20
Moment y (kN.m)	912.00	903.50	502.70	463.60	189.00	153.60	47.00
Moment z (kN.m)	575.90	569.70	1086.10	1083.40	99.80	111.10	242.40
Displacement							
Lateral (xy) (mm)	13.57	13.47	13.94	13.75	1.79	1.60	2.27
Vertical (z) (mm)	2.28	2.71	2.06	2.31	2.48	2.50	2.57
Resultant (mm)	13.68	13.64	13.97	14.00	2.79	2.75	3.39

Under liquefaction conditions, the axial load has no significant increase, but moment force y and z-direction significantly increase. On the other hand, the shear forces y and z directions have decreased since the shear forces are partially converted into moment forces due to the increased length of the free-standing pile. Meanwhile, for MIDAS GTS NX modeling, both static state and liquefaction state are performed only for extreme loading conditions, as the previous conclusions from MIDAS Civil are considered. The result shows axial force increases, shear force a few decreases, and moment forces increase significantly consistently. A significant increase in lateral displacement can recognize through the addition of axial and lateral forces during liquefaction. The results of the MIDAS GTS NX output can be seen in Table 9 below:

IOP Conf. Series: Earth and Environmental Science 1184 (2023) 012007 doi:10.1088/1755-1315/1184/1/012007

Table 9. MIDAS GTS NX outpun in static and liquefaction state

Output	Static State Load Combination				Liquefaction state Load Combination			
	Ext. 1	Ext. 2	Ext. 3	Ext. 4	Ext. 1	Ext. 2	Ext. 3	Ext. 4
Forces								
Axial (kN)	2674.57	3056.46	2456.75	2729.76	2962.96	3356.16	2678.31	3009.11
Shear y (kN)	298.03	332.81	414.54	414.88	176.88	158.60	281.80	281.35
Shear z (kN)	365.52	368.64	242.58	246.02	257.36	248.81	174.93	151.76
Moment y (kN.m)	651.74	669.40	430.57	413.71	949.97	951.39	577.13	504.68
Moment z (kN.m)	570.11	594.82	759.16	739.17	631.64	626.26	1013.04	989.87
Displacement								
Lateral (xy) (mm)	7.14	7.19	5.19	4.80	15.27	15.22	12.74	12.33
Vertical (z) (mm)	5.22	5.85	4.80	5.02	5.80	6.50	5.41	5.38
Resultant (mm)	8.84	9.24	6.98	6.95	16.04	16.39	13.59	13.43
Bearing Capacity								
Pile Force (kN)	284.47	278.66	275.19	251.00	303.76	288.53	296.67	265.72
Pile Tip (kN)	594.25	581.42	575.35	525.87	634.25	604.53	620.18	556.14

5.2. Beam diagram output in Midas Civil

The beam diagram output shows that the axial load value gradually decreases under static conditions, while the decrease occurs after 6.0 m depth under liquefaction conditions. Under static conditions, the shear value also gradually decreases from the surface, while under liquefaction conditions, the decrease occurs after 6.0 m depth. Meanwhile, there is a significant increase in moment force output induced by the length of the freestanding pile emerging during liquefaction. At the osculation between liquefied and non-liquefied layers, conditions of the maximum moment in y and z-directions must also be taken into account, since the value of the moment reaches a second peak. After passing a depth of > 9.0 m, the moment value gradually decreases until the end of pile. In general, the liquefaction conditions show an increase in axial, shear, and moment values compared to the static state until it reaches a non-liquefied layer. The results are shown in Figures 12 and 13. Modeling with MIDAS Civil has limitations for calculating the foundation as a soil structure interaction only, even though it can generate internal forces for pile material design.

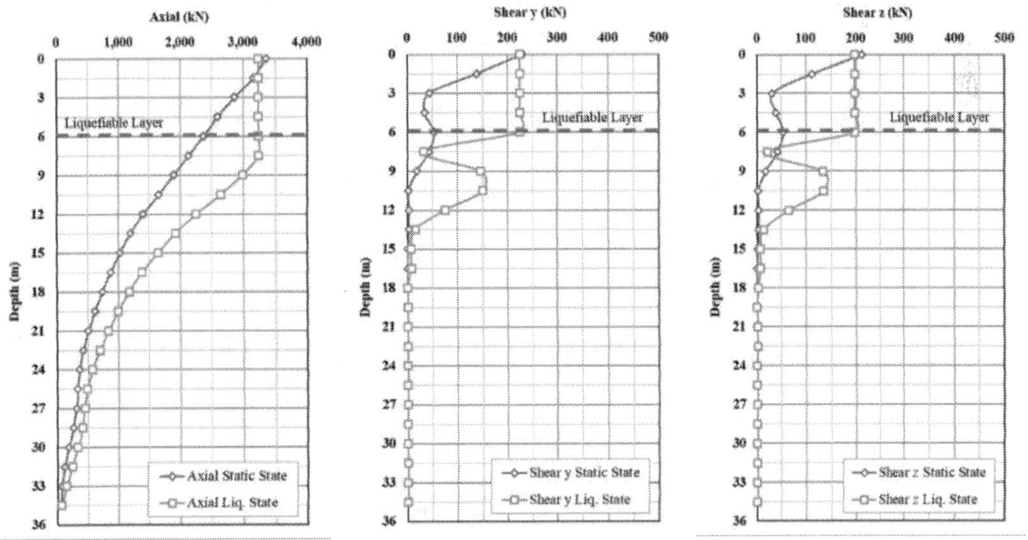

Figure 12. Axial and shear force output using MIDAS Civil

IOP Conf. Series: Earth and Environmental Science 1184 (2023) 012007 doi:10.1088/1755-1315/1184/1/012007

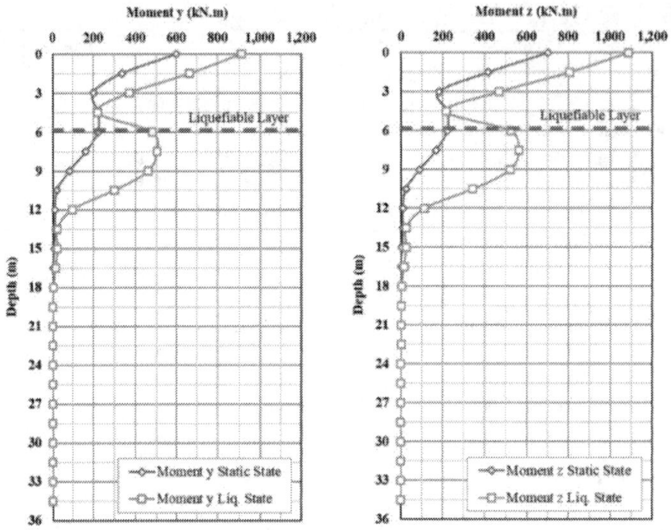

Figure 13. Moment force output using MIDAS Civil

5.3. Beam diagram output in Midas GTS NX

MIDAS GTS NX uses numerical simulation methods in soil layer modeling. This study uses MIDAS GTS NX and aims to determine the bearing capacity of the soil by considering the load that occurs on the pile. The output of the internal forces at the pile has the same consistency as with MIDAS Civil. The axial force in the liquefaction state increases slightly, but the shear force decreases. This is caused by the lateral force being converted into significant moment force, which rises at the top of pile. The output results are shown in Figures 14 and 15. MIDAS Civil and GTS NX results show a relatively consistent similarity in values. However, in MIDAS GTS NX, the pile is considered elastic such that the axial, shear, and moment forces propagate until the bottom of pile.

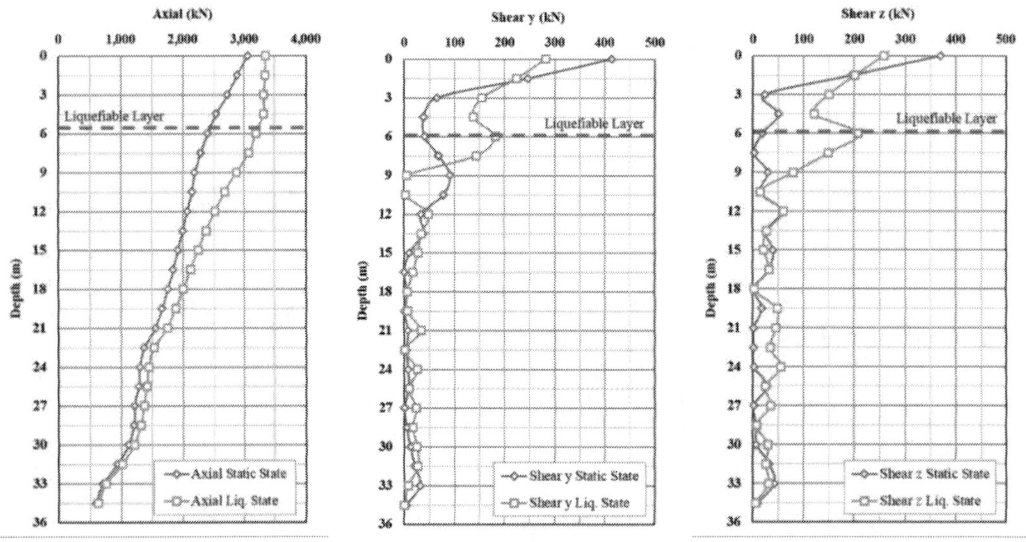

Figure 14. Axial and shear force output using MIDAS GTS NX

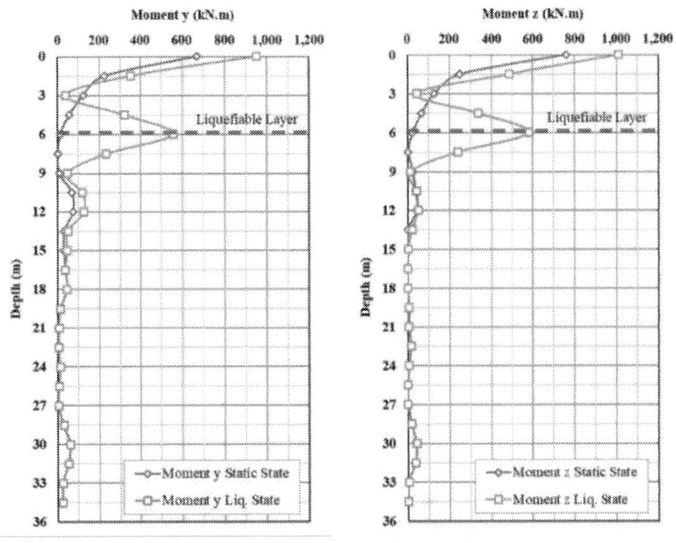

Figure 15. Moment force output using MIDAS GTS NX

5.4. Axial bearing assessment

One of the advantages of using MIDAS GTS NX is that it can generate an axial bearing capacity based on the option of pile force as skin resistance (kN/m) and pile tip as end bearing (kN) (see Figure 16). The resulting axial bearing capacity is a reaction of load applied to the pile and then transferred to each layer of soil. Considering the design maximum axial load from MIDAS Civil modeling and the results of pile driving analysis (PDA) tests on the pile in the field using a safety factor value = 2.0 (this value was taken according to Likins (2004) [27]) shows that the axial load carrying capacity is still slightly above the design load = 3456.02 kN and below the PDA test value = 4480.50 kN as shown in Figure 17. The value of axial bearing capacity in the static and liquefaction state indicates the value tends not to change significantly. Based on these results, MIDAS GTS NX can be used to estimate the axial bearing capacity when designing a new bridge or evaluating the existing bridge.

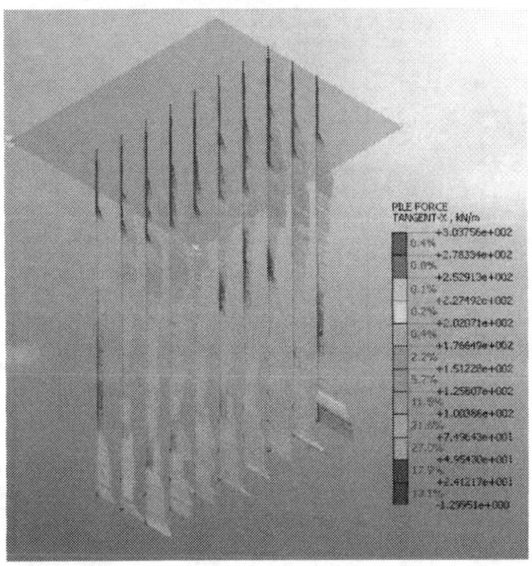

Figure 16. MIDAS GTS NX Output for Pile force

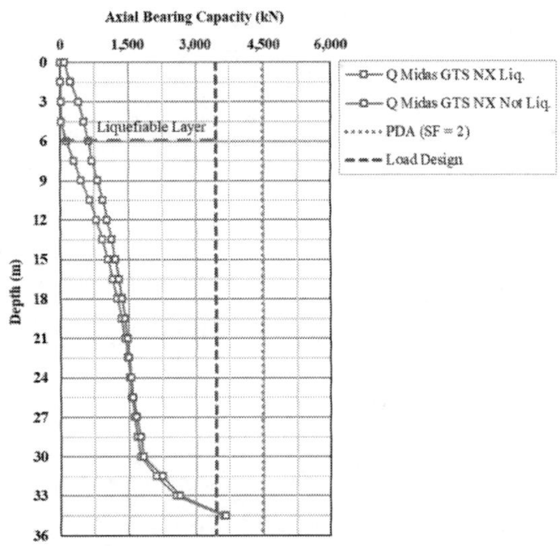

Figure 17. Axial bearing capacity vs. Load Design and PDA Test

5.5. Lateral bearing assessment

MIDAS GTS NX has the advantage that it can estimate the displacement value for both the pile and the surrounding soil layer, as shown in Figure 18. In a static state, the lateral displacement generated by MIDAS GTS NX is larger than that of MIDAS Civil (see Figure 19). In the meantime, in a liquefaction state, MIDAS GTS NX and MIDAS Civil have value consistently. However, the difference can be seen in the behavior when sticking to the bottom of soil. For MIDAS Civil at a depth of > 9.0 m, the displacement is close to 0, while for MIDAS GTS NX, the depreciation is gradual till the bottom. This confirms that the interface relationship between the pile and the ground is considered an elastic non-linear with a fixity point near the bottom of the pile (the pile is considered wedged at the base of pile) on MIDAS GTS NX. In contrast, this study shows that MIDAS Civil has a fixity point value near the liquefaction layer. This conforms to Minno et al. (2015) that the MIDAS finite element allows for more realistic stress and strain distribution in the soil as plasticity is considered according to the Mohr-Coulomb failure criterion [28].

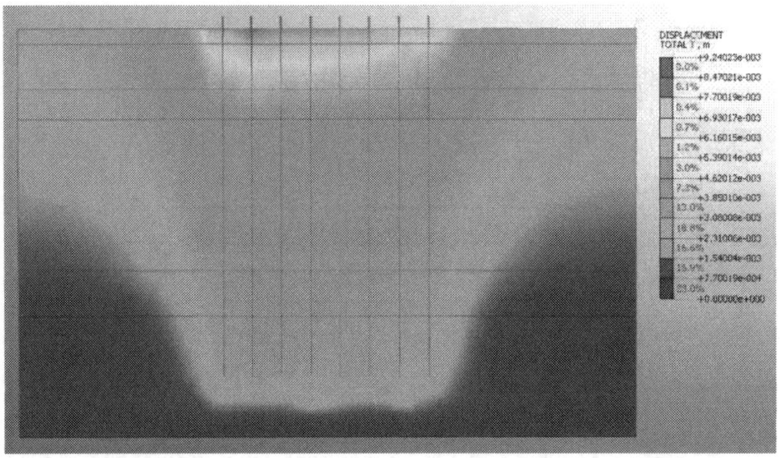

Figure 18. MIDAS GTS NX Output for Displacement Total

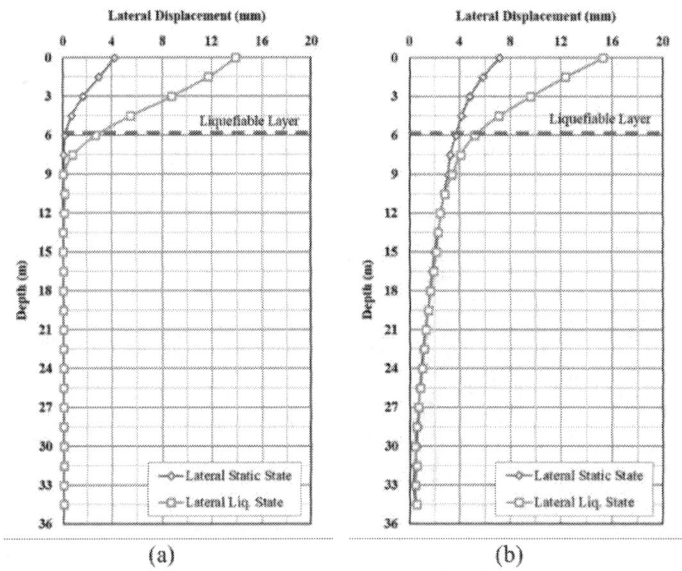

Figure 19. Lateral displacement output by MIDAS; (a) Civil, and (b) GTS NX

Based on the allowable displacement value of 25.4 mm, the MIDAS output indicates that the Kretek 2 bridge is stable with a maximum lateral displacement of 15.3 mm under static and liquefaction conditions. Referring to Xiao et al (2016), the displacement value of 25.4 mm shows that the damage occurring to the bridge can still be tolerated. This result is also supported by the lateral static load test with a 100% load of 245.25 kN, giving a total displacement value of 4.17 mm, and a 200% load of 490.50 kN, giving a total displacement value of 14.98 mm. Based on those results, the pile performance remains reliable, and the bearing capacity can still support the lateral loads.

However, MIDAS Civil and GTS NX are software that complete each other. MIDAS Civil is needed for modeling the bridge superstructure and substructure with obtaining the behavior of the bridge under combined loads. Meanwhile, MIDAS GTS NX is very useful for modeling the soil layer structure and its impact on the pile foundation to get axial bearing capacity value. Both can produce nearly similar lateral displacement values (differences $\leq$ 4.0mm) in a liquefaction state for lateral capacity. For further investigations, modeling with particular sand materials such as UBCSand and the time history dynamic analysis is required so that model can be nearly close to actual condition.

6. Conclusion

The bearing capacity assessment of bored piles can be performed with MIDAS Civil and GTS NX. MIDAS Civil provides a stiffer pile output with a fixity point value close to the joint of the liquefied and non-liquefied layers. In contrast, the MIDAS GTS NX offers a more elastic pile with a pinched condition at the bottom of pile. The Midas results compared to the pile driving analysis and the static load test in the field showed that the pile is capable of supporting axial and lateral loads. Based on the results of the foundation modeling, during liquefaction, there is an increase in axial and moment force but a decrease in shear force. While lateral displacement is significantly increased, foundation performance remains stable. Despite this, MIDAS Civil and GTS NX are software that complete each other. Midas Civil emphasizes superstructures modeled with foundations as pile springs for soil structure interactions. Midas GTS NX emphasizes modeling soil layers with 3D numerical simulations and piles as a soil structure interaction so that the value of the axial bearing capacity (skin resistance and end bearing) shows up. Also, both can give similar lateral displacement values in the liquefaction state. The results of bearing capacity assessment show that Kretek 2 bridge foundation is sufficient to carry the load in the static and liquefaction state.

7. References

[1] Dash S R Bhattacharya S and Blakeborough A, 2010 Bending-buckling interaction as a failure mechanism of piles in liquefiable soils *Soil Dyn. Earthq. Eng.* 30, 1–2 p. 32–39.
[2] Imran I Santoso B Pramudito A and Zamad M K, 2019 *Simulation of Palu IV Bridge Collapse using near-fault ground motions* .
[3] Badan Geologi, 2019 *Atlas Zona Kerentanan Likuefaksi Indonesia* Bandung: Kementerian Energi Sumber Daya Mineral.
[4] Zakariya A Rifa'i A and Ismanti S, 2022 Ground Motion and Liquefaction Study at Opak River Estuary Bantul in *5th International Conference on Eartquake Engineering and Disaster Mitigation*, September 2022.
[5] Muntafi Y Nojima N and Jamal A U, 2020 Damage Probability Assessment of Hospital Buildings in Yogyakarta, Indonesia as Essential Facility due to an Earthquake Scenario *J. Civ. Eng. Forum* 6, 3 p. 225.
[6] Sari E K, 2017, Perancangan Fondasi Tiang Bor pada Jembatan Kretek 2 Bantul, Yogyakarta, Universitas Gadjah Mada.
[7] Hardiyatmo H C, 2018 *Analisis dan Perancangan Fondasi II* 4th ed. Yogyakarta: Gadjah Mada University Press.
[8] Meyerhof G G, 1976 Bearing Capacity and Settlement of Pile Foundations *ASCE J Geotech Eng Div* 102 p. 197–228.
[9] Reese L C and Wright S J, 1977, Drilled Shaft Manual, Volume I, Construction Procedures and Design for Axial Loading, Washington, DC.
[10] Poulos H G and Davis E H, 1980 *Pile Foundation Analysis and Design* Canada: Rainbow-Bridge

Book Co.
[11] O'neill M W and Reese L C, 1989, New Design Method for Drilled Shaft From Common Soil and Rock Tests, in *Foundation Engineering - Current Principles and Practices*, p. 1026–1039.
[12] Kardogan P S O Isik N S Onur M I and Bhattacharya S, 2019 A Study on the Laterally Loaded Pile Behaviour in Liquefied Soil Using P-Y Method *IOP Conf. Ser. Mater. Sci. Eng.* 471, 4.
[13] Hansen J B, 1961 The Ultimate design of piles against transversal loads *Geotek. Inst. (Danish Geotech. Institute)* 12 p. 16.
[14] Broms B B, 1964 Lateral Resistance of Piles in Cohesionless Soils *J. Soil Mech. Found. Div.* 90, SM2 p. 123–156.
[15] Evan L T and Duncan J M, 1982, Simplified Analysis of Laterally Loaded Piles, Berkeley.
[16] Badan Standarisasi Nasional Indonesia, 2017 *SNI 8460:2017 Persyaratan Perancangan Geoteknik* 8460 .
[17] Xiao M Qiu T Khosrojerdi M Basu P and Withiam J L, 2016 Synthesis and Evaluation of the Service Limit State of Engineered Fills for Bridge Support *Fed. Highw. Adm.* FHWA-HRT-1, February p. 151.
[18] Rostami R B. Mickovski S Hytiris N and Bhattacharya S, 2021 The Dynamic Behaviour of Pile Foundations in Seismically Liquefiable Soils: Failure Mechanisms, Analysis, Re-Qualification *Earthquakes - From Tectonics to Build.* December.
[19] Huded P and Dash S R, 2021 Influence of p-y curves on buckling capacity of pile foundation *Int. Sch. Sci. Res. Innov.* 15, 8 p. 238–242.
[20] Elgridly E A Fayed A L and Ali A A A F, 2022 Efficiency of pile groups in sand soil under lateral static loads *Innov. Infrastruct. Solut.* 7, 1.
[21] Bowles J E, 1997 *Foundation Analysis and Design International Fifth Edition* Fifth Illnois: The McGraw-Hill Companies, Inc.
[22] Jeongsik P, 2018, Nonlinear 3-Dimensional Analysis of Dynamic p-y Curve in Sand, Yonsei University.
[23] Lombardi D Dash S R Bhattacharya S and Wood D M, 2016 Construction of Simplified Design p-y curves for liquefied soils September.
[24] Badan Standardisasi Nasional, 2016 *SNI 1725:2016 Pembebanan untuk Jembatan* 1725 .
[25] Badan Standarisasi Nasional Indonesia, 2016 *SNI 2833:2016 Perencanaan Jembatan Terhadap Beban Gempa* 2833rd ed. Jakarta.
[26] Asrurifak M, 2021, Laporan Final Site Specific Response Spectra (SSRS) Jembatan Kretek II D.I. YOGYAKARTA, Bantul.
[27] Likins G, 2004 Pile Testing - Selection and Economy of Safety Factors 40743, July 2004 p. 239–252.
[28] Minno M Persio R and Petrella F, 2015 Finite element modeling of a piled raft for a tall building on cohesionless soil *Geotech. Eng. Infrastruct. Dev. - Proc. XVI Eur. Conf. Soil Mech. Geotech. Eng. ECSMGE 2015* 7, March p. 4019–4024.

Acknowledgments
The authors thank the Unit Work of National Road Implementation (PJN) Yogyakarta Special Region Province Territory, National Road Implementation Agency (BBPJN) of Central Java – Yogyakarta Special Region for soil investigation data support.

IOP Conf. Series: Earth and Environmental Science 1184 (2023) 012008 doi:10.1088/1755-1315/1184/1/012008

Settlement Analysis in Potentially Liquifiable Soil at Kretek 2 Bridge Area

F Nurdiansyah[1, 2], A D Adi[1] and S Ismanti[1]

[1] Department of Civil and Environment Engineering, Faculty of Engineering, Universitas Gadjah Mada, Yogyakarta 55281, Indonesia

[2] Directorate General of Highways, The Ministry of Public Works and Housing, South Jakarta 12110, Indonesia

agusdadi1@ugm.ac.id

Abstract. The Kretek 2 Bridge is one of the important infrastructures built on the southern coast road of Java. Referring to the Map of Indonesia Liquefaction Vulnerability Zone in 2019, Kretek 2 Bridge was identified as a high liquefaction susceptibility zone. The result of liquefaction potential analysis on the slab-on-pile section indicates that the soil layer has potential to liquefy if an earthquake occurs. Based on the Liquefaction Severity Index, the slab-on-pile section of the Kretek 2 Bridge has a high severity index at BH-10, moderate severity index at BH-11, and low severity index at BH-01 and BH-12. One of liquefaction effects is a ground settlement. Settlement due to liquefaction is a vertical soil deformation in the soil layer caused by compaction due to the earthquake. This study was conducted by calculating the settlement of sand deposits in slab-on-pile section using an equation proposed by Zhang et al. in 2002 and Yoshimine in 2006. The results of settlement analysis using Zhang's method to predict settlement in the study area, BH-10 having the most significant settlement at 29.803 cm. Using the settlement analysis method by Yoshimine, the highest settlement found in BH-10, which is 30.401 cm.

1. Introduction

Indonesia is located in the most active earthquake region in the world. Geographically, Indonesia is the meeting point of 4 large tectonic plates, as seen in **Figure 1**. The Eurasian Plate moves southeast direction, Indo-Australian Plate moves northward, and both the Pacific and Filipino Plates move westward [1]. These tectonic plates cause geological activities such as earthquakes, volcanic formations, or submarine trenches. One of the disasters that may occur due to an earthquake is liquefaction. Liquefaction is a condition where the soil changes from solid to liquid due to earthquake vibration or cyclic loading. Cyclic loadings from earthquakes cause an increase in pore water and cause a loss of soil strength. Historically, liquefaction occurred during the Yogyakarta Earthquake with a magnitude of 6.3 sourced from the movement of the Opak Fault. These liquefaction phenomena, such as lateral spreading and sand boils (**Figure 2**), were detected in some locations at Yogyakarta when the earthquake occurred.

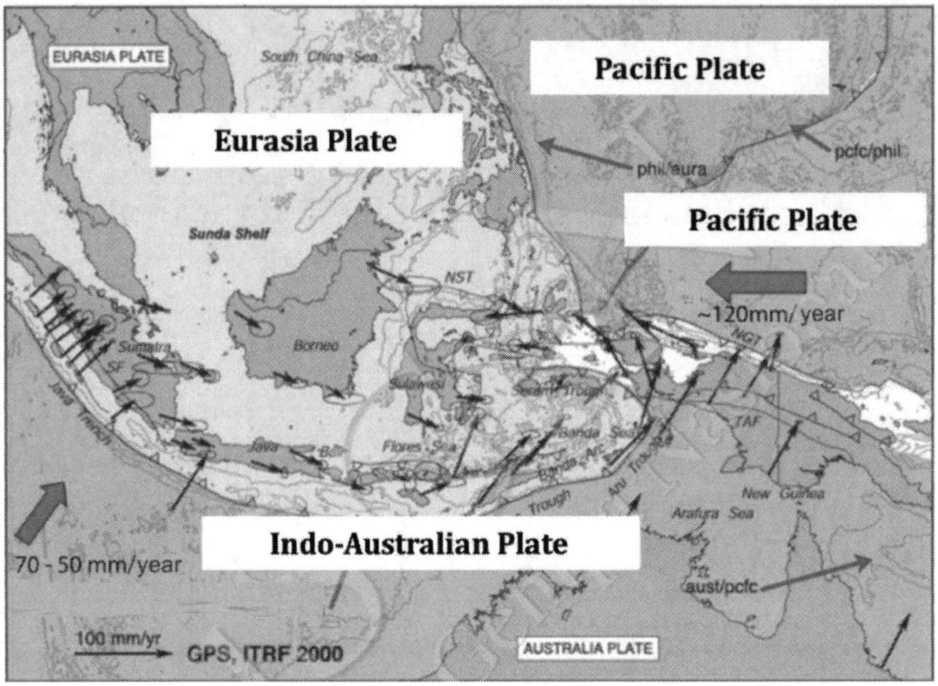

Figure 1. Active tectonic map in Indonesia [1]

(a) (b)

Figure 2. Liquefaction phenomena were found in Yogyakarta Earthquake in 2006 [3]. (a) sand boiling in an open well, (b) lateral spreading

The Kretek 2 Bridge will be one of the vital infrastructures to improve connectivity in Java's southern coast road.The Kretek 2 Bridge located in Kretek District, Bantul Regency, Yogyakarta. The bridge began construction in 2021 and is planned for completion by the end of 2022. According to the Map of the Indonesia Liquefaction Vulnerability Zone (see **Figure 3**), the Kretek 2 Bridge is within the high liquefaction vulnerability zone [2].

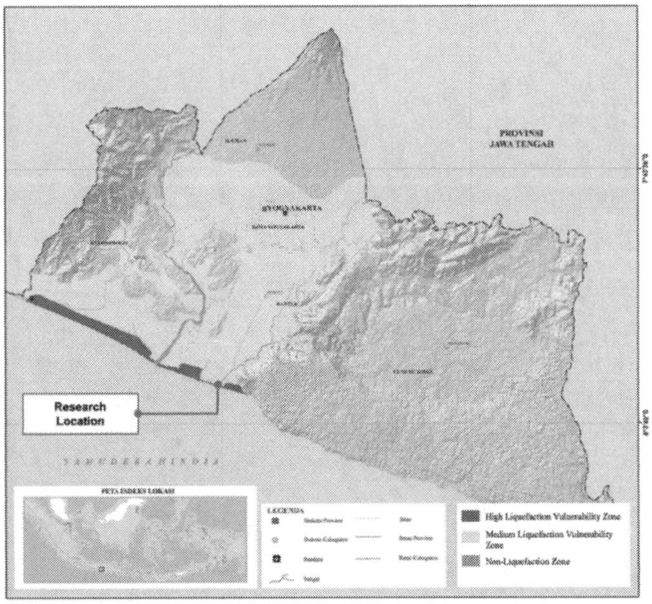

Figure 3. Yogyakarta liquefaction zone in Map of Indonesia Liquefaction Vulnerability Zone [2]

The Kretek Bridge 2 can be divided into four sections (see **Figure 4**). The first section is the Samas' approach road sections on the west side, the second part located at the northwest of the Kretek 2 Bridge and the main focus of this study is the slab-on-pile section, the third part is the main bridge section, and the last section is the embankment and retaining wall in the parangtritis' side. This research only discusses the estimation of ground settlement that occurs due to liquefaction in the slab-on-pile section and the structure of slab on pile is not discussed in this study.

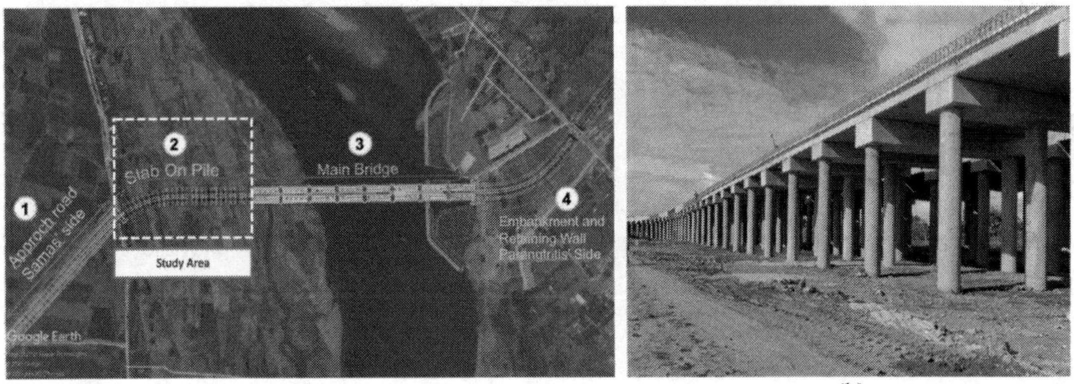

(a) (b)

Figure 4. (a) Study area for slab-on-pile section of Kretek 2 Bridge (b) slab-on-pile construction

The Kretek 2 Bridge was built to cross the Opak River. Based on soil investigation report, the soil layer at study area dominated by sandy soil and has a low N-SPT value in the topsoil (**Figure 5**). In BH-10 the soil layer is dominated by sandy soil with a loose – very dense density to a depth of -30 meters, and there is a layer of clay with a dense consistency at a depth of -31.5 meters to -37.5 meters. BH-11 and BH-12 are dominated by sandy soil with a dense – very dense density to a depth of -39 meters. BH-01 is dominated by granular soil (sand) to a depth of -24 meters, from a depth of -25.5 a dense layer of clay is found to a depth of -39 meters. This type of soil is very vulnerable to liquefaction.

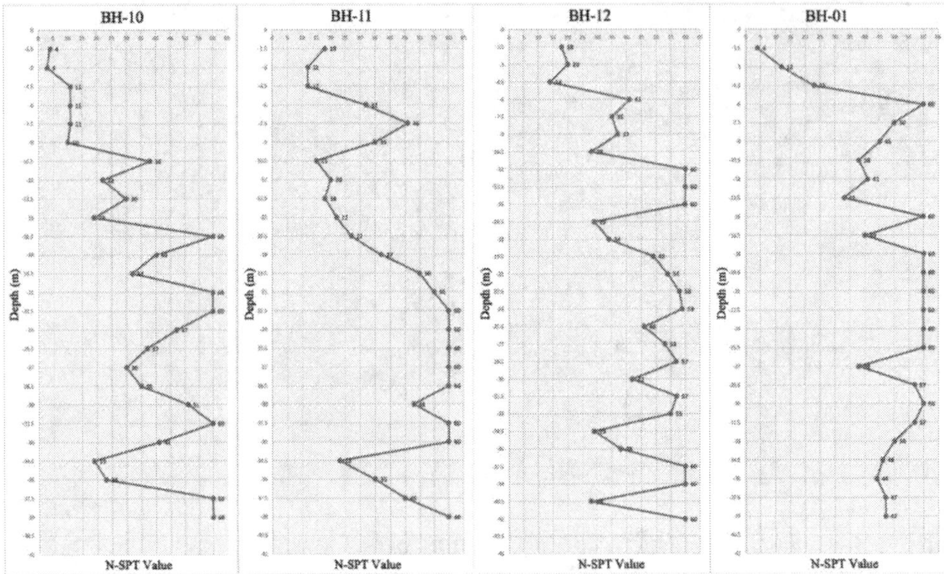

Figure 5. N-SPT value at study area

Based on the Geological Map of the Yogyakarta Special Region Province, the location of the Kretek 2 Bridge consists of the sedimentary formations of the Merapi Young Volcano, whose soil layer is dominated by sand and sandy clay. The Merapi Young Volcano deposit is composed of river and coastal alluvial deposits[4]. Based on lithostratigraphy, geological conditions on the Kretek 2 Bridge are classified as quaternary deposits that are less than 500 years old [5]. Quaternary deposits have a high to very high liquefaction susceptibility [6].

Based on the soil investigation report in the slab-on-pile section using data from BH-10, BH-11, BH-12, and BH-01, the groundwater level depth is between 1.2 - 1.5 meters below the surface ground, as shown in **Table 1**.

Table 1. Groundwater table at slab-on-pile section of Kretek 2 Bridge

Boreholes	GWT level (meter)	Liquefaction Susceptibility
BH-10	1.2	Very High
BH-11	1.5	Very High
BH-12	1.5	Very High
BH-01	1.2	Very High

From the groundwater table shown in **Table 1**, groundwater table with depths of less than 3 meters is categorized as a very high level of liquefaction susceptibility [7].

Preliminary analysis of grain size distribution in the study area adopted from Tsuchida (1970) study to define the grain size distribution that is vulnerable to liquefaction. The grain size distribution in the study area is shown in **Figure 6**. The grain size distribution in the study area indicates that the soil layer of slab-on-pile section has potential for liquefaction

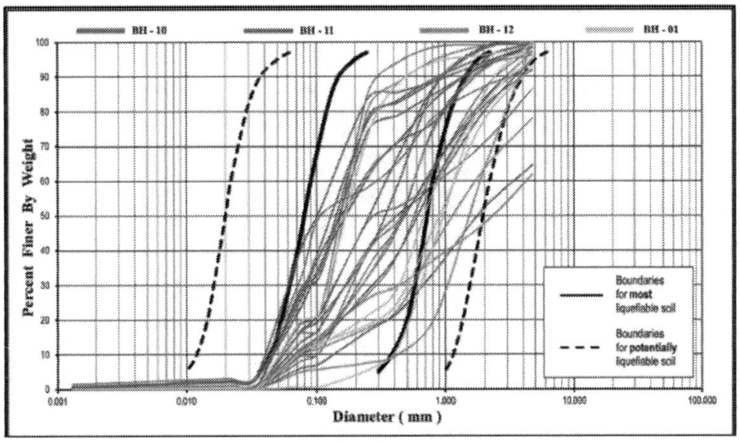

Figure 6. Comparison between grain size distribution in the study area and grain size distribution of liquifiable soil proposed by Tsuchida (1970)

This study attempts to study the potential for liquefaction in the slab-on-pile segment of the Kretek 2 Bridge using Liquefaction Severity Index (LSI) and identify the liquefaction-induced settlement that can occur. Based on the result of this study, we can classify the hazard level of settlement that can be used as a reference in mitigation.

2. Methods

The procedure presented in this article utilize the following steps:
Step 1. Asses the liquefaction potential based on Idriss and Boulanger's method [8] and Evaluate the liquefaction severity using LSI method by Sonmez and Gokceoglu [9].
Step 2. Calculate liquefaction-induced settlement using Zhang *et al.* method [10] and Yoshimine *et al.* method [11]

2.1. Liquefaction Potential Analysis

The liquefaction potential in soils is often evaluated using simplified procedure method originally proposed by Seed and Idriss [12] based on the Standard Penetration Test (SPT). In this study, the method used to analyze the liquefaction potential is the simplified procedure method developed by Idriss and Boulanger in 2008 [8]. The value of safety factor against liquefaction in the liquefaction potential analysis is calculated based on the ratio between cyclic resistance ratio (CRR) and cyclic shear ratio (CSR). The CRR value is the value of the soil resistance to cyclic loads caused by the earthquake. The CRR value is calculated using the corrected SPT blow counts, correction factor for fines content, and correction for effective overburden pressure. The CRR value is calculated using **Equation 1**

$$CRR_{M=7.5,\sigma'_{vc}=1} = \exp\left(\frac{(N_1)_{60cs}}{14.1} + \left(\frac{(N_1)_{60cs}}{126}\right)^2 - \left(\frac{(N_1)_{60cs}}{23.6}\right)^3 - \left(\frac{(N_1)_{60cs}}{25.4}\right)^4 - 2.8\right) \quad (1)$$

Where, $CRR_{M=7.5,\sigma'_{vc}=1}$ is the CRR value in the moment magnitude (M) = 7,5 and overburden pressure (σ'_{vc}) = 1 atm. Research indicates that other corrections, such as earthquake magnitude, overburden pressure, and static shear stress, can also be made to the CRR for any eartquake moment Magnitude. **Equation 2** calculates the CRR value for the any moment magnitude and other overburden pressures according to the conditions of any location.

$$CRR_{M,\sigma'_{vc}} = CRR_{M=7.5,\sigma'_{vc}=1} \cdot MSF \cdot K_\sigma \quad (2)$$

where, MSF is the correction factor of magnitude, $K_\sigma$is the correction factor of overburden pressure with a maximum value of 1.1 [8]. Coefficient $C_\sigma$with a maximum value of 0.3 [7], $\sigma'_v$is the effective stress, and P_a is the overburden pressure of 101 kPa.

Cyclic shear ratio (CSR) is a value that indicates the ratio of soil stress caused by cyclic loading. The value of CSR was formulated by Seed and Idriss [12] and developed by Idriss and Boulanger [8]. CSR value is obtained using **Equation 3**.

$$CSR = 0.65 \cdot r_d \left(\frac{a_{max}}{g}\right)\left(\frac{\sigma_v}{\sigma'_v}\right) \quad (3)$$

with, 0.65 is a assumed value that is equivalent to 65% of the maximum induced stress, r_d= depth reduction factor, a_{max} is maximum horizontal acceleration at ground surface due to earthquake, g is the acceleration of gravity , σ_v and σ'_v are the total and effective overburden stress. The value of the safety factor against liquefaction can be calculated using the following equation.

$$FS_{Liq} = \frac{CRR_{M,\sigma'_{vc}}}{CSR} \quad (4)$$

Based on **Equation 4**, the FS_{Liq} defined as the ratio of CRR that will cause liquefaction of the soil to the CSR developed in the soil by the earthquake loads. The value of FS_{Liq}<1 and FS_{Liq} > 1 indicates that the soils are classified as liquefiable and non-liquefiable, respectively, while the FS_{Liq} = 1 represents limiting equilibrium.

2.2. Liquefaction Severity Index

Liquefaction Severity Index (LSI) was used to evaluate the severity of liquefaction events proposed by Sonmez and Gokceoglu [9]. The LSI value is obtained by using probability numbers and the soil depth function. The LSI value calculated using **Equation 5 - Equation 8**

$$LSI = \int_0^{20} P_L(z).w(z)dz \quad (5)$$

$$P_L(z) = \frac{1}{1+\left({FS}/{0.96}\right)^{4.5}} \text{ for } FS \leq 1.411 \quad (6)$$

$$P_L(z) = 0 \text{ for } FS > 1.411 \quad (7)$$

$$w(z) = 10 - 0.5z \quad (8)$$

where, *LSI* is *liquefaction severity index*, $P_L(z)$ is the probability of liquefaction with safety factor consideration, dan $w(z)$ is depth weight factor and z is depth (≤ 20m). This method limits the maximum FS value to 1.411, if FS value greater than 1.411 the $P_L(z)$ value is 0.

The LSI value indicates the level of liquefaction potential in the studied area and is classified into six categories. The classification of LSI value can be seen in **Table 2.**

Table 2. Liquefaction Severity Index Classification

LSI Range	*Description*
85 < LSI ≤ 100	*Very high*
65 < LSI ≤ 85	*High*
35 < LSI ≤ 65	*Moderate*
15 < LSI ≤ 35	*Low*
0 < LSI ≤ 15	*Very low*
LSI = 0	*Non-liquified*

2.3. Liquefaction-induced settlement analysis

2.3.1. Zhang et al. method

One of the effects of liquefaction is a ground settlement. Ground settlement occurs due to soil compaction due to earthquakes which causes the cavities in the loose sand layer to fill each other. A ground settlement has the potential to cause damage to the structure above. According to Zhang *et al.* [10], liquefaction-induced settlement is a vertical deformation in the soil layer caused by densification and compaction of granular soil, which was initially loose but solidified due to the earthquake. During the reconsolidation of sandy soils, liquefaction-induced settlement can be observed through the volumetric strain that occurs [13]. Zhang *et al.*[10] propose an equation to calculate the amount of ground settlement due to liquefaction. Liquefaction-induced settlement using this method uses the correlation between the value of FS_{Liq}, $qc_{1N_{cs}}$, dan ε_v. Because in this study the existing data is N-SPT value, it is necessary to convert the value of N-SPT to the value of CPT data ($qc_{1N_{cs}}$) using equation proposed by Andrus *et al.* [14]. The NSPT value is converted to CPT value using **Equation 9**

$$(qc_{1N})_{cs} = \left(\frac{(N_1)_{60cs}}{0.488}\right)^{\frac{1}{0.7879}} \tag{9}$$

where, $(N_1)_{60cs}$ is the corrected value of N-SPT for fines content, $(qc_{1N})_{cs}$ is the corrected value of CPT for fines content

After the value of $qc_{1N_{cs}}$ is obtained, the next step is to estimate the volumetric strain value (ε_v). The estimated volumetric strain value was obtained using the correlation between the value of FS_{Liq} and $qc_{1N_{cs}}$ [10] as seen in the **Table 3**.

Table 3. An empirical equation of volumetric strain due to liquefaction [10]

Safety Factor (FS_{Liq})	$(qc_{1N})_{cs}$	*volumetric strain* (ε_v)
$\leq$ 0.50	$33 \leq (qc_{1N})_{cs} \leq 200$	$102(qc_{1N})_{cs}^{-0.82}$
0.6	$33 \leq (qc_{1N})_{cs} \leq 147$	$102(qc_{1N})_{cs}^{-0.82}$
	$147 \leq (qc_{1N})_{cs} \leq 200$	$2411(qc_{1N})_{cs}^{-1.45}$
0.7	$33 \leq (qc_{1N})_{cs} \leq 110$	$102(qc_{1N})_{cs}^{-0.82}$
	$110 \leq (qc_{1N})_{cs} \leq 200$	$1701\,(qc_{1N})_{cs}^{-1.42}$

Safety Factor (FS_{Liq})	$(qc_{1N})_{cs}$	*volumetric strain* (ε_v)
0.8	$33 \leq (qc_{1N})_{cs} \leq 80$	$102(qc_{1N})_{cs}^{-0.82}$
	$80 \leq (qc_{1N})_{cs} \leq 200$	$1690\ (qc_{1N})_{cs}^{-1.46}$
0.9	$33 \leq (qc_{1N})_{cs} \leq 60$	$102(qc_{1N})_{cs}^{-0.82}$
	$60 \leq (qc_{1N})_{cs} \leq 200$	$1430\ (qc_{1N})_{cs}^{-1.48}$
1.0	$33 \leq (qc_{1N})_{cs} \leq 200$	$64\ (qc_{1N})_{cs}^{-0.93}$
1.1	$33 \leq (qc_{1N})_{cs} \leq 200$	$11\ (qc_{1N})_{cs}^{-0.65}$
1.2	$33 \leq (qc_{1N})_{cs} \leq 200$	$9.7(qc_{1N})_{cs}^{-0.69}$
1.3	$33 \leq (qc_{1N})_{cs} \leq 200$	$7.6(qc_{1N})_{cs}^{-0.71}$
2.0	$33 \leq (qc_{1N})_{cs} \leq 200$	0

According Zhang *et al.* [10] liquefaction-induced settlement can be estimated using **Equation 10**

$$S = \sum_{i=1}^{n} \varepsilon_v \,\Delta Z_i \tag{10}$$

where, S is post -liquefaction settlement in cm, ε_v is volumetric strain (%), and ΔZ is soil thickness per layer in cm

2.3.2. Yoshimine et al. method
Another method to estimating liquefaction-induced settlement is using Yoshimine *et al.* [11]. Yoshimine *et al.* in 2006 proposed the liquefaction-induced settlement in the soil by considering the possible shear deformation. The Yoshimine's liquefaction-induced settlement method is the result of the development by Ishihara and Yoshimine [13]. They observed that the volumetric strain that occurs during post-liquefaction sand sample reconsolidation is directly related to the maximum shear strain that rises during undrained cyclic loading and to the initial relative density of the sand. The liquefaction-induced settlement method by Yoshimine *et al.* calculated using following equation.

$$\gamma_{\lim} = 1.859\left(1.1 - \sqrt{\frac{(N_1)_{60cs}}{46}}\right)^3 \geq 0 \tag{11}$$

$$F_\alpha = 0.032 + 0.69\sqrt{(N_1)_{60cs}} - 0.13(N_1)_{60cs} \tag{12}$$

$$\gamma_{\max} = 0 \qquad \text{if } FS_{liq} \geq 2 \tag{13.a}$$

$$\gamma_{\max} = \min\left(\gamma_{\lim}\ 0.035(2 - FS_{liq})\left(\frac{1 - F_\alpha}{FS_{liq} - F_\alpha}\right)\right) \qquad \text{if } 2 > FS_{liq} > F_\alpha \tag{13.b}$$

$$\gamma_{\max} = \gamma_{\lim} \qquad \text{if } FS_{liq} \leq F_\alpha \tag{13.c}$$

$$\varepsilon_v = 1.5 \exp\left(-0.369\sqrt{(N_1)_{60cs}}\right) \cdot \min\,(0.08; \gamma_{\max}) \tag{14}$$

$$S_{v-1D} = \int_0^{20} \varepsilon_v . dz \tag{15}$$

where, $\gamma_{\lim}$ is minimum shear strain limit, F_α is limiting value of FS_{liq}, $\gamma_{\max}$ is maximum shear strain, ε_v is volumetric strain, and S_{v-1D} is settlement due to liquefaction-induced one-dimensional reconsolidation

Ishihara and Yoshimine [13] classified the relation between damage extent and settlement estimation shown in **Table 4**.

Table 4. Relation between damage extent and approximate settlements

Extent of damage	Settlements (cm)	Phenomena on the ground surface
Light to no damage	0 - 10	Minor cracks
Medium damage	10 - 30	Small cracks, oozing of sand
Extensive damage	30 - 70	Large cracks, spouting of sands, large offsets, lateral movement

3. Results and Discussions

3.1. Liquefaction potential analysis result

The results of the liquefaction potential analysis using the Idriss and Boulanger simplified procedure methods [8] at four boreholes (BH-10, BH-11, BH-12 , and BH-01) shows that the slab-on-pile section of Kretek 2 Bridge has potential liquifiable soil if earthquake occurs. The results of liquefaction potential analysis shown in the **Table 5** - **Table 8**

Table 5. Results of Liquefaction potential analysis in BH-10

Depth (m)	SPT Blow	N_{60}	$(N_1)_{60cs}$	CSR	CRR	FS_{Liq}	Description
1.5	4	3.73	10.94	0.391	0.161	0.411	Liquified
3	3	2.98	9.37	0.506	0.146	0.288	Liquified
4.5	11	12.19	17.08	0.551	0.222	0.403	Liquified
6	11	12.19	17.35	0.569	0.220	0.386	Liquified
7.5	11	12.19	17.64	0.573	0.218	0.381	Liquified
9	10	11.67	17.90	0.570	0.217	0.382	Liquified
10.5	38	44.33	49.41	0.562	2.000	3.561	Non-Liquified
12	22	25.67	30.01	0.550	0.554	1.006	Non-Liquified
13.5	30	35.00	38.04	0.537	2.000	3.726	Non-Liquified
15	19	22.17	24.90	0.522	0.319	0.611	Liquified
16.5	60	70.00	68.15	0.507	2.000	3.944	Non-Liquified
18	40	46.67	46.22	0.492	1.995	4.056	Non-Liquified
19.5	32	37.33	36.34	0.477	1.461	3.064	Non-Liquified

IOP Conf. Series: Earth and Environmental Science 1184 (2023) 012008 doi:10.1088/1755-1315/1184/1/012008

Table 6. Results of Liquefaction potential analysis in BH-11

Depth (m)	SPT Blow	N_{60}	$(N_1)_{60cs}$	CSR	CRR	FS_{Liq}	Description
1.5	18	16.80	31.65	0.347	0.787	2.269	Non-Liquified
3	12	11.90	22.06	0.464	0.301	0.650	Liquified
4.5	12	13.30	19.99	0.514	0.261	0.508	Liquified
6	32	35.47	42.22	0.536	2.000	3.734	Non-Liquified
7.5	46	50.98	56.43	0.543	2.000	3.682	Non-Liquified
9	35	40.83	42.13	0.542	2.000	3.687	Non-Liquified
10.5	15	17.50	16.99	0.536	0.202	0.376	Liquified
12	20	23.33	24.39	0.527	0.314	0.595	Liquified
13.5	18	21.00	21.28	0.515	0.250	0.485	Liquified
15	22	25.67	24.71	0.502	0.311	0.619	Liquified
16.5	27	31.50	29.17	0.488	0.465	0.952	Liquified
18	37	43.17	40.11	0.474	1.955	4.125	Non-Liquified
19.5	50	58.33	52.93	0.460	1.903	4.138	Non-Liquified

Table 7. Results of Liquefaction potential analysis in BH-12

Depth (m)	SPT Blow	N_{60}	$(N_1)_{60cs}$	CSR	CRR	FS_{Liq}	Description
1.5	18	16.80	28.79	0.347	0.539	1.556	Non-Liquified
3	20	19.83	29.07	0.460	0.557	1.210	Non-Liquified
4.5	14	15.52	20.15	0.508	0.263	0.517	Liquified
6	41	45.44	50.07	0.529	2.000	3.780	Non-Liquified
7.5	35	38.79	43.04	0.536	2.000	3.731	Non-Liquified
9	37	43.17	45.69	0.535	2.000	3.738	Non-Liquified
10.5	28	32.67	32.74	0.529	0.829	1.568	Non-Liquified
12	60	70.00	67.22	0.519	2.000	3.853	Non-Liquified
13.5	60	70.00	65.41	0.507	2.000	3.942	Non-Liquified
15	60	70.00	63.81	0.494	2.000	4.046	Non-Liquified
16.5	29	33.83	30.55	0.481	0.545	1.135	Non-Liquified
18	34	39.67	35.04	0.467	1.104	2.366	Non-Liquified
19.5	49	57.17	50.37	0.453	1.880	4.153	Non-Liquified

Table 8. Results of Liquefaction potential analysis in BH-01

Depth (m)	SPT Blow	N_{60}	$(N_1)_{60cs}$	CSR	CRR	FS_{Liq}	Description
1.5	4	3.73	8.57	0.387	0.140	0.361	Liquified
3	12	11.90	18.24	0.492	0.240	0.488	Liquified
4.5	23	25.49	33.38	0.532	1.046	1.965	Non-Liquified
6	60	66.50	76.96	0.548	2.000	3.650	Non-Liquified
7.5	50	55.42	61.67	0.551	2.000	3.629	Non-Liquified
9	45	52.50	56.31	0.547	2.000	3.656	Non-Liquified
10.5	38	44.33	46.46	0.539	2.000	3.713	Non-Liquified
12	41	47.83	48.41	0.527	2.000	3.794	Non-Liquified
13.5	33	38.50	38.03	0.514	2.000	3.891	Non-Liquified
15	60	70.00	65.85	0.500	2.000	4.002	Non-Liquified
16.5	40	46.67	43.75	0.485	1.988	4.098	Non-Liquified
18	60	70.00	63.10	0.470	1.931	4.106	Non-Liquified
19.5	60	70.00	61.93	0.456	1.879	4.121	Non-Liquified

Based on the results the liquefaction potential analysis, FS_{liq}<1 dominantly found in BH-10 and BH-11, while in BH-12 there was only 1 liquefied layer and BH-01 there were 2 layers of liquefied soil. The parameter used in the Liquefaction Severity Index (LSI) calculation is the FS_{liq}, which has been obtained from the simplified procedure method. Calculation results of the LSI value at each borehole are shown in **Table 9**.

Table 9. Results of Liquefaction Severity Index

Borehole	Number of liquefied soil layers	LSI Value	LSI Classification
BH-10	7	71.35	High
BH-11	7	43.55	Moderate
BH-12	1	15.11	Low
BH-01	2	25.88	Low

The results in **Table 9** also show that if the number of liquefied soil layers increases, the LSI value will also be higher. The number of liquefied layers in BH-10 and BH-11 were 7 layers each, but the LSI value in BH-10 was higher. This is because in BH-10 the value of $FS_{\text{liq}} < 1$ mostly occurs at a depth of <10 meters.

3.2. Liquefaction-induced settlement analysis result

The results of liquefaction-induced settlement analysis using 2 methods (Zhang, 2002 and Yoshimine, 2006) showed that the highest settlement was in BH-10. Using the Zhang method the settlement in BH-10 was 29.80 cm, referring to the **Table 4** (relation between damage extent and approximate settlements) by Ishihara and Yoshimine [13], the level of damage at point BH-10 is medium-damage. , Using the Yoshimine method, the liquefaction-induced settlement in BH-10 was 30.40 cm and classified into extensive damage. The result of liquefaction-induced settlement shown in the **Table 10**.

Table 10. Results of liquefaction-induced settlement

Borehole	LSI Value	Zhang et al. 2002		Yoshimine et al. 2006	
		Approximate Settlement (cm)	Extent of damage	Approximate Settlement (cm)	Extent of damage
BH-10	71.35	29.80	Medium damage	30.40	Extensive damage
BH-11	43.55	17.97	Medium damage	20.82	Medium damage
BH-12	15.11	3.60	Light to no damage	5.53	Light to no damage
BH-01	25.88	13.66	Medium damage	10.85	Medium damage

Based on the results of the analysis in **Table 9** and **Table 10**, it is shown that the higher the liquefaction potential value, settlement value will be greater. This also indicates that the occurrence of liquefaction can cause ground settlement.

4. Conclusions

The analysis of liquefaction potential was determined using simplified procedure and LSI method based on soil investigation report on four boreholes in the slab-on-pile section of Kretek 2 Bridge, which resulted in different severity in each boreholes. The results using the LSI method show that BH-10 is in the high category with an LSI value of 71.35, BH-11 is in the moderate category with an LSI value of 43.55, BH-12 is in the low category with an LSI of 15.11 and BH-01 is in the low category with an LSI value of 25.88. Based on the LSI value obtained, slab-on-pile section of Kretek 2 Bridge has the potential to liquefy if an earthquake occurs. Based on the LSI method proposed by Sonmez and Gokceoglu, the smaller the FS_{liq} value obtained from simplified procedure method, the higher the LSI value.

Liquefaction events can lead to ground settlement. The results of this study indicate that the higher the LSI value, the greater the approximate ground settlement. The highest liquefaction-induced settlement occurs in BH-10, which has an LSI value of 71.35 and is in the high category. Using the method of Zhang 2002, the settlement value on BH-10 is 29.80 cm. Using the method of Yoshimine 2006, the settlement value of BH-10 is 30.40 cm.

Based on the results of this study, the higher the LSI value, ground settlement value will be greater. This also indicates that the occurrence of liquefaction can cause ground settlement and have a dangerous impact on the bridge structure. Further analysis is needed to determine the effect of liquefaction and liquefaction-induced settlement to the bearing capacity of the pile foundation at the slab-on-pile section of Kretek 2 Bridge

References

[1] Tim Pusat Studi Gempa Nasional, 2017 *Peta Sumber Dan Bahaya Gempa Indonesia Tahun 2017* Jakarta: Pusat Penelitian dan Pengembangan Perumahan Pemukiman, Badan Penelitian dan Pengembangan Kementerian Pekerjaan Umum dan Perumahan Rakyat.

[2] Badan Geologi, 2019 *Atlas Zona Kerentanan Likuefaksi Indonesia* Jakarta: Kementerian Energi dan Sumber Daya Mineral.

[3] Mase L Z, 2017 Experimental Liquefaction Study of Southern Yogyakarta Using Shaking Table *J. Tek. Sipil ITB* **24**, 1 hal. 11–18.

[4] Rahardjo W Sukandarrumidi dan Rosidi H, 1995, Peta Geologi Lembar Yogyakarta. Pusat Penelitian dan Pengembangan Geologi, Bandung.

[5] Surono, 2009 Litostratigrafi Pegunungan Selatan Bagian Timur Daerah Istimewa Yogyakarta dan Jawa Tengah *J. Geol. dan Sumberd. Miner.* **19**, 3 hal. 209–221.

[6] Youd T L dan Perkins D M, 1978 Mapping Liquefaction-Induced Ground Failure Potential *ASCE J Geotech Eng Div* **104**, 4 hal. 433–446.

[7] Youd T L Tinsley J C Perkins D M King E J dan Preston R F, 1979 Liquefaction Potential Map of San Fernando Valley, California *Geol. Surv. Circ. (United States)* 807 hal. 37–48.

[8] Idriss I M dan Boulanger R W, 2008 *Soil Liquefaction During Earthquakes* Earthquake Engineering Research Institute (EERI).

[9] Sonmez H dan Gokceoglu C, 2005 A liquefaction severity index suggested for engineering practice *Environ. Geol.* **48**, 1 hal. 81–91.

[10] Zhang G Robertson P K dan Brachman R W I, 2002 Estimating liquefaction-induced ground settlements from CPT for level ground *Can. Geotech. J.* **39**, 5 hal. 1168–1180.

[11] Yoshimine M Nishizaki H Amano K dan Hosono Y, Feb 2006 Flow deformation of liquefied sand under constant shear load and its application to analysis of flow slide of infinite slope *Soil Dyn. Earthq. Eng.* **26** hal. 253–264.

[12] Seed H B dan Idriss I M, 1971 Simplified Procedure for Evaluating Soil Liquefaction Potential *J. Soil Mech. Found. Div. ASCE 97, SM9* hal. 1249–1273.

[13] Ishihara K dan Yoshimine M, Mar 1992 Evaluation of Settlements in Sand Deposits Following Liquefaction During Earthquakes *Soils Found.* **32**, 1 hal. 173–188.

[14] Andrus R D Piratheepan P Ellis B S Zhang J dan Juang C H, 2004 Comparing liquefaction evaluation methods using penetration-VS relationships *Soil Dyn. Earthq. Eng.* **24**, 9–10 hal. 713–721.

IOP Conf. Series: Earth and Environmental Science 1184 (2023) 012009 doi:10.1088/1755-1315/1184/1/012009

The Effect of High Pore Water Pressure on Liquefaction Potential in Sidoarjo Mud Volcanoes (LUSI)

D Wijanarko[1,2], A D Adi[1] and A Rifa'i[1].

[1]Department of Civil and Environment Engineering, Faculty of Engineering, Universitas Gadjah Mada, Yogyakarta 55281, Indonesia
[2]Directorate General of Water Resources, Ministry of Public Works and Housing, Jakarta 12110, Indonesia

agusdadi1@ugm.ac.id, dedywijanarko @mail.ugm.ac.id

Abstract. LUSI until now is still active and holds the potential for subsequent disasters due to the influence of high pore water pressure in the affected area. The presence of high pore water pressure will cause a decrease in effective stress and decrease in soil / rock strength. This type of subsequent disaster is closely related to the process of liquefaction which has the potential to occur with the addition of earthquake events. Analysis of liquefaction potential with empirical analysis on standard penetration test (SPT) based evaluation, with input of soil properties parameters, seismicity, and groundwater level parameters. The groundwater level parameter in the calculation will show a normal pore water pressure value in accordance with the depth of the water-saturated layer. Input of high pore water pressure parameters so far has not been done. CPT-u measurement data for pore pressure value input consists of 13 points of measurements was on the northern embankment, with the lowest level of stability. The input of these parameters will provide variations in the differences in the level of potential liquefaction in the study area. The result is changes occur in the decrease in the safety factor value ranging from 0.1-0.4 and the change in the liquefaction potential index (LPI) value from originally low to high, and originally high to very high. The anomaly input of pore pressure in CPT-u measurement results ranges from 40%- 70% higher than normal pore pressure in the sand layer. Analysis of the potential for liquefaction in the study area showed that the results of the high- very high liquefaction potential in the northern and eastern research areas, and the low Liquefaction potential was in the southern and western areas.

1. Introduction

The mud flow in Porong Sidoarjo was reported on May 29, 2006, is continuing and there is no sign of stoppings. Facts at the research site regarding the presence of high pore pressure [1], as well as the impact of subsequent disasters, being a very important discussion regarding the potential dangers of Liquefaction in the LUSI area. Liquefaction is the change in the condition of the soil (granular) of sand saturated with water to liquid due to the increase in the pressure of the water, the price of which becomes equal to the total pressure due to the occurrence of dynamic loads, so that the effective stress of the soil becomes zero [2].

The conditions for the occurrence of liquefaction refer to the definition and the historical record of liquifaction that has occurred, have been fulfilled and occurred in LUSI including: the presence of granular and unconsolidated sand soils, shallow groundwater levels, the age of young sediment deposits, the presence of land slopes and are closely related to seismic activity. Geological conditions of the study area in the back arc basin of the Kendeng Zone, located north of the active magmatic arc. From the results of the anomaly gravity survey, it shows that the area with sediment is very thick, due to the rapid sedimentation process, so it has the potential to produce rock formations with high pressure, it also appears that the mud volcanoes along the Java Island are in this zone [3].

2. Research Methode

The main cause of liquefaction is a rise in pore pressure due to the earthquake which causes a decrease in effective stress. Starting from these two parameters, this research was carried out by combining methods for calculating the potential for liquefaction which in general have been used based on the formulation of Idris and Boulanger. From the formulation, the author tried to enter the value of the pore water pressure parameter of the measurement results directly, to replace the value of the calculation result based on the formula. This is done considering that Lusi conditions are different and there is a large pore water pressure anomaly below the surface. Empirical Analysis of soil Liquefaction, SPT based evaluation, Idriss and Boulanger 2008 (IB) method [4]. To determine the safety factor of the potential liquefaction based on the comparison of the Cyclic Resistant Ratio (CRR) value to the Cyclic stress ratio (CSR) value. Liquefaction potential index (LPI) has been used to find out the limits of the potential liquefaction that will occur. [5] Developed the relationship between the thickness of the liquefaction and non-liquefaction layers at a certain acceleration of the soil surface, to determine whether the liquefaction that occurs can give rise to identifiable manifestations on the surface. Drilling data and SPT measurements in this study consist of 26 bore holes: 7 drilling points in 2015-2016, 16 drilling points in 2019, and 3 drilling points in 2021. To input pore water value, use the CPT-u measurement 13 CPT-u points in 2014 and 2021, the focus of measurements was on the northern embankment, with the lowest level of stability (**Figure 1**), total area of the research site in the sidoarjo mud affected area, East Java, covering an area of 1.167ha.

Figure 1. CPT-u Measurement Locations in the study area were conducted in 2014 and in 2021

3. Liquefaction Potential Analysis.

3.1. Empirical Analysis of soil Liquefaction, SPT based evaluation, IB 2008 method

The method to determine the safety factor of the potential liquefaction based on the comparison of the Cyclic Resistant Ratio (CRR) value to the Cyclic stress ratio (CSR) value. To find soils that are susceptible to liquefaction, one needs to analyze the values of the CSR and CRR at each specific depth at the study site. Analysis of the potential for liquefaction requires calculation of the safety factor (FS) against liquefaction using the CSR and CRR values of the ground in situ. Liquefaction may occur if the CSR ratio caused by an earthquake is greater than the CRR ratio of the ground in situ.

3.1.1. Cyclic Resistant Ratio (CRR)

The soil's ability to withstand liquefaction (CRR) can be determined based on test result data. The CRR value depends on test result data such as CPT or SPT test results because it is generally correlated with in situ parameters, such as CPT resistance penetration value, SPT number of strokes, or shear wave velocity, Vs [4]. (Equation 1)

$$CRR_{M=7.5,\sigma'_{vc}=1} = \exp\left(\frac{(N_1)_{60cs}}{14.1} + \left(\frac{(N_1)_{60cs}}{126}\right)^2 - \left(\frac{(N_1)_{60cs}}{23.6}\right)^3 - \left(\frac{(N_1)_{60cs}}{25.4}\right)^4 - 2.8\right) \quad (1)$$

$CRR_{M=7.5,\sigma'_{vc}=1}$ is the CRR value in the moment magnitude (M) = 7,5 and overburden pressure (σ'_{vc}) = 1 atm

3.1.2. Cyclic Stress Ratio (CSR)

Liquefaction that occurs in saturated sand soils can be caused by seismic stress ratio (SSR) or commonly called CSR caused by earthquakes. The magnitude of CSR is influenced by maximum earthquake acceleration, earth acceleration, total soil stress, effective soil stress, and depth factor reduction. The formula still uses assumptions on earthquakes with a magnitude of 7.5 and overburden pressure (σ'_{vc}) = 1 atm, so it needs a correction factor by entering parameters (MSF) is the correction factor of magnitude, K_σ is the correction factor of overburden pressure with a maximum value of 1.1 (Equation 2) [4].

$$CSR_{M,\sigma'_v} = 0,65 \frac{\sigma_v}{\sigma'_v} \frac{\alpha_{max}}{g} \frac{1}{MSF} \frac{1}{K_\sigma} \quad (2)$$

3.1.3. Safety of Factor (FS)

If the value of soil resistance to liquefaction CRR is greater than the loading that occurs or CSR then the soil is safe from liquefaction. However, if the value of CRR is smaller than CSR, the land is not safe from liquefaction.

The higher the safety factor against liquefaction the more resistant the soil is to liquefaction. However, Soils that have a safety factor slightly greater than 1.0 can be liquefied during earthquakes. For example, if the lower layer melts, the resulting upward flow of water may lead to a layer above it melting even though its safe factor is slightly greater than 1. [6] Proposed a safety factor value against potential liquefaction of 1.3. Soil conditions with a safety factor value of less than 1.3 will have the possibility of liquefaction, if the soil has a safety factor equal to 1.3 then the soil condition is in critical condition, and soil with a safety factor of more than 1.3 will be safe from liquefaction.

3.2. Liquefaction Potential Index (LPI)

Analysis using LPI can be carried out based on liquefaction analysis with the simplified procedure method. This method is used to find out the limits of the potential liquefaction that will occur [5] developed a relationship between the thickness of the liquefied soil and the non-liquefied soil at a given soil level acceleration, to determine whether the liquefaction that occurs can give rise to identifiable manifestations on the surface.

Assumptions in the development of liquefaction level analysis using the LPI method include soil depth that has the potential for liquefaction from the surface to a depth of 20 meters. The LPI assumes that the damage to liquefaction is proportional to the following conditions: Liquefied layer thickness with a maximum depth of 20 meters, Distance of liquefied layer to soil, number of layers with a safety factor value of less than one (FS<1). Soil Type Parameters [5] states that liquefaction events that have occurred during earthquakes have been found in deposits containing deposits of fine to medium sand and sand containing a low plasticity ratio. Thus, the type of soil susceptible to liquefaction is nonplastic (noncohesive) soil. It is estimated that noncohesion soils that are prone to liquefaction are clean sands, noncohesive silty sands, noncohesive silt and gravel.

4. Analysis Result

The From the formulation above (Equation 1, Equation 2), a replacement of the pore water pressure value of the calculation result is carried out, replaced by entering the pore water pressure value of the measurement results directly at the research site (**Figure 2**). To obtain analysis results that are more in line with the conditions at LUSI. We plot the pore pressure value data from measurements in the existing formula until the value of the safety factor change is obtained.

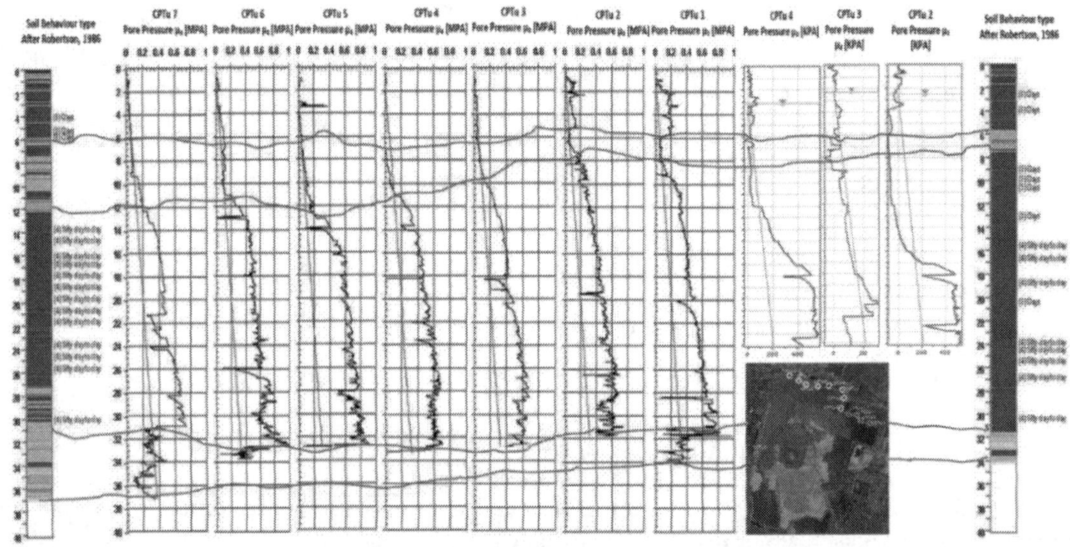

Figure 2. Pore pressure value in the study area

The correlation between measurements is focused on the presence of loose sand layers that have the potential to the liquefaction process, the distribution pattern of this sand layer greatly determines the level of liquefaction potential of each area the thicker and the shallower the position of the sand layer will give value of the potential liquefaction higher. The results of the CPT-u readings on the sand layer show an anomaly pattern of high pore water pressure of 40-70% of the normal pore water pressure condition, and the closer to the surface the anomalous to be smaller and at the top of the sand layer will provide normal pore water pressure readings. This pattern is different from the pattern of clay or silt layers, in this layer anomaly the pore water pressure is able to reach 2-3 times the normal pore water pressure

The earthquake event used, the Tuban earthquake in 1950 with a magnitude of 6.6 and a depth of 45 km, it's based on the earthquake catalog from BMKG, earthquake events were obtained around the study area, where the area of the seismicity study area was made ± 110 km from the location of the embankment [7].The maximum peak soil acceleration value present at the study site based on Indonesia Earthquake Hazard Map in 2017 [8]. According to the peak acceleration map in bedrock (SB), the acceleration at the study site was 0.15 g - 0.2 g, The value of the site coefficient is seen from the SE site class for the study site of 1.9, then the maximum acceleration value on the surface (amax) is 0,28-0.38 g.

The calculation results based on the idriss and boulanger methods and LPI calculations, it appears that the results change due to the input of high pore water pressure anomaly conditions parameters at the study site. Changes occur in the decrease in the safety factor value ranging from 0.1-0.4 (**Table. 1,** and **Table. 2**) And the change in the LPI value from originally low to high, and originally high to very high. The anomaly input of pore water pressure in CPT-u measurement results ranges from 40%- 70% higher than normal pore water pressure in the sand layer.

Table1. Safety factor analysis results BH1-BH5, compared to FS results without Pore Pressure and FS' inputs with pore presure parameter

NO	DEPTH	BH1		BH2		BH3		BH4		BH5	
		FS	FS'	FS	FS'	FS	FS'	FS	FS'	FS	FS'
1	0-2	-	-	-	-	0,38	0,38	0,73	0,73	0,62	0,62
2	2-4	0,72	0,72	-	-	0,34	0,34	0,35	0,35	0,52	0,35
3	4-6	0,46	0,46	0,31	0,31	0,32	0,32	0,38	0,38	0,35	0,12
4	6-8	0,46	0,46	0,31	0,20	0,28	0,16	0,28	0,17	1,13	1,13
5	8-10	0,27	0,20	0,31	0,06	0,27	0,05	0,26	0,05	-	-
6	10-12	0,29	0,12	-	-	-	-	-	-	-	-
7	12-14	-	-	-	-	-	-	-	-	-	-
8	14-16	-	-	-	-	-	-	-	-	-	-
9	16-18	-	-	-	-	-	-	-	-	-	-
10	18-20	-	-	-	-	-	-	-	-	-	-

Table2. Safety factor analysis results BH11-DB01, compared to FS results without Pore Pressure and FS' inputs with pore presure parameter

NO	DEPTH	BH11		BH12		BH14		BH15		DB01	
		FS	FS'	FS	FS'	FS	FS'	FS	FS'	FS	FS'
1	0-2	-	-	-	-	-	-	-	-	0,49	0,49
2	2-4	-.	-	-	-	0,62	0,35	-	-	0,30	0,30
3	4-6	-	-	1,00	0,83	-	-	-	-	0,91	0,91
4	6-8	0,62	0,62	0,97	0,68	-	-	0,67	0,28	0,45	0,45
5	8-10	0,56	0,42	-	-	-	-	-	-	0,45	0,31
6	10-12	0,60	0,32	-	-	-	-	-	-	0,29	0,29
7	12-14	-	-	-	-	-	-	-	-	-	-
8	14-16	-	-	-	-	-	-	-	-	-	-
9	16-18	-	-	-	-	-	-	-	-	-	-
10	18-20	-	-	-	-	-	-	-	-	-	-

Comparing the results of the analysis of liquefaction potential by entering the parameters of high pore pressure and with normal pore pressure. The results showed that a change from the original low liquefaction potential to a high liquefaction potential (in the west and south areas), and from the beginning the high liquefaction potential to a very high potential (in the northern and eastern areas), as shown on the **Figure 3** and **Figure 4**.

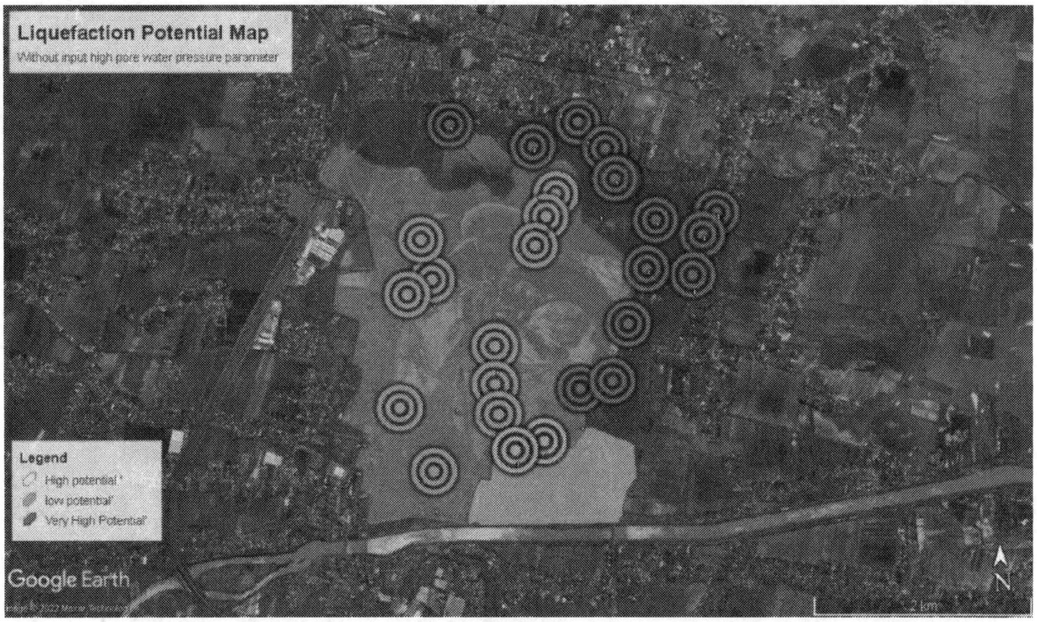

Figure 3. Potential Liquefaction based on (LPI), without input high pore presure parameter

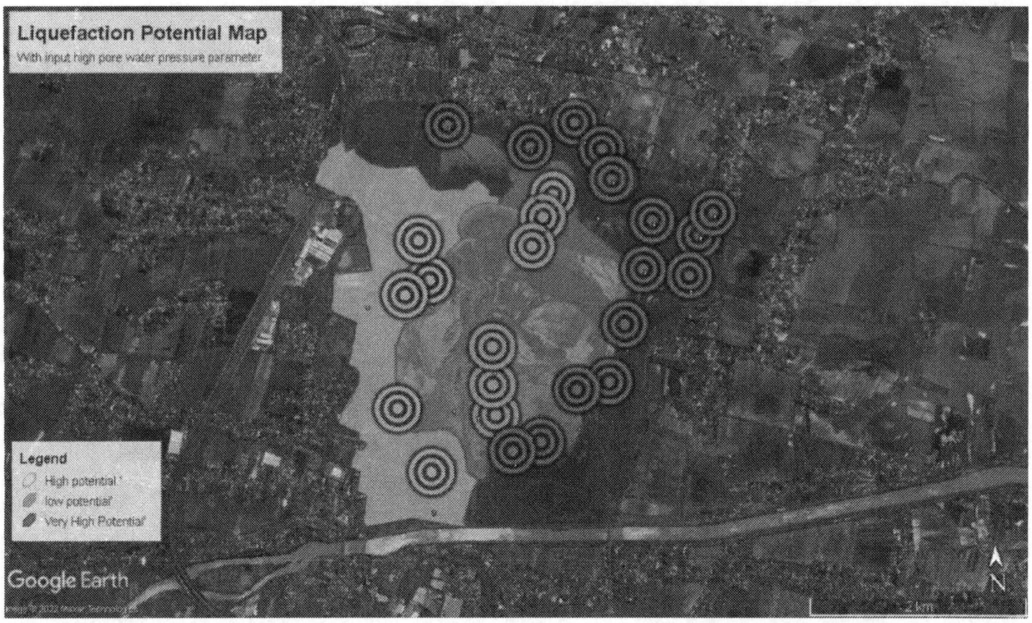

Figure 4. Potential Liquefaction based on LPI, with input parameter hight pore presure peremeter

Indications of liquefaction in the shallow layers are also found in the form of sand boils and soft-sediment deformation structures, occurs in the southern part of the area.Occurs in areas with high Liquefaction potential (LPI value > 5) this pattern corresponds to the results of previous studies by [9] Damage due to liquifaction occurs in soils with LPI > 5.
In the area inside the mud embankment gives a low liquefaction potential result, this result is largely determined by the composition of the constituent area in the embankment composed of mud material / LUSI material about 84.47% consists of fine-grained soil dominated by clay size particles [10]. From this composition will give a low liquefaction potential value to the area around the center of the mudflow.

5. Discussion

The research was conducted with Empirical Analysis of Soil Liquefaction potential with SPT based evaluation, of course, many other methods can also be done as a comparison.
Changes in the results of the analysis with the input of the existing pore water pressure parameter are very dominant in affecting the change in the results of the analysis, but it is necessary to input other parameters that have not been covered much by the existing formula, for example, such as the input shear strength parameters such as cohesion and the angle of internal friction need to be carried out further research
The problem in determining the potential of Liquefaction is that proving the level of accuracy of the analysis results is difficult to verify.

6. Conclusion

Analysis of the potential for liquefaction in the study area showed that the results of the high- very high liquefaction potential in the northern and eastern research areas, and the low Liquefaction potential was in the southern and western areas.
The distribution of high-very high liquefaction potential is influenced by the distribution pattern of loose sand at shallow depths with low fine content, SPT values of less than 15, groundwater level conditions of less than 1m, and the dominant parameter is the high pore water pressure on the sand layer.
The calculation results based on the Idriss and Boulanger methods and LPI calculations, it appears that the results change due to the input of high pore water pressure anomaly conditions parameters at the study site. Changes occur in the decrease in the safety factor value ranging from 0.1-0.4 And the change in the LPI value from originally low to high, and originally high to very high. The anomaly input of pore pressure in CPT-u measurement results ranges from 40%- 70% higher than normal pore water pressure in the sand layer
The effects of liquefaction process are often found in the phenomenon of sand boils, soft sediment deformation structures, and subsidence in the research area

7. Acknowledgments

The authors are grateful to those who made the work possible for Ministry of Public Works and Housing and Pusat Pengendalian Lumpur Sidoarjo which have support the data and all parts of this article.

References

[1] Tanikawa W, Sakaguchi M, Wibowo H T, Shimamoto T, Tadai O, 2010. Fluid transport

properties and estimation of overpressure at the Lusi mud volcano, East Java Basin. Eng. Geol. 116, 73e85

[2] Seed R B and Idriss I M, 1971 J. Soil Mech. Found. Div. ASCE97, 1249

[3] Untung M and Sato Y, 1978, Gravity and geological studies in Java, Indonesia: Geological Survey of Indonesia and Geological Survey of Japan,Special Publication 6, 207 p

[4] Idriss I M and Boulanger R W, 2008 Soil Liquefaction During Earthquakes Earthquake Engineering Research Institute (EERI).

[5] Ishihara K and Yoshimine M, "Evaluation of Settlements in Sand Deposits Following Liquefaction During Earthquakes," Soils and Foundations. 32(1), 173-188 (1992).

[6] Martin G R. & Lew M (eds). 1999. Recommended procedures for implementation of DMG Special Publication 117 guidelines for analysing and mitigating liquefaction in California. Southern California Earthquake Center, University of Southern California, Los Angeles, CA.

[7] BMKG, Badan Meteorologi, Klimatologi, dan Geofosika, [Online]. Available: https://dataonline.bmkg.go.id/data_gempa_bumi. (Accessed March 2022).

[8] National Center of Earthquake Studies, 2017, Earthquake Source and Hazard Maps (Research and Development Agency of the Ministry of Housing and Public Works, Bandung).

[9] Toprak S and Holzer T L, 2003 J. Geotech. Geoenviron. Eng. ASCE 129(4), 315

[10] L. Handoko, A. Rifai, N. Yosufuku, & R. Ishikura, "Physical properties and mineral content of Sidoarjo mud volcano", Procedia Engineering 125, 324 – 330, (2015).

[11] Ministry of Energy and Mineral Resources, 2019, Atlas Zona Kerentanan Likuefaksi Nasional (Geological Agency, Bandung,).

Failure Investigation of Braced Excavation in Soft Clays: Case Study on the Collapse of Nicoll Highway

Ari Surya Abdi[1], **Chang-Yu Ou**[2]
[1]Department of Civil & Construction Engineering, National Taiwan University of Science and Technology, No. 43, Section 4, Keelung Rd., Da'an District, Taipei City 106, Taiwan
[2]Department of Civil & Construction Engineering, National Taiwan University of Science and Technology, No. 43, Section 4, Keelung Rd., Da'an District, Taipei City 106, Taiwan
Email: arisuryaabdi.asa@gmail.com

Abstract. Deep excavation in clay normally causes large soil movement, which may induce basal heave failure. In this study, the three-dimensional finite element analysis with the strength reduction method was employed to evaluate the typical failure mechanism of braced excavation in soft clays. The collapse of Nicoll Highway was selected as a failure case history and further conducted by considering the full elastoplastic structural behavior. The results showed that the computed wall displacements were successfully verified through the observation data. It was also found that the excavation failure was initiated by the yielding of the strut-waler connection, which was similar to that of the field observation. Finally, the proposed finite element model adopted in this study can be used as a design recommendation to evaluate the failure mechanism and further prevent possible braced excavation failure in soft clays.
Keywords: basal heave, clays, deep excavation, elastoplastic structural, finite element

1. Introduction

The bracing system is commonly adopted in deep excavations to ensure the diaphragm wall can withstand construction. In general, the bracing system consists of the center post, horizontal steel strut, and steel waler. The horizontal steel struts are installed perpendicular to the diaphragm wall to resist the earth pressure acting behind the wall, while the steel waler is applied to transfer the force acting on the wall to the horizontal strut. Furthermore, the function of center posts is to support the strut to not fall out due to their own mass [1].

As a matter of fact, excavation activities in clay often resulted in a large soil movement behind the wall. Consequently, the stress of the structural support system may exceed its strength material and induce excavation failure. According to previous studies [2–5], the yield of the diaphragm wall and the structural bracing system could trigger basal heave failure. For example, the collapse of the Nicoll Highway excavation case was initially caused by the buckling of the strut-waler connection, followed by the failure of the diaphragm wall and construction area [6]. Hence, an appropriate design of the structural system should be considered properly in braced excavations.

In this study, the collapse of the Nicoll Highway excavation case was investigated to evaluate the typical braced excavation failure mechanism. A three-dimensional finite element method (3D FEM) was used to model the actual excavation geometry accurately. The full elastoplastic structural behavior was used to let the structural model could yield during the analyses. Furthermore, the strength reduction method was adopted in the stability analysis to assess the factor of safety against basal heave at the final excavation stage.

2. Numerical Methods

2.1. Collapse of Nicoll Highway Excavation Case

The Nicoll Highway excavation case was part of the MRT Circle Line project located in Singapore. The project was occupied a 211-m-long, while the final excavation level was planned to go up to 11th stage (GL – 33.6 m). However, the collapse occurred at the 10th stage (GL – 30.6 m), while the failure was extended from Type M3 area to M2 area, as illustrated in Figure 1. According to the COI report [6], the buckling of strut-waler connection was initially found during the 7th excavation stage. Prior to collapse, the buckling of strut-waler connection was also found at the ninth level of strut (10th excavation stage), which further caused the excavation failure, resulting the crater with a diameter of about 100 m.

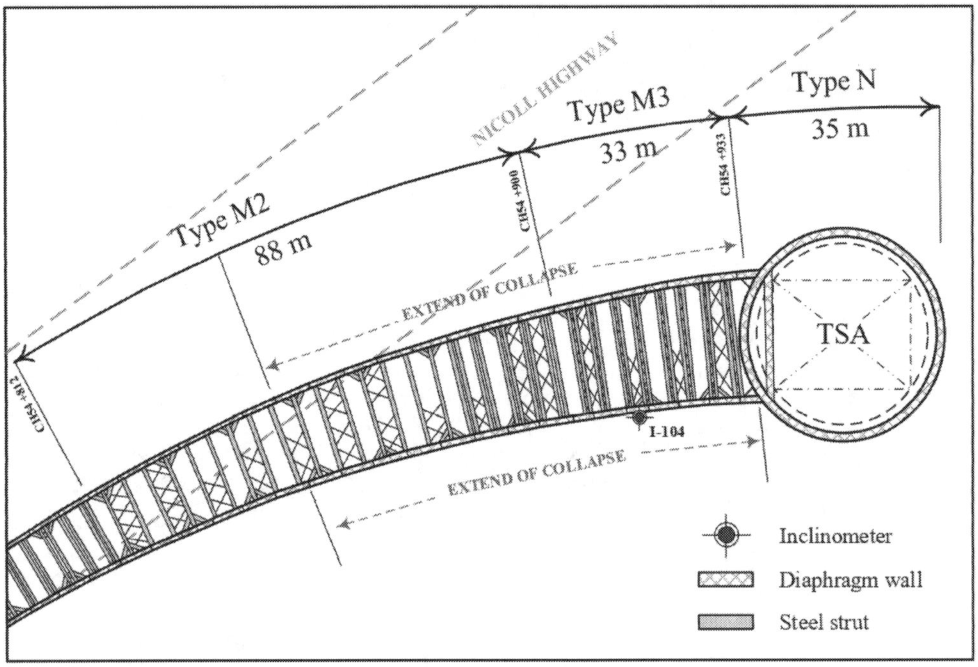

Figure 1. Detail of the collapsed area in the Nicoll Highway excavation case.

The soil layer dominantly consists of marine clay, while the old alluvium (defined as hard stratum) is found approximately GL – 41.5 m below the ground surface. The ground surface was covered by backfill with a unit weight (γ_t) of 19 kN/m^3, while the unit weight of marine clay ranged from 16 to 16.8 kN/m^3. The unit weight of the upper fluvial deposit ranged from 19 to 20 kN/m^3, whereas the unit weight of estuarine deposits was 16.8 kN/m^3. The unit weight of the lower fluvial deposit and old alluvium is found to be 20 kN/m^3. In addition, the groundwater level was located at Gl -2.4 m.

The excavation was supported by a diaphragm wall with a length and thickness are 44.3 m and 0.8 m, respectively. The bracing system was adopted to restrain the excessive wall displacement, consisting of nine strut levels arranged by an average horizontal spacing of 4 m. Two continuous jet grout piles (upper and lower JGP parts) were implemented near the final excavation level to support the diaphragm wall during excavation. The soil condition, detailed structural support system, and excavation profile are detailed in Figure. 2.

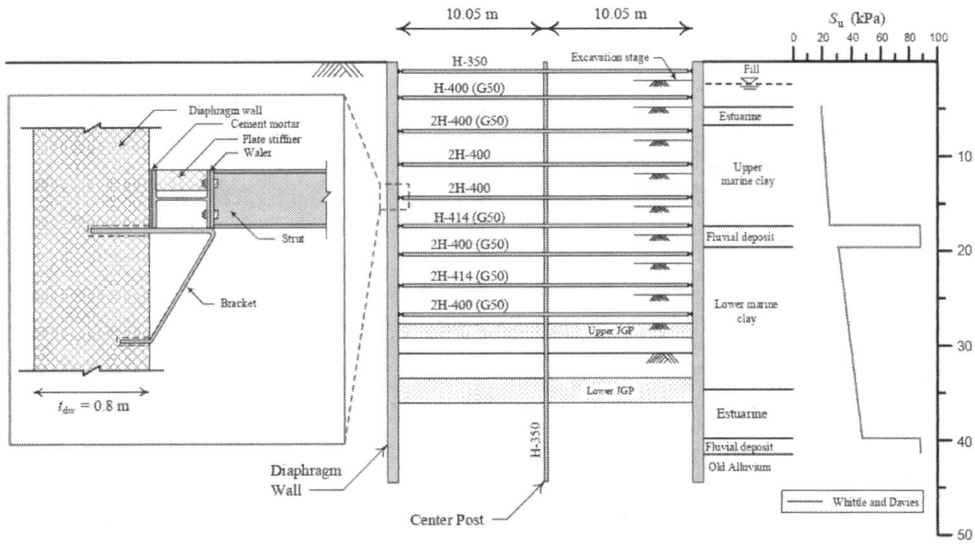

Figure 2. Detail of structural support system, excavation profile, and soil condition in the Nicoll Highway excavation case (s_u was obtained from Whittle and Davies [7]).

2.2. Determination of soil and structural properties

The three-dimensional finite element method (3D FEM) was performed by using the commercial geotechnical software PLAXIS 3D v.21 [8]. The soil was modeled by a 10-node tetrahedral element, while the Mohr-Coulomb soil model was used to simulate the soil strength characteristic near failure. In the analyses, the backfill layer was performed by drained material, in which the cohesion (c'), internal friction angle (ϕ'), and Poisson's ratio (v) were assumed to be 0.1 kPa, 30 °, and 0.3, respectively. Young's modulus of the backfill layer (E') was estimated based on the N-SPT value ($E' = 2000N$), where the N value of backfill was approximately 5. On the other hand, the undrained material was adopted to simulate the soft marine clay with $c_u = s_u$ (see Figure. 2), $\phi = 0$, and $v_u = 0.495$. The value of $E_u = 200s_u$ was chosen for the typical marine soft clay in Singapore. The design of undrained shear strength and Young's modulus of the JGP layer were 300 kPa and 150 MPa, respectively [6].

The plate element was used to model the diaphragm wall, while the horizontal strut and center post were modeled as beam elements. The nominal value for Young's modulus of diaphragm wall concrete was estimated by $E = 4700\sqrt{f_c'}$ (f_c' = 40 MPa), following the suggestion by ACI 318-95 [9]. A reduction factor of 0.8 was used for wall stiffness to consider the cracks caused by the large bending [10,11]. On the other hand, the yield stress (σ_y), Poisson's ratio (v_s), and Young's modulus of steel material were assumed to be 250 MPa, 0.3, and 2.04×10^8 kPa, respectively. The plastic axial stress (N_p) and plastic bending moment (M_p) were estimated based on σ_y value, cross-section area (A), and section modulus of material (S). A reduction of 10% in the strut and center post parameters (i.e., E and σ_y) was applied to represent the improper installation during construction.

Furthermore, because the buckling could occur in the connection between the horizontal strut and waler, the strut-waler connection was considered in the analyses. According to Abdi and Ou [2], the strut-waler connection could be represented as a short beam element, which is typically in the range of 0.3 m to 0.4 m (depending on the steel waler dimension). Under such a case, it is expected that the yield of the strut-waler connection is mainly due to the axial force, while the bending moment is relatively small. The reduction of σ_y value for the strut–wall connection parameter (α = 0.7) was considered to represent the possible buckling failure caused by the large compression.

Moreover, a 12-node interface element was employed to model the interaction behavior between the wall (including the diaphragm and buttress wall) and the surrounding soil. The roughness of the interface element is controlled by the strength reduction factor (R_{inter}), which is related to the friction ratio between the wall adhesion and the soil cohesion ($\alpha = s_u/c_w$). R_{inter} value is set to be 0.67, following the typical roughness of concrete surface [1].

The 3D FEM modeling was focused on the Type M3 area (largest subsided area). The horizontal boundary was extended far from the excavation area to minimize the effects of the boundary restraints. The nodes along the side and bottom boundaries were constrained by the roller and fixed boundary, respectively. A very fine mesh with local refinement was adopted in the analyses to provide an accurate numerical solution. Figures. 3(a) and (b) show the typical finite element mesh and detail of structural modeling in the analyses.

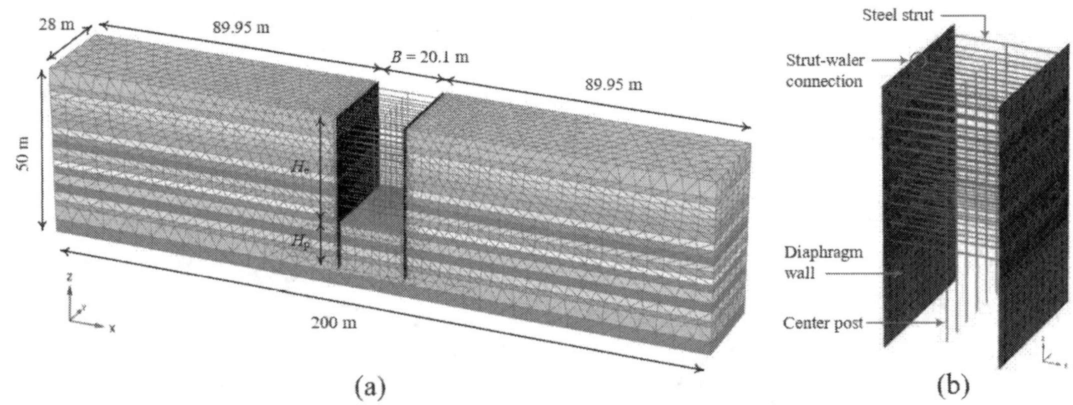

Figure 3. Typical excavation geometry used in the 3D FEM analyses: (a) finite element mesh; (b) detail of structural modeling.

2.3. The Concept of Strength Reduction Method

In general, the safety factor is usually defined as the ratio of the collapse load over the working load. Indeed, this concept could be adopted to assess the safety factor for foundation engineerings, such as footings or piles. However, for the slope and excavation case, the failure was initially induced by the load from the overburden pressure. In such a case, the failure surface is highly affected by the soil strength. Thus, other methods to determine the safety factor could be adopted, such as the strength reduction method. The strength reduction method is generally performed by reducing the soil strength until the failure occurs (i.e., divergence condition) [2,12–17]. In this analysis, the corresponding strength reduction is defined as the factor of safety against basal heave ($FS_{fem} = SR_{max}$). Figures. 4(a) and (b) illustrate the reduction process of the Mohr-Coulomb (MC) failure envelope. Thus, the concept of the strength reduction method in relation to the MC failure criterion can be expressed as:

$$\tau = c + \sigma_n \tan\phi \text{ (MC failure criterion)} \quad (1)$$

$$c_r = \frac{c}{SR};\ \phi_r = \tan^{-1}\left(\frac{\tan\phi}{SR}\right) \text{ (during strength reduction)} \quad (2)$$

IOP Conf. Series: Earth and Environmental Science 1184 (2023) 012010 doi:10.1088/1755-1315/1184/1/012010

$$c_f = \frac{c}{FS_{\text{fem}}};\ \phi_f = \tan^{-1}\left(\frac{\tan\phi}{FS_{\text{fem}}}\right) \text{ (at failure [maximum strength reduction])} \quad (3)$$

where τ is the shear strength; σ_n is the normal stress along the shear plane; c is the cohesion; ϕ is the internal friction angle; SR is the strength reduction value; FS_{fem} is the factor of safety at maximum strength reduction; and parameters with the subscript "f" refer to the reduced strength at failure.

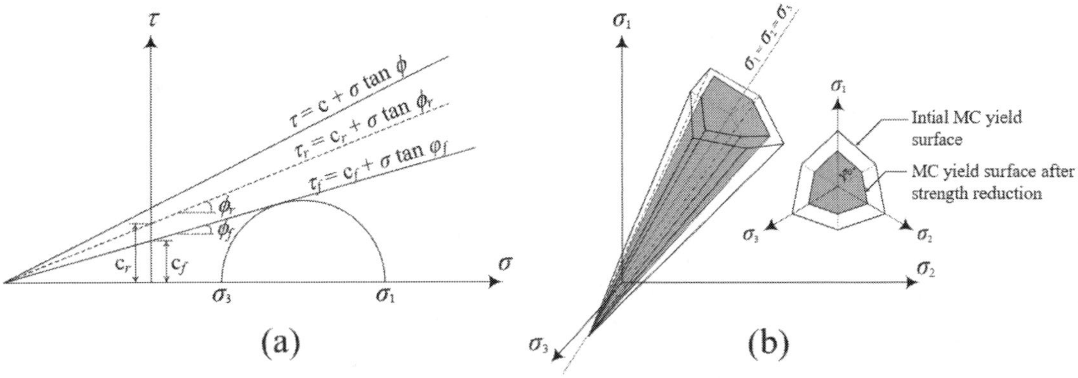

Figure 4. The strength reduction concept for the MC soil model: (a) linear envelope in the Mohr diagram; (b) pyramidal surface in principal stress space and cross-section in the equipressure plane.

In addition, the strength reduction is also applied to the interface strength element during the stability analyses. The tolerated error was set to be 0.01, following the default from the PLAXIS for the convergence criteria of global error. The arc-length control procedure was used in the analysis to obtain reliable collapse loads during strength reduction. Furthermore, a very fine mesh with local refinement in the interesting area was applied to provide accurate numerical solutions.

3. Results and Discussions

3.1. Excessive Wall Displacement

The inclinometer data located in the Type M3 area (I-104, see Figure. 1) was used to validate the 3D FEM results. The design of maximum wall displacement is supposed to be around 14.5 cm. However, excessive wall displacement occurred during construction. According to the COI report [6], the maximum wall displacement at the 10th excavation stage reached 35 cm, which was located near the excavation level. This indicates that the demolishing of the upper JGP induced excessive wall movement and further caused the failure of the diaphragm wall.

Figure. 5 shows the comparison between the computed wall displacement (FEM results) and field measurement (I-104). For the field measurement in stage 2, the wall displacement is like a cantilever shape (i.e., the largest wall displacement occurred near the ground surface). As the excavation activity goes deeper (stage 10), the trend of wall displacement along the depth is more like a curvilinear shape, in which the largest wall displacement was found near the final excavation level. This is due to the horizontal struts that restrain the upper part of the diaphragm wall. Furthermore, the results showed a close agreement between the FEM results and field measurement for the initial stage (2nd excavation stage) and final stage (10th excavation stage). This clearly implies that the FEM results were validated through the observation data. Hence, the 3D FEM analysis adopted in this study could be reasonably used to further assess the failure mechanism of braced excavation.

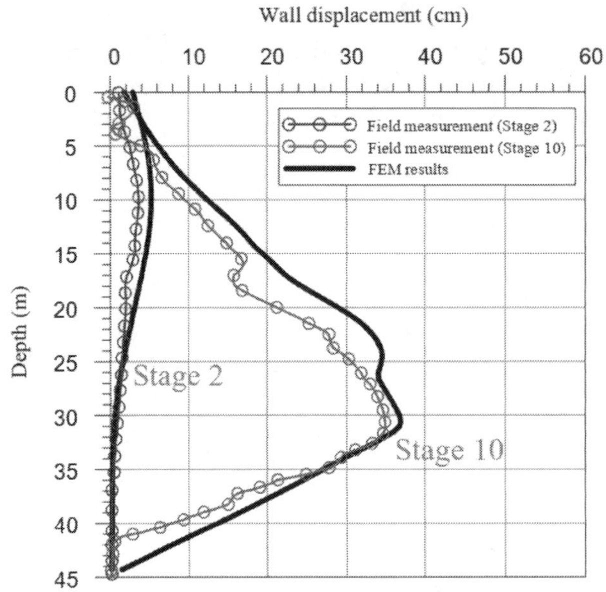

Figure 5. Comparison of computed wall displacement and field measurements.

3.2. Basal Heave Failure

The stability analysis was performed at the final excavation stage (10th excavation stage). The soil strength was successively reduced until the convergence solutions no longer remained (i.e., divergence condition), as mentioned in the preceding section. According to the stability results, it is found that the maximum strength reduction is 1.120 at the divergence condition, which was defined as the factor of safety against basal heave. This indicates that the final excavation stage in the FEM analysis was prior to the collapse, which was similar to the field observation.

Figure. 6(a) shows the displacement vector at divergence condition. As shown in the figure, the soil behind the wall was moving upward, passing through the lower part of the diaphragm wall, causing the large upward soil movement inside the excavation. The soil movement further caused the upward movement of the center post, resulting in the large bending of horizontal struts. Indeed, this phenomenon is similar to that of the field observation, where the center post was moving upward when the excavation failure occurred.

In addition to the stability analysis, the generation of the plastic point was further investigated, as shown in Figure 6(b). It is found that the plastic points were concentrated below the final excavation surface and extended far behind the diaphragm wall. It is due to the large soil movement that caused the shear failure and formed the typical failure surface. Besides, the range of the extended failure area from the excavation center was found to be around 49 m, which was similar to the actual failure condition (the failure area was extended with a radius of 50 m). Hence, it is concluded that the 3D FEM modeling adopted in this study should be considered for the proper design of braced excavation in soft clays.

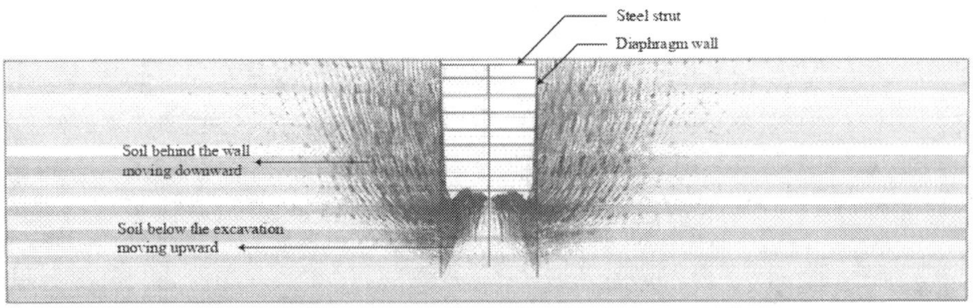

(a) Displacement vector at divergence condition

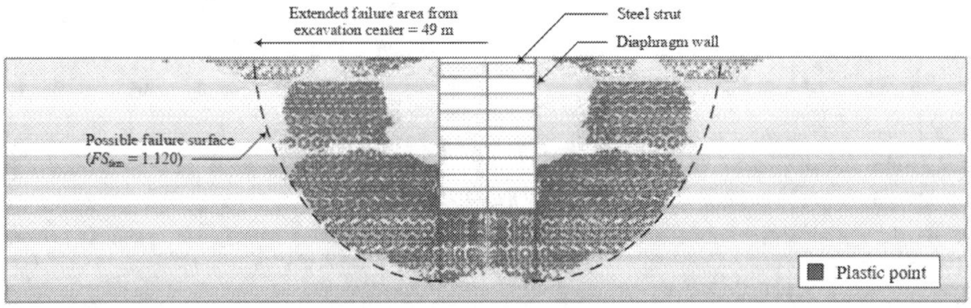

(b) Plastic point generation at divergence condition

Figure 6. The stability results at divergence condition: (a) displacement vector; (b) plastic point generation.

3.3. Internal Force of Structural Support System

The internal force curve of the structural support system (horizontal steel struts and strut-waler connection) was presented to evaluate the plastic behavior bracing system from the initial excavation stage until the failure occurred, as plotted in Figure. 7. The boundary line (*BL*) herein has an implication of the ultimate combination between the plastic axial force (N_p) and the plastic bending moment (M_p). Under such conditions, when the stress acting on the structural element exceeds the BL value, the plastic condition of the structural material has been reached.

Figure 7(a) shows the detailed schematics of the structural level and excavation stage. It should be noted that the number inside the circle indicates the excavation stage. In the internal force curve (see Figure 7[b]), the sequence number of the excavation stage was started after installing the strut and connection. For example, because the 1st level of strut and connection were installed after completing the 1st excavation stage, the internal force development was started from the 2nd excavation stage. On the other hand, the 2nd level of strut and connection was initially developed from the 3rd excavation stage.

As shown in Figure 7(b), the strut-waler connection is mainly subjected to the axial force, while the bending moment is relatively small. It is because the joint between the diaphragm walls and the strut-waler connection is assumed to be hinged, while its length is typically short. Hence, the internal force is only developed in the N_p axis. The axial force in the strut-waler connection was generally decreased during the initial excavation stage and relatively stable at a certain stage. For the horizontal steel strut, the internal force developed in the N_p and M_p axis, which was different from that of the strut-waler connection. It is due to the large upward movement of the center post, causing the bending in the horizontal steel strut, as mentioned earlier. The bending moment gradually increased at the certain stage,

IOP Conf. Series: Earth and Environmental Science 1184 (2023) 012010 doi:10.1088/1755-1315/1184/1/012010

particularly at the divergence condition. Nevertheless, the internal force of 1st and 2nd level of strut and strut-waler connections did not reach the plastic condition.

The internal force curve of 4th, 5th, and 6th are shown in Figure 7(c). It is found that the 6th level of connection reached the boundary line, implying that the strut-waler connection was subjected to buckling failure. Furthermore, the 9th level of connection also achieved the limiting value of axial force, as indicated in Figure 7(d). This further caused the rapid numerical solution to reach the divergence condition, resulting in a lower factor of safety (FS_{fem} = 1.120). Besides, these findings are similar to that of the field observation (the buckling occurred at 6th and 9th levels of the strut-waler connection). Hence, the proper design of a structural support system should be highlighted when designing braced excavation.

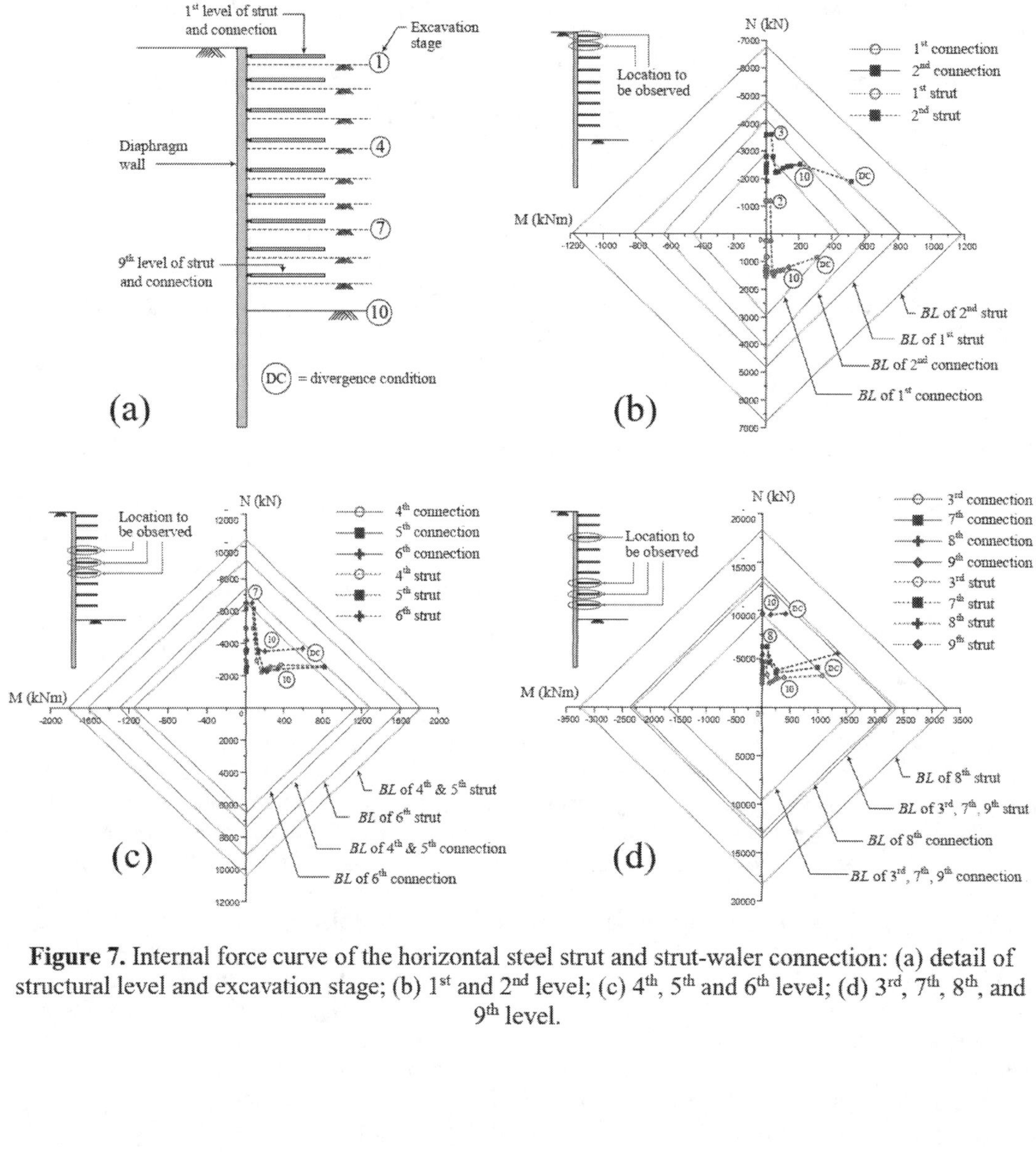

Figure 7. Internal force curve of the horizontal steel strut and strut-waler connection: (a) detail of structural level and excavation stage; (b) 1st and 2nd level; (c) 4th, 5th and 6th level; (d) 3rd, 7th, 8th, and 9th level.

4. Conclusion

The three-dimensional finite element analysis with the strength reduction method was employed to evaluate the excessive wall displacement and failure mechanism in the collapse of the Nicoll Highway excavation case. The structural support systems, such as the diaphragm wall, center post, and horizontal steel strut, were modeled as elastoplastic structural behavior to represent the actual failure condition. In addition, the strut-waler connection model was also proposed to consider the possible buckling failure in the connection between strut and waler. Furthermore, the following conclusions can be drawn:

1. The close agreement between the computed wall displacement and field measurement can be obtained by considering the proposed FEM model.
2. A similar failure mechanism was obtained between the FEM model and the actual failure condition, where the excavation failure was initiated by the yielding of the strut-waler connection, followed by the failure of the horizontal strut and surrounding soil.
3. The proposed FEM model can be used as a design recommendation to evaluate the failure mechanism and further prevent possible basal heave failure in future construction.

References

[1] Ou C-Y 2006 Deep Excavation Theory and Practice *Taylor Fr.*
[2] Abdi A S and Ou C-Y 2022 A Study of the Failure Mechanism of Braced Excavations Using 3D Finite-Element Analysis *Int. J. Geomech.* **22** 1–14
[3] Do T N, Ou C Y and Chen R P 2016 A study of failure mechanisms of deep excavations in soft clay using the finite element method *Comput. Geotech.* **73** 153–63
[4] Ukritchon B, Whittle A J and Sloan S W 2003 Undrained Stability of Braced Excavations in Clay *J. Geotech. Geoenvironmental Eng.* **129** 738–55
[5] Choosrithong K and Schweiger H F 2020 Numerical Investigation of Sequential Strut Failure on Performance of Deep Excavations in Soft Soil *Int. J. Geomech.* **20** 04020063
[6] COI C of I 2004 Report of the Committee of Inquiry into the incident at the MRT circle line worksite that led t the collapse of Nicoll Highway
[7] Whittle A J and Davies R V 2006 Nicoll Highway collapse: evaluation of geotechnical factors affecting design of excavation support system *Int. Conf. Deep Excav.* **28** 30
[8] Brinkgreve R, Engin E and Swolfs W 2013 PLAXIS 3D 2019 reference manual 470
[9] ACI A C I 2014 Building Code Requirements for Structural Concrete *Am. Concr. Inst.* 524
[10] Lim A, Hsieh P G and Ou C Y 2016 Evaluation of buttress wall shapes to limit movements induced by deep excavation *Comput. Geotech.* **78** 155–70
[11] Lim A, Ou C Y and Hsieh P G 2019 An innovative earth retaining supported system for deep excavation *Comput. Geotech.* **114** 103135
[12] Goh A T C, Zhang W G and Wong K S 2019 Deterministic and reliability analysis of basal heave stability for excavation in spatial variable soils *Comput. Geotech.* **108** 152–60
[13] Goh A T C 2017 Deterministic and reliability assessment of basal heave stability for braced excavations with jet grout base slab *Eng. Geol.* **218** 63–9
[14] Faheem H, Cai F, Ugai K and Hagiwara T 2003 Two-dimensional base stability of excavations in soft soils using FEM *Comput. Geotech.* **30** 141–63
[15] Faheem H, Cai F and Ugai K 2004 Three-dimensional base stability of rectangular excavations in soft soils using FEM *Comput. Geotech.* **31** 67–74
[16] Do T N and Ou C Y 2020 Factors affecting the stability of deep excavations in clay with consideration of a full elastoplastic support system *Acta Geotech.* **15** 1707–22
[17] Do T-N, Ou C-Y and Lim A 2013 Evaluation of Factors of Safety against Basal Heave for Deep Excavations in Soft Clay Using the Finite-Element Method *J. Geotech. Geoenvironmental Eng.* **139** 2125–35

The Potential of Bentonite and Chitosan Mixtures as Clay Liner Base Material

T Hidayat[1] Y F Arifin[1, 2]

[1]Civil Engineering Master's Program, University of Lambung Mangkurat, Indonesia
[2]Wetland Based Material Research Center, University of Lambung Mangkurat

y.arifin@ulm.ac.id

Abstract. A polymer-enhanced bentonite-sand mixture has been widely used as a clay liner. This research focuses on the potential use of chitosan as a natural polymer mixed with bentonite as a clay liner base material. The bentonite used is a type of calcium bentonite, which is easily obtained in the market. Meanwhile, chitosan is produced from shrimp shell waste and is also easily available in the market. The process of mixing chitosan with a percentage of 2–6% of bentonite was carried out using a certain method, which is described in detail in the article. The materials' physical, chemical, and engineering characteristics, such as Atterberg limits, permeability, and compressive strength, were examined after mixing. The results show that the mixing process was successful for the finding of minerals and functional groups based on the XRD and FTIR results. Promising results of the study include larger clusters, higher plasticity, decreased permeability, and higher compressive strengths of the mixtures after chitosan was added to bentonite.

1. Introduction

Considerable use has been made of bentonite as a material for constructing clay liners. It is the most favorable material for clay liners due to its water absorption and impermeability. In addition to being pricey, however, bentonite has a significant swelling potential, therefore it is usually mixed with other materials, such as sand, when it is used. Its application as a clay liner has been widely studied in relation to varying proportions of sand and bentonite [1]. The addition of sand decreases the mixture's swelling characteristics without altering its water resistance [2].

To improve its performance, bentonite is mixed with a polymer. In civil engineering, polymer and bentonite mixtures are not used separately. Mixing this material with sand improves its field compactability, swelling pressure, and shear strength. This substance is also known as polymer-enhanced bentonite-sand combination (PEBSM). Until now, the PEBSM has been widely used in the field. Some of the benefits of this material include almost no difference in water retention during the drying and wetting processes, a large ability to expand so that it has very low permeability, and no cracks when conditions are dry [3]. Field investigation reveals that the PEBSM layer prevented arsenic migration from tailings to the cover soil above it [4].

In recent years, natural polymers such as chitosan have been utilized for a wider variety of applications than synthetic polymers. Chitosan is an extract from the shells of crustaceans such as shrimp and crabs. Chitosan is widely produced in Indonesia because of its abundance as a material derived from shrimp and crab processing factories. As a natural polymer, chitosan has good characteristics, such as being biodegradable, non-toxic, and able to absorb heavy metals [5,6].

The success of using chitosan as an additive lies in the preparation process [7]. Several methods were used to combine chitosan and clay to achieve the best results for each application. Aguilar et al. [8] used a concentration of 0.5% mass with a volume ratio of chitosan and 1% acetic acid solution. It was found that approximately 1% chitosan was needed to obtain a material that could protect the soil surface from being easily eroded and 3% to increase the compressive strength of the sample. For a more effective preparation, Chen et al. [6] mixed 1 g of chitosan with 2% acidic acid for 2 hours, then added sodium tripolyphosphate solution, stirring for 4 hours. This process is continued by filtering and washing with water until a solution with a pH of 7 is obtained. This mixture was used to keep heavy metals like Cd and Pb in waste.

A mixture of chitosan and soil has also begun to be used for construction [8,9]. Aguilar et al. [8] conducted research on the possibility of using chitosan mixed with clay as an erosion-resistant material by utilizing the ability of the mixture to prevent water from entering the soil. Taytak et al [9] also use a mixture of chitosan with bentonite and kaolinite as the basic material for clay liners. The addition of chitosan to a mixture of bentonite and kaolinite can increase the density and reduce the permeability of the mixture.

This research discusses the compacted bentonite-chitosan mixture as a clay liner base material. Tests were carried out in the laboratory to determine the technical behavior of the bentonite-chitosan mixture based on its physical and hydraulic properties. This research is expected to be able to provide an overview and explain the results of a mixture of bentonite and chitosan as a potential material for a base for clay liners.

2. Material Used

In this investigation, calcium bentonite, which is commercially available in Indonesia, was utilized. The bentonite has a specific gravity of 2.71 and a water content of 14.17%. The soil used consists of 1.4% fine sand, 8.3% silt, and 90.3% clay. The liquid limit, plastic limit, shrinkage limit, and plasticity index of bentonite are 351.71, 44.68, 41.89, and 307.03%, respectively. The composition of exchangeable cations is 0.34 meq/g Na^+, 18.70 meq/g Ca^{2+}, 0.20 meq/g Mg^{2+}, and 0.58 meq/g K^+. Chitosan is made from shrimp shell waste that has been ground into a powder and is easily available on the market.

3. Methods and Procedures

Sample preparation

The samples used were bentonite and chitosan mixed with several treatments, following the methods used by several researchers [8,9]. In this study, three concentrations were prepared, namely 2, 4, and 6% chitosan in the mixture. Following are the procedures that were taken to prepare the composition of 2% chitosan:

1. Create an acidic solution with a concentration of 2% (v/v).
2. Add 2 g of chitosan powder to the solution prepared in Step 1.
3. Mix 98 g of bentonite in the solution made in Step 2 for two hours with a stirrer.
4. Mix the ingredients in step 3 with 100 mL of sodium tripolyphosphate solution (5.0 wt%) and stir for 4 hours.
5. The mixture in Step 4 is then filtered and washed with distilled water to pH 7.
6. The mixture is then put in the oven at 60°C until the powder is dry and can be used.
7. The above weights and percentages can be changed depending on the amount of mixture desired by means of multiples of each material.

Tests of bentonite-chitosan mixtures

Several tests were undertaken to determine the efficacy of the bentonite and chitosan mixture. Included in these tests are FTIR, SEM, and XRD [10,11]. Other engineering properties tested include the liquid limit and plastic limit tests, the permeability test, and the unconfined compression test. For the

IOP Conf. Series: Earth and Environmental Science 1184 (2023) 012011 doi:10.1088/1755-1315/1184/1/012011

permeability test and the unconfined compression test, the samples were statically compacted with a hydraulic jack at a dry unit weight of 16 kN/m^3 and a moisture content of 10%. A type of falling head permeability test was carried out for these samples.

4. Results and Discussions

Characterization of the Bentonite-Chitosan Mixtures

The XRD results of bentonite and bentonite-chitosan mixtures with chitosan percentages of 2, 4, and 6% are shown in Figure 1. In general, these results show an XRD pattern at 2θ from 10° to 70°. To analyze the pattern in more detail, the pattern is divided into 2θ-patterns at 18°–24° and 24°–30°, as shown in Figures 1(a) and 1(b). In Figures 1(a) and 1(b), the change in shape of the 2θ peak is shown at 20.9° and 26.7°, respectively. Changes in the peak of the XRD results in this range are due to the interaction of chitosan with the bentonite surface [12][13].

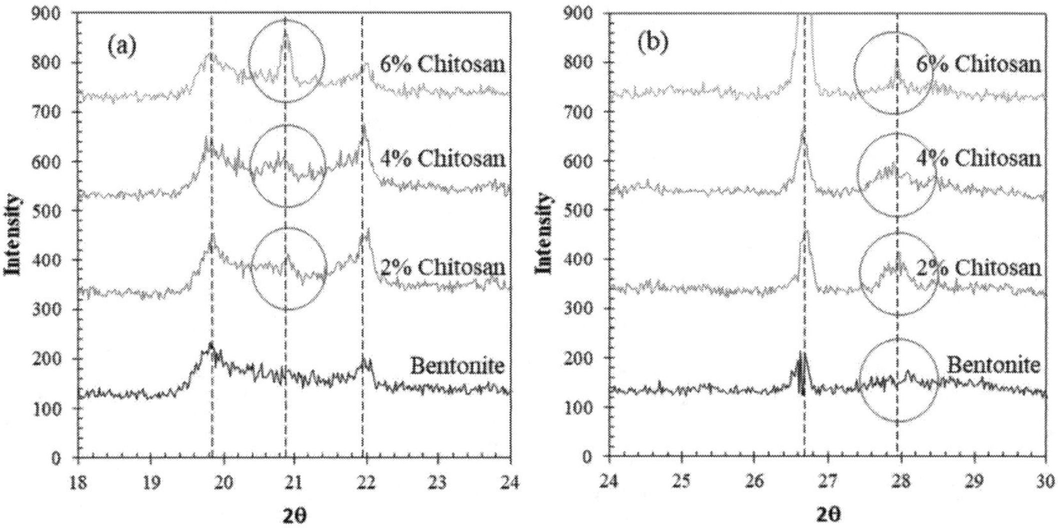

Figure 1. XRD results of the bentonite-chitosan mixtures at 2θ of (a) 18°–24° and (b) 24°–30°

Figure 2 shows the results of the FTIR test sample of a mixture of bentonite and chitosan. Some changes in the peaks of the FTIR results are seen at wavelengths of 3606, 2921, 2877, 1867, 1645, and 1589 cm^{-1}. Some of the wavelength peaks that approach these wavelengths are related to the interaction of chitosan and bentonite, such as 3615 cm^{-1} for OH stretching [13], 2910 cm^{-1} for C-H stretching (Hristodor et al., 2014), 2914 cm^{-1} for C-H stretching [12], 1642 cm^{-1} for amide group [12], 1640 cm^{-1} for N-H bending [14], and 1589 cm^{-1} for N-H main amine bending [14]. Stretching and bending are related to changes or molecular vibrations, each of which means a change in bond length (stretch) and a change in bond angle (bend).

Figures 3(a)–3(d) show the SEM results of bentonite samples and bentonite-chitosan mixtures with a chitosan percentage of 2–6%. As shown in the figure, the bentonite sample aggregate ranges from 1–2μm. This size increases with the addition of chitosan, which has a size of more than 5μm. As indicated by the FTIR and XRD data, the inclusion of chitosan in the internal pores increases the attraction force between the particles. This increase resulted in increased water absorption on the clay surface. The interaction between chitosan and bentonite, which results in a greater increase in crystal structure, was also observed by other researcher [12].

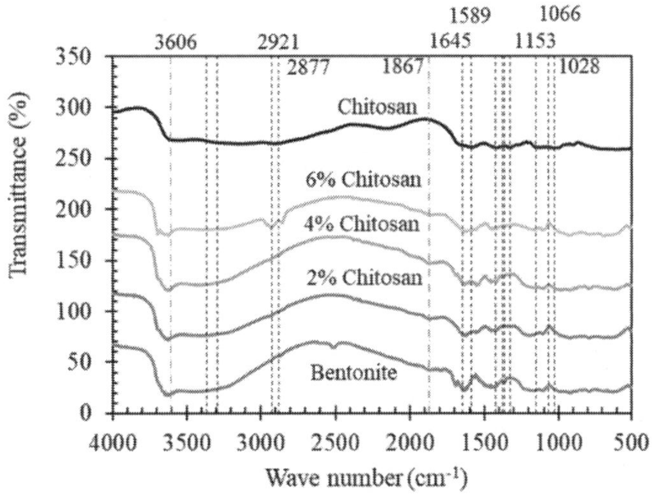

Figure 2. FTIR results of bentonite, chitosan, and bentonite-chitosan mixtures

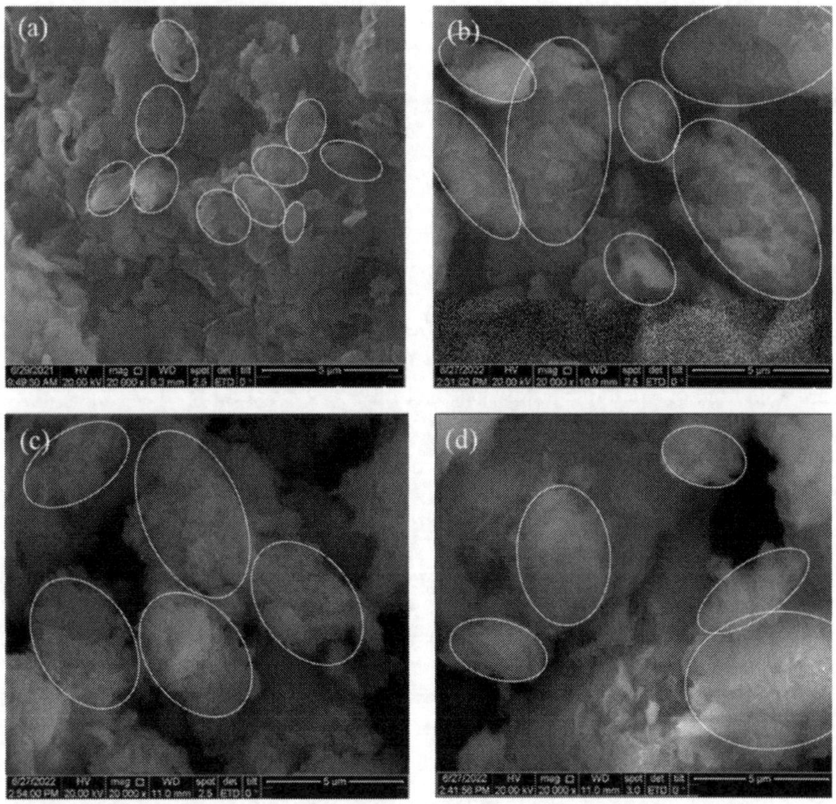

Figure 3. SEM results of bentonite and bentonite-chitosan mixtures: (a) bentonite, (b) 2% chitosan, (c) 4% chitosan, and (d) 6% chitosan

The liquid and plastic limits as a function of chitosan percentage are depicted in Figures 4(a) and 4(b). As shown in Figure 4(a), the sample's liquid limit increased from 371% (liquid limit bentonite) to

404% at 6% chitosan. The addition was almost linear at 6.6% per 1 percent chitosan. This also occurs at the plastic limit, where the value increases from 42% before adding chitosan to 48% after adding 6% chitosan, or increasing 1% per one percent addition of chitosan. The increase in the plastic limit is much smaller than the liquid limit.

One other important parameter that is usually considered to see the effect of adding additives to the soil is the plasticity index. The plasticity index is the difference between the liquid limit and the plastic limit. Analyzing the results in Figure 4, the plasticity index of the sample increased from 329% to 356%, or an increase of 4.5%, with the addition of one percent chitosan.

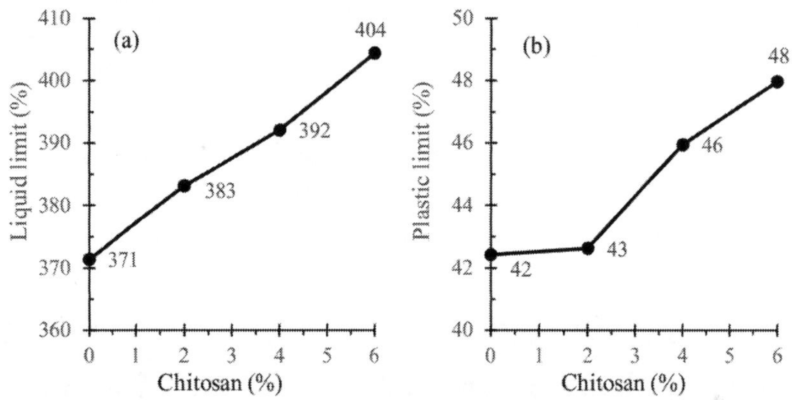

Figure 4. (a) Liquid limit and (b) plastic limit as a function of chitosan

Permeability of compacted bentonite-chitosan mixtures

Figure 5 shows the permeability of bentonite and the bentonite-chitosan mixture. Permeability decreased from 2.6×10^{-10} cm/s (bentonite) to 2.0×10^{-10} cm/s. This behavior is in accordance with the results of FITR, XRD, SEM, and plasticity tests. The increase in the size of the clusters shows the strong attraction between the bentonite particles. When it expands, the pores are closed, and the permeability becomes low. An increase in the value of soil plasticity, as shown in Figure 4, is also the expected result of this study. This parameter is closely related to hydraulic properties, including permeability. Generally, samples with high plasticity have low permeability.

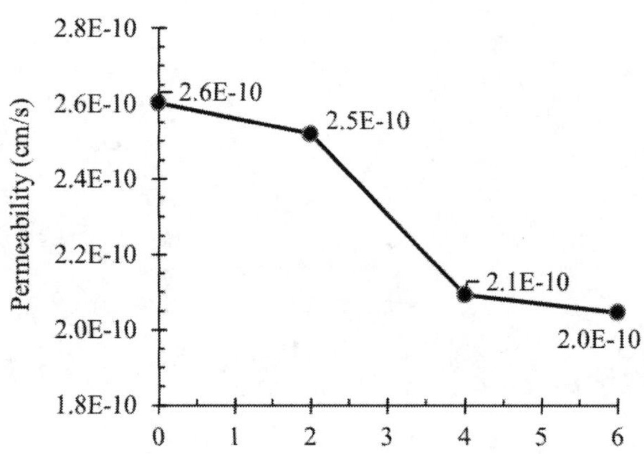

Figure 5. Permeability of the bentonite-chitosan mixtures as a function of chitosan content

Compressive strength of compacted bentonite-chitosan mixtures

Figure 6 shows the effect of the percentage of chitosan on the compressive strength of a mixture of bentonite and chitosan. As seen in the figure, qu increases from 10.61 kPa (bentonite) to 13.57 kPa at 6% chitosan. This relationship is almost linear, with an increase in compressive strength of 0.49 kPa with the addition of 1% chitosan. In general, the increase in soil plasticity is associated with cohesive soils that have high cohesion values. The addition of chitosan in the soil increases the bond between the particles, as shown in the SEM results, and is in accordance with the results of this compressive strength.

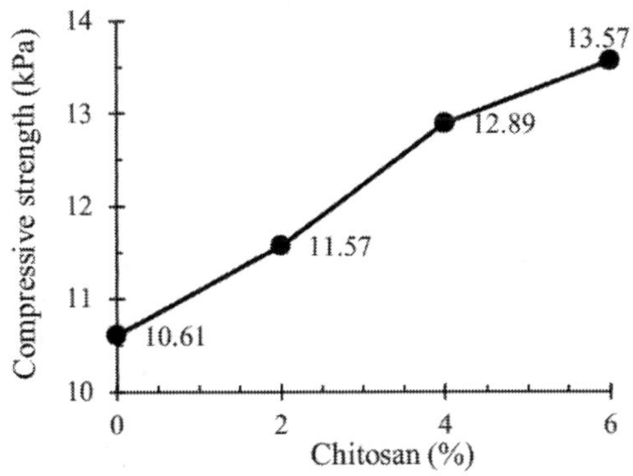

Figure 6. Compressive strength as a function chitosan content

5. Conclusions

The effectiveness of a mixture of bentonite and chitosan as a base material for clay liners has been described and analyzed through this research. In the XRD and FTIR analyses, the addition of chitosan to bentonite led to promising results, as evidenced by the appearance of new peaks due to the creation of minerals and functional groups. The cluster size of the bentonite and chitosan mixture was also larger than that without chitosan, as evidenced by the SEM results. The addition of 2-6% chitosan also resulted in an increase in the liquid limit, plastic limit, and plasticity index. This increase is very beneficial because of the linear relationship between plasticity and soil permeability. This was demonstrated by a decrease in the permeability of the mixture from 2.6×10^{-10} to 2.0×10^{-10} cm/s. Chitosan also adds more benefits by increasing the compressive strength from 10.61 to 13.57 kPa. Based on the result, further research could be done on the mixtures of bentonite, chitosan, and sand.

References

[1] Alzamel M, Fall M and Haruna S 2022 Swelling ability and behaviour of bentonite-based materials for deep repository engineered barrier systems: Influence of physical, chemical and thermal factors *J. Rock Mech. Geotech. Eng.* **14** 689–702

[2] Alzamel M, Haruna S and Fall M 2022 Saturated hydraulic conductivity of bentonite–sand barrier material for nuclear waste repository: effects of physical, mechanical thermal and chemical factors *Environ. Earth Sci.* **81**

[3] Agus S S and Arifin Y F 2018 Compacted Polymer-Enhanced Bentonite-Sand Mixture – Behaviour and Potential Applications *20th SEAGC-3rd AGSEA Conference in conjunction with 22nd Annual Indonesia National Conference on Geotechnical Engineering* ed M Irsyam, I Towhata, R Nazir, B Hsiung, P P Raharjo and P T Simatupang (Jakarta) pp 415–20

[4] Hosney M S and Rowe R K 2017 Performance of polymer-enhanced bentonite–sand mixture for covering arsenic-rich gold mine tailings for up to 4 years *Can. Geotech. J.* **54** 588–99

[5] Kusumawati N 2009 Pemanfaatan Limbah Kulit Udang Sebagai Bahan Baku Pembuatan Membran Ultrafiltrasi *Inotek* **13** 113–20

[6] Chen L, Wu P, Chen M, Lai X, Ahmed Z, Zhu N, Dang Z, Bi Y and Liu T 2018 Preparation and characterization of the eco-friendly chitosan/vermiculite biocomposite with excellent removal

capacity for cadmium and lead *Appl. Clay Sci.* **159** 74–82

[7] Kotal M and Bhowmick A K 2015 Polymer nanocomposites from modified clays: Recent advances and challenges *Prog. Polym. Sci.* **51** 127–87

[8] Aguilar R, Nakamatsu J, Ramírez E, Elgegren M, Ayarza J, Kim S, Pando M A and Ortega-San-Martin L 2016 The potential use of chitosan as a biopolymer additive for enhanced mechanical properties and water resistance of earthen construction *Constr. Build. Mater.* **114** 625–37

[9] Taytak B, Pulat H and Yukselen-Aksoy Y 2012 Improvement of engineering properties of soils by biopolymer additives *3rd Int. Conf. New Dev. Soil Mech. Geotech. Eng.* 28–30

[10] Lewandowska K, Sionkowska A, Kaczmarek B and Furtos G 2014 Characterization of chitosan composites with various clays *Int. J. Biol. Macromol.* **65** 534–41

[11] Giannakas A, Grigoriadi K, Leontiou A, Barkoula N M and Ladavos A 2014 Preparation, characterization, mechanical and barrier properties investigation of chitosan-clay nanocomposites *Carbohydr. Polym.* **108** 103–11

[12] Thakur G, Singh A and Singh I 2016 Chitosan-montmorillonite polymer composites: Formulation and evaluation of sustained release tablets of aceclofenac *Sci. Pharm.* **84** 603–17

[13] Hristodor C, Vrinceanu N, Pui A, Novac O, Copcia V E and Popovici E 2012 Textural and Morphological Characterization of Chitosan/Bentonite Nanocomposite *Environ. Eng. Manag. J.* **11** 573–8

[14] Lertsutthiwong P, Noomun K and Khunthon S 2012 Influence of chitosan characteristics on the properties of biopolymeric chitosan – montmorillonite *Prog. Nat. Sci. Mater. Int.* **22** 502–8

Characterization of sprayable mulch film from yam nagara starch for agricultural application

RA Ghani[1,3], M Elma[1,2,3*], RA Lestari[1,2], A Alsiren[1,3], Nasruddin[1,3], and A Rahma[2,3]

[1]Department of Chemical Engineering, Faculty of Engineering, Lambung Mangkurat University, Banjarmasin 70123, Indonesia
[2]Doctoral Program of Environmental Science, Postgraduate Program, Lambung Mangkurat University, Banjarmasin 70123, Indonesia
[3]Materials and Membranes Research Group (M[2]ReG), Lambung Mangkurat University, Banjarbaru 70714, Indonesia

E-mail: melma@ulm.ac.id

Abstract. The conventional mulch film generally produced from plastic to control of weed in agriculture activities. Nevertheless, the used of conventional mulch film has impact to pollute the environment, due to non-degradable. Therefore, mulch film from natural material derived yam nagara starch can be alternative to substitute the conventional mulch film that more eco-friendly. The aims of this study are to fabricate and characterization of mulch film from yam nagara starch which prepared by sprayed technique. The yam nagara starch prepared from local sweet potato varieties (*Ipomoea batatas L*) of Nagara, South Kalimantan. Fabrication of mulch film was conducted by mixed yam nagara starch with varied composition (0.25-0.75 wt%), hydrolysed keratin of chicken feather, corn gluten meal (CGM), acetate acid, and carbon black. The final solution of yam nagara mulch film was applied on top soil as plant medium using compressor spray gun and left to dry for 24 h. Furthermore, the sprayable mulch films were characterized by several parameters i.e., FTIR, water solubility, water uptake, tensile strength and elongation. Sprayable mulch film from yam nagara starch were successfully fabricated for agriculture application. The results show the sprayable mulch film with composition of 0.75 wt% yam nagara starch obtain a good plastic-like characteristics and may be affordable to be degraded due to easily to can soluble with water and high-water uptake properties. In other hand, sprayable mulch film from yam nagara starch in this work affordable to restrains weeds of the chilli plants for agriculture application.

1. Introduction

Today, the human population is increasing along with its demand which affects industrial and agricultural sector [1-9]. Besides that, especially Indonesia as agraris country has a large area of land even though it's an archipelago form [10-14].Weed growth in agricultural activities is often an obstacle to the growth and production of cultivated plants. This is due to competition between cultivated plants and weeds to obtain the necessary growth support sources such as light, space, nutrients and humidity. Weed control is still a major problem in agricultural activities. In general, weed management is done by using plastic mulch [4, 15]. Currently, the use of plastic mulch is used in vegetable production, but the use of plastic mulch in horticulture causes serious impacts in the form of waste [14, 16]. Plastic mulch used is generally made of polyethylene and synthetic polymers derived from petroleum (non-renewable resources)[17-19]. These materials are not biodegradable and need to be removed with special measures [15, 17, 20-24]. Therefore, it is necessary to control weeds without waste by using biodegradable mulch. One of the materials for making biodegradable mulch is starch [25]. This material is biodegradable, renewable, and inexpensive, making it an attractive source for the production of biodegradable mulch[26]. In addition, starch raw materials are easy to obtain. The starch material is mostly produced from sweet potatoes that are often found in South Kalimantan, namely Yam Nagara (Ipomoea batatas L). Yam Nagara has a starch content of 45 to 70% [27].

So far, the utilization of cassava is only used by the community South Kalimantan as an alternative food for rice and snacks. Wrong One potential of yam that can be utilized is to make it into a spray-based eco-friendly plastic mulch. Recently, sprayable polymer systems biodegradable mulch film for soil coating is the focus of research that is increasingly in demand [25]. Previously the use of plastic mulch conventionally placed on the ground produces air space between the ground and plastic.

A layer of mulch that can be sprayed directly onto the soil and during the drying process, a layer is formed following the unevenness of the soil surface [28]. The advantages of sprayable biodegradable mulch film are its efficient use, its end-of-life can be removed naturally without high costs and no negative effects on the environment [29-31].

Previously, Gloeb [15] has conducted research on the effectiveness of weed control from bio-based sprayable mulch films in special crop systems. This research used corn starch as a material for making sprayable biodegradable mulch film. However, for now there is still no application using cassava starch in Indonesia. Therefore, this study investigate how the characteristics of environmentally friendly plastic mulch from Nagara yam starch use spray for agricultural cultivation.

2. Materials and Methods

2.1 Chemicals and materials

Keratin hydrolyzate from chicken farm waste in the form of chicken feathers, sodium hydroxide (NaOH; 0.8M) as a mixture for making keratin hydrolyzate, yam Nagara starch which is mashed using a 200 mesh sieve, glycerol, corn gluten meal (CGM), acetic acid (CH3COOH; 100 %) to adjust the solution to the standard pH of the soil, carbon black as a coloring agent and aquades

2.2 Pretreatment

Yam Nagara starch is processed through the manufacture of Yam Nagara starch. Then the yam flour was washed using distilled water and soaked for 24 hours. Then rinsed and drained. The resulting starch precipitate was then dried in an oven at 70 °C for 24 hours. The dried Nagara yam starch was mashed and sieved with a size of 200 mesh.

2.3 Exprerimental setup

The manufacture of sprayable biodegradable mulch film was started by mixing cassava starch with a concentration (0,25%, 0,5%, 0,7%) with 700 mL of hydrolyzed keratin and the addition of 303 mL of distilled water in a beaker. Keratin hydrolyzate was previously obtained by soaking raw chicken feathers in 0.8 M NaOH for 48 hours. The soaking results are separated by filtration. The hydrolyzed keratin filtrate was then mixed with the Nagara yam starch. The mixture was heated and homogenized using hotplates at 70° for 30 minutes. Then while the starch-keratin mixture is heated, glycerol is added with and Corn Gluten Meal (CGM) each 20 grams. Then added CH3COOH little by little until homogeneous and the pH of the mixture reached 6.8 - 7.5. Then 2% carbon black was added to the mixture.

2.4 *Application of Sprayable Biodegradable Mulch Film*

The sprayable biodegradable mulch film solution was evenly applied to an area of 60 x 23 cm2 using a spray gun F75G-2 with a nozzle size of 1.5 mm and a tube volume capacity of 300 mL. Then the sprayable biodegradable mulch film solution that has been sprayed is allowed to dry and form a film for 24 hours.

IOP Conf. Series: Earth and Environmental Science 1184 (2023) 012012 doi:10.1088/1755-1315/1184/1/012012

2.5 Analysis Characteristic

Then the Sprayable Biodegradable Mulch Film will be analyzed by Scanning Electron Microscope (SEM) and Fourier Transform Infrared Spectroscopy (FTIR), mechanical properties, and biodegradation. Then the Sprayable Biodegradable Mulch Film was also tested for its water resistance and water solubility.

Figure 1. (a) Sprayable biodegradable mulch film, (b) spray gun used for the spray ability tests, (c) application of sprayable biodegradable mulch film

3. Result and Discussions

3.1 Characteristics of Sprayable Biodegradable Mulch Film from Yam Nagara Starch

The use of the Michelson interferometer gives the FTIR method an advantage over conventional infrared spectroscopy and other spectroscopic methods. Among them is the molecular structure information can be obtained precisely and accurately (has a high resolution). Another advantage of this method is that it can be used to identify samples in various phases (gas, solid or liquid). FTIR is used to determine the functional group of a compound. In Fig. 2 shown SBMF has several components of functional groups, namely O-H, C-H, amide II, C-C and C-O. The absorption spectra of the 3271 cm^{-1} regions show the stretching vibration range of the O-H group bond and in the 2930 cm^{-1} spectra it shows the antisymmetric and symmetric stretching vibrations of the C-H bond [32]. The amide II bond (NH bending) appears in the absorption spectra of the 1552 cm^{-1} region [33]. Furthermore, at the peak of the 1407 cm^{-1} spectra is C-C stretching [34]. Meanwhile at 1022 cm^{-1}, the C-O stretching range of starch (C-O-H and C-O-H) is seen [35].

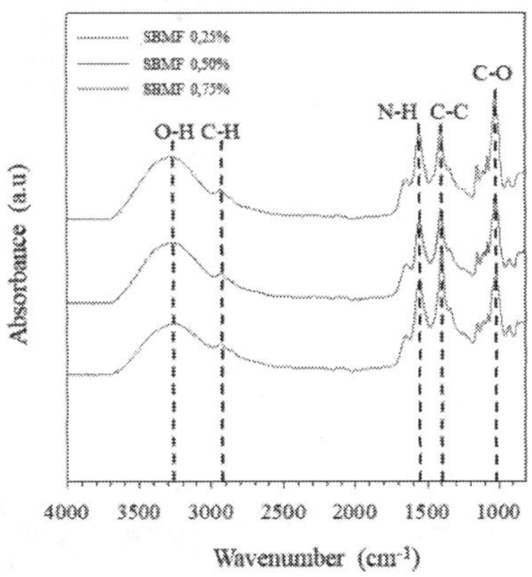

Figure 2. FTIR test results from BMF

The solubility test in water was determined by the composition of the mixture of environmentally friendly spray-based plastic mulch (SBMF) produced. Solubility test aims to determine the occurrence of bonds in the polymer and the level or regularity of the bonds in the polymer and to estimate the stability of the mulch film against water. The solubility of the mulch film in water was determined by the solubility test in water, namely the percentage of swelling of the mulch film in the presence of water [36].

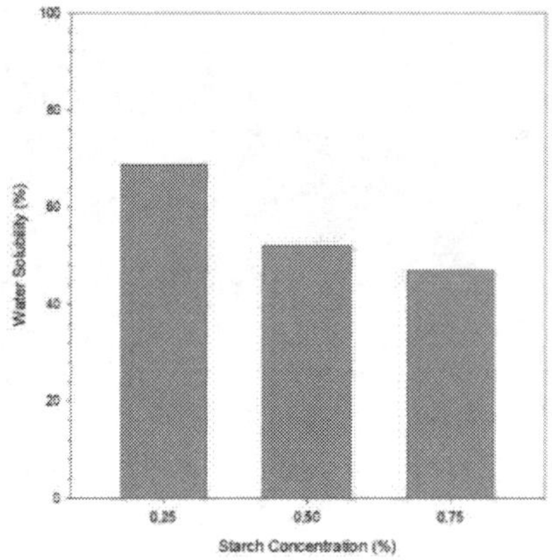

Figure 3. Effect of starch concentration on solubility in water

Fig 3. shows a decreasing trendline as the concentration of cassava starch increases. The highest solubility was found at 0.25% starch concentration and the lowest was 0.75%. This shows the water solubility of 0.25% sweet potato starch indicating the ease of the film to dissolve in water. This is because with the increase in starch, the glycerol: starch ratio will decrease, the hydrophilic groups in the mulch film constituents decrease. The lower the hydrophilic group in the constituent materials, the lower the solubility of the mulch film. Glycerol plays a role in increasing the solubility of starch-based mulch films. This is because glycerol is hydrophilic, so the higher the concentration of glycerol will cause the solubility of the mulch film to increase. Solubility in water is required to improve the integrity of the film and its resistance to water.

This test is carried out to determine the occurrence of bonds in the polymer and the degree or regularity of the bonds in the polymer which is determined by the percentage addition of polymer weight after water absorption occurs. The properties of plastic film resistance to water are determined by the percentage of swelling of the film by the presence of water, the lower the water absorption value, the better the mulch film properties, while the higher the water absorption, the mulch film properties will be easily damaged [9].

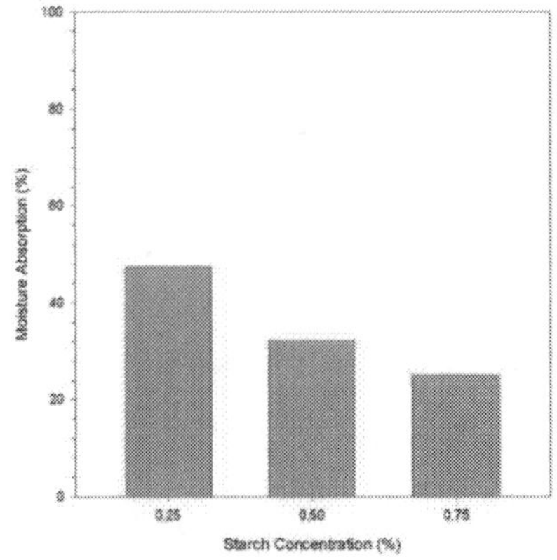

Figure 4. Effect of starch concentration on water absorption

Fig 4. shows that in this study the addition of 0.75% starch had the best water resistance compared to the others, which was indicated by the smallest water absorption value of 25.24%. The results of the water resistance test of environmentally friendly spray-based plastic mulch (SBMF) obtained an average value ranging from 25.24% to 47.49%. The lowest value of SBMF water resistance was indicated by the addition of 0.75% starch, which was 25.24% and the highest value was found in the addition of 0.25% starch, which was 47.49%. As for the addition of 0.50%, the water absorption value is 32.35%.

The value of the percent of water absorbed will be inversely proportional to the water resistance of the plastic, the smaller the value of the percent of water absorbed by the plastic, the greater the water resistance of the plastic. The addition of glycerol as a plasticizer affects the value of water resistance to the film where the addition of glycerol increases the permeability to gases, water vapor, and solutes [11]. Another factor causing the high value of water absorption in the film is caused by its constituent components. Starch has hydrophilic properties which have a major influence on water absorption. The high value of water resistance is influenced by the large content of OH- hydroxyl groups derived from starch that has not been completely modified.

The biodegradation test was carried out to determine how quickly the mulch film was degraded by microorganisms in an environment [37, 38]. The media used is soil because in the soil there are many types of microorganisms (fungi, bacteria and algae) and in large quantities, so that it will support the degradation process that will be carried out.

(a)

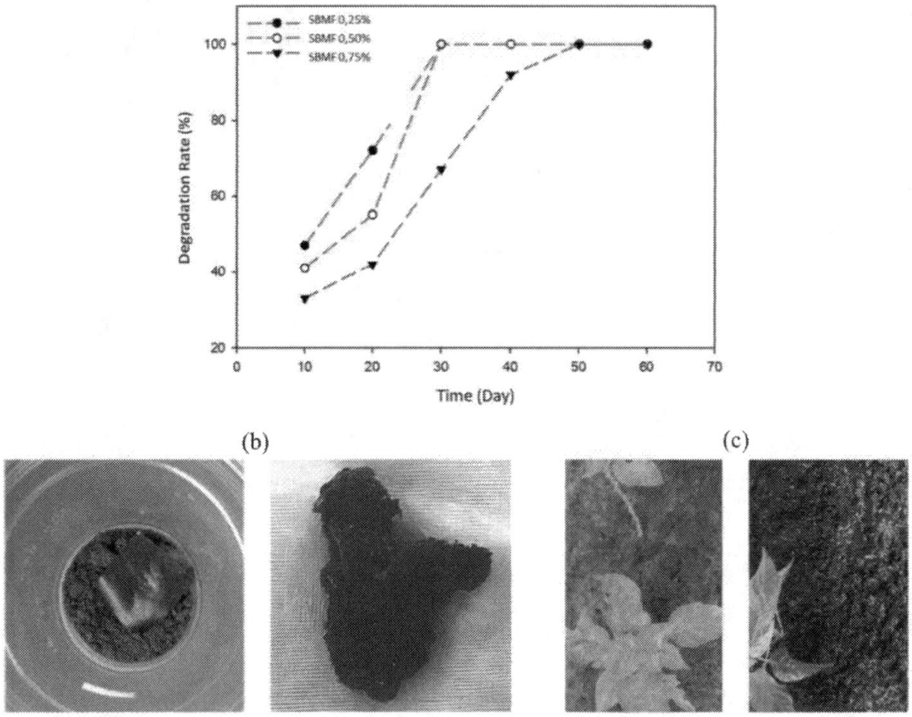

0 Day 40 Days Bare Soil SBMF

Figure 5. (a) Results of biodegradation analysis from SBMF, (b) biodegradation test 0 days (left) and 40 days (right) on 0.75% SBMF sample, (c) comparison of soil covered with SBMF (right), and in bare soil (left)

The sample with the best degradation was obtained in the 0.75% SBMF sample because the perfect degradation time was obtained on the 50th day while the 0.25% SBMF and 0.75% SBMF samples were perfectly graded earlier on the 30th day of the sample. The percentage of reduced film weight increased as the amount of starch added was reduced. According to previous reports [25], the biodegradability rate of degradable mulch films is usually around 32% in 100 days, whereas conventional LDPE plastic films show almost no degradation over the same time period. Therefore, the degradation rate of the SBMF sample was faster than that of ordinary degradable mulch film, which indicates that the biodegradability rate is less effective.

Morphological analysis using Scanning Electron Microscopy (SEM) can determine film homogeneity, cracks, and microscopic mixing surface smoothness [39-44]. The biodegradable mulch film which was analyzed for morphology was the sample with the addition of 0.75% starch because it had the best resistance and solubility in water.

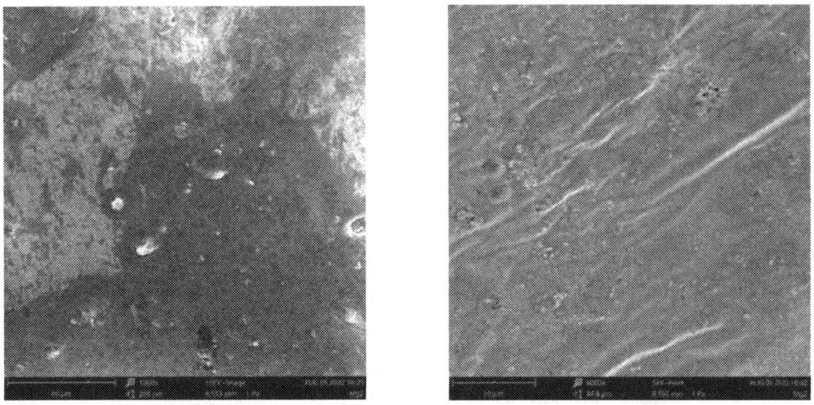

(a) (b)

Figure 6. Morphology of Biodegradable Mulch Film (a) 0.75% SBMF with 1000x Magnification (b) 0.75% SBMF with 6000x Magnification

Fig 6. shows the surface (a) of a 0.75% SBMF sample with 1000 times magnification which looks fibrous, uneven, and still has lumps. In addition, it is seen that there is an uneven accumulation of layers due to the presence of cavities. SEM results of 0.75% SBMF sample with a magnification of 6000 times strengthen the presence of pores in the sample. This happens because the solution is not mixed homogeneously perfectly. Solutions that are not mixed homogeneously can be caused by the mixing process using only a magnetic stirrer. A large amount of carbon black, which is not homogeneously dispersed in the mixture on the surface of the coating becomes an accessible place for water deposition [45-47]. In this case water cannot interact directly with the polymer matrix, but acts as a plasticizer so that the mechanical response is lower than the expected value. The large inhomogeneity of the dispersed carbon black particles in the polymer matrix is clearly demonstrated. The SEM results are supported by the physical properties on the surface of the 0.75% SBMF film which is still rough to the touch.

Analysis of mechanical properties includes tensile strength and elongation. Tensile strength shows the quality of the physical character of the film in its resistance to the environment [48, 49]. As for elongation, to determine the elongation ability of the film, the higher the elongation value, the more flexible and plastic the film is.

Table 1. Mechanical properties test results

No.	Parameter	Result	Unit
1	Tensile Strength	0,374	N/mm^2
2	Elongation	0,112	%

Elasticity is the most important evaluation index for the mechanical properties of a mulch film, which is usually characterized by low tensile strength and high elongation at break [50, 51]. If the tensile strength of the mulch film is large, this means that the film is stiff and less prone to deformation, so that plant seedlings under the mulch film cannot break the film on their own as they grow [49, 52]. Mechanical parameters carried out on the tested material regarding the spray black coating sample, it is very difficult to carry out mechanical analysis due to the presence of micro-holes, so that the material itself is torn before testing, and the spray layer thickness measurement is incorrect, due to the very large amount of carbon black in it, even 2% to 1.5% polysaccharide component. It is clear from Table 1. that the elongation at break and elongation in the 0.75% BMF sample are 0.374 N/mm^2 and 0.112%, respectively, so that the elasticity of this film is relatively poor. The tensile strength and elongation values are very low compared to Elma, *et al.* [27] with the same material source, Nagara yam. The study obtain 1,84 N/mm^2 film tensile strength. This may differ from the evidence that, when polymeric materials are filled with carbon black, usually the tensile strength increases [53-58], while the elongation value decreases. In our case, two synergistic effects can justify the different behavior of which is the problem of thickness determination, and the large water absorption ability of the polymer layer on the other hand.

4. Conclusion

The lowest SBMF water resistance value is shown in the addition of 0.75% starch which is equal to 25.24% and the highest value is found in the addition of 0.25% starch that is equal to 47.49%. Whereas for addition of 0.50%, the water absorption value is 32.35%. The more the smaller the percentage of water absorbed by the mulch film, the larger the the water resistance of the mulch film . Solubility value in The highest water content was found at 0.25% starch concentration and the lowest was 0.75%. This indicates water solubility at a starch concentration of 0.25% shows the ease of the film to dissolve in water. Solubility in water is needed to increase the film integrity and resistance against water. The elongation at break and elongation in the 0.75% SBMF sample were 0.374 N/mm2 and 0.112%, respectively, so that the film elasticity it is relatively poor with very high tensile and elongation strength values low. The sample with the best degradation in the 0.75% SBMF sample because perfect degradation time obtained on the 50th day while on 0.25% SBMF and 0.75% SBMF samples completely degraded earlier on the 30th day of the sample. In the application, there is no weed growth during the degradation period, so the use of sprayable biodegradable mulch film is considered to be able to overcome weed growth for plants and does not produce plastic waste for the environment.

Acknowledgement

This Work is supported by PT. Indofood Sukses Makmur Tbk. through the Indofood Research Nugraha program for the period 2021-2022. Authors thanks to Pertamina Patra Niaga Integrated Terminal (IT) Banjarmasin, PT. Pertamina, Banjarmasin- Indonesia for financial support during experiments. Also, authors thanks to Materials and Membranes Research Group for the facilities provided. Muthia thanks to National Battery Research Institure (NBRI) for the 2022 Grant Prize Award funded by collaboration between NBRI and Queen Marry University of London as well as British Council. Finally, Muthia thanks to the Tesis, Dissertation and Applied Research Universities 2021-2023 Grant, Directorat Jenderal of Higher Education, Research and Technology the Ministry of Education Republic of Indonesia.

References

[1] Elma, Muthia, Ghani, Rhafiq Abdul, Rahma, Aulia, Alyanti, Alya Dita, and Dony, Novrian, "Banana Peels Pectin Templated Silica Ultrafiltration Membrane in Disk Plate Configuration Applied for Wetland Water Treatment," *Journal of Advanced Research in Fluid Mechanics and Thermal Sciences,* vol. 100, no. 1, pp. 77-88, (2022).

[2] Elma, Muthia, Rahma, Aulia, Kusumawati, Uun, Pratama, Reza Satria Kelik, and Alyanti, Alya Dita, "Single Vs Multichannel Silica-Pectin Ultrafiltration Membranes for Treatment of Natural Peat Water," *Journal of Advanced Research in Fluid Mechanics and Thermal Sciences,* vol. 100, no. 2, pp. 33-46, (2022).

[3] Syauqiah, I, Elma, M, Mailani, DP, and Pratiwi, N, "Activated carbon from Nypa (Nypa fruticans) leaves applied for the Fe and Mn removal," in *IOP Conference Series: Materials Science and Engineering*, 2020, vol. 980, no. 1: IOP Publishing, p. 012073.

[4] Adam, Mohd Ridhwan *et al.*, "Ammonia removal by adsorptive clinoptilolite ceramic membrane: Effect of dosage, isothermal behavior and regeneration process," *Korean Journal of Chemical Engineering,* vol. 38, no. 4, pp. 807-815, (2021).

[5] Rahman, Sazila Karina, Rampun, Erdina Lulu Atika, Rahma, Aulia, and Elma, Muthia, "Deconvolution of carbon silica templated thin film using ES40 and P123 via rapid thermal processing method," *Materials Today: Proceedings,* vol. 31, pp. 75-78, (2020).
[6] Lestari, Riani Ayu *et al.*, "Organo Silica Membranes for Wetland Saline Water Desalination: Effect of membranes calcination temperatures," in *E3S Web of Conferences*, 2020, vol. 148: EDP Sciences, p. 07006, https://doi.org/10.1051/e3sconf/202014807006.
[7] Elma, Muthia, Rakhman, Arief, and Hidayati, Rahmi, "Silica P123 Membranes for desalination of wetland saline water in South Kalimantan," in *IOP Conference Series: Earth and Environmental Science*, 2018, vol. 175, no. 1: IOP Publishing, p. 012007, https://doi.org/10.1088/1755-1315/175/1/012007.
[8] Syauqiyah, Isna, Elma, Muthia, Putra, Meilana D, Rahma, Aulia, Pratiwi, Amalia E, and Rampun, Erdina LA, "Interlayer-free silica-carbon template membranes from pectin and P123 for water desalination," in *MATEC Web of Conferences*, 2019, vol. 280: EDP Sciences, p. 03017, https://doi.org/10.1051/matecconf/201928003017.
[9] Elma, Muthia, "Optimum ratio between waste cooking oil and coconut oilas raw material for biodiesel production," *International Journal on Advanced Science, Engineering and Information Technology,* (2017).
[10] Elma, Muthia, Pratiwi, Amalia Enggar, Rahma, Aulia, Rampun, Erdina Lulu Atika, and Handayani, Noni, "The Performance of Membranes Interlayer-Free Silica-Pectin Templated for Seawater Desalination via Pervaporation Operated at High Temperature of Feed Solution," in *Materials Science Forum*, 2020, vol. 981: Trans Tech Publ, pp. 349-355.
[11] Assyaifi, Zaini L *et al.*, "Photocatalytic–pervaporation using membranes based on organo-silica for wetland saline water desalination," *Membrane Technology,* vol. 2021, no. 7, pp. 7-11, (2021).
[12] Elma, Muthia *et al.*, "Organo-silica membrane for brine water pervaporation," in *IOP Conference Series: Earth and Environmental Science*, 2020, vol. 473, no. 1: IOP Publishing, p. 012129.
[13] Elma, Muthia *et al.*, "Long-Term Performance and Stability of Interlayer-Free Mesoporous Silica Membranes for Wetland Saline Water Pervaporation," *Polymers,* vol. 14, no. 5, p. 895, (2022), https://doi.org/10.3390/polym14050895.
[14] Rahma, Aulia, Elma, Muthia, Rampun, Erdina Lulu Atika, Pratiwi, Amalia Enggar, Rakhman, Arief, and Fitriani, Fitriani, "Rapid thermal processing and long term stability of interlayer-free silica-P123 membranes for wetland saline water desalination," *Journal of Advanced Research in Fluid Mechanics and Thermal Sciences,* vol. 71, no. 2, pp. 1-9, (2020), https://doi.org/10.37934/arfmts.71.2.19.
[15] Gloeb, Eliott, "Weed Control Efficacy of Bio-Based Sprayable Mulch Films in Specialty Crop Systems," Agronomy and Horticulture Department, University of Nebraska - Lincoln, 201, 2020. [Online]. Available: https://digitalcommons.unl.edu/agronhortdiss/201
[16] Vox, G, Santagata, G, Malinconico, M, Immirzi, B, Mugnozza, G Scarascia, and Schettini, E, "Biodegradable films and spray coatings as eco-friendly alternative to petro-chemical derived mulching films," *Journal of Agricultural Engineering,* vol. 44, no. s2, (2013).
[17] Elma, Muthia *et al.*, "Hollow fiber membrane applied for Sasirangan wastewater desalination integrated with photocatalysis and pervaporation set-up," *Materials Today: Proceedings,* vol. 51, pp. 1298-1302, (2022), https://doi.org/10.1007/s13726-019-00700-y.
[18] Hidayah, Nur *et al.*, "Physicochemical properties of membrane adsorber from palm empty fruit bunch (PEFB) by acid activation," *Membranes,* vol. 11, no. 12, p. 917, (2021).
[19] Elma, Muthia, Rampun, Erdina Lulu Atika, Rahma, Aulia, Pratiwi, Amalia Enggar, Abdi, Chairul, and Rossadi, Raisa, "Effect of two stages adsorption as pre-treatment of natural organic matter removal in ultrafiltration process for peat water treatment," in *Materials Science Forum*, 2020, vol. 988: Trans Tech Publ, pp. 114-121, https://doi.org/10.4028/www.scientific.net/MSF.988.114.
[20] Pradhana, EA *et al.*, "The functionalization study of PVDF/TiO2 hollow fibre membranes under vacuum calcination exposure," in *Journal of Physics: Conference Series*, 2021, vol. 1912, no. 1: IOP Publishing, p. 012035.
[21] Sumardi, Anna *et al.*, "Designing a mesoporous hybrid organo-silica thin film prepared from an organic catalyst," *Membrane Technology,* vol. 2021, no. 2, pp. 5-8, (2021).

[22] Elma, Muthia *et al.*, "Physicochemical properties of mesoporous organo-silica xerogels fabricated through organo catalyst," *Membranes,* vol. 11, no. 8, p. 607, (2021), https://doi.org/10.3390/membranes11080607.

[23] Elma, Muthia *et al.*, "Carbon templated strategies of mesoporous silica applied for water desalination: A review," *Journal of Water Process Engineering,* vol. 38, p. 101520, (2020), https://doi.org/10.1016/j.jwpe.2020.101520.

[24] Elma, Muthia and Riskawati, Nur, "Silica membranes for wetland saline water desalination: performance and long term stability," in *IOP Conference Series: Earth and Environmental Science*, 2018, vol. 175, no. 1: IOP Publishing, p. 012006, https://doi.org/10.1088/1755-1315/175/1/012006.

[25] Adhikari, Raju, Bristow, Keith L, Casey, Philip S, Freischmidt, George, Hornbuckle, John W, and Adhikari, Benu, "Preformed and sprayable polymeric mulch film to improve agricultural water use efficiency," *Agricultural Water Management,* vol. 169, pp. 1-13, (2016).

[26] Raharjo, Yanuardi *et al.*, "Dialysis Membranes for Acute Kidney Injury," *Membranes,* vol. 12, no. 3, p. 325, (2022).

[27] Elma, Muthia, Pradana, Eggy A, Sihombing, Natalia, Thala'ah, Rizky Noor, Rahma, Aulia, and Rampun, Erdina LA, "Edible Film Fabrication derived from Nagara Starch and Red Ginger Essential Oil as a Package of Instant Noodles Seasoning," in *IOP Conference Series: Earth and Environmental Science*, 2022, vol. 999, no. 1: IOP Publishing, p. 012024.

[28] Schettini, E, Vox, G, Malinconico, M, Immirzi, B, and Santagata, G, "Physical properties of innovative biodegradable spray coating for soil mulching in greenhouse cultivation," in *International Conference on Sustainable Greenhouse Systems-Greensys2004 691*, 2004, pp. 725-732.

[29] Giaccone, Matteo *et al.*, "Biodegradable mulching spray for weed control in the cultivation of containerized ornamental shrubs," *Chemical and Biological Technologies in Agriculture,* vol. 5, no. 1, pp. 1-8, (2018).

[30] Nabilah, NN Ain, Rahmawati, Ratri, Bilad, MR, Pratiwi, AE, Elma, Muthia, and Jusoh, Norwahyu, "Finned spacer panel system for fouling control in Chlorella vulgaris harvesting," in *IOP Conference Series: Materials Science and Engineering*, 2020, vol. 736, no. 6: IOP Publishing, p. 062009.

[31] Razak, Nik Nurul Ain Nabilah *et al.*, "Finned spacer for enhancing the impact of air bubbles for membrane fouling control in Chlorella vulgaris filtration," *Bioresource Technology Reports,* vol. 11, p. 100429, (2020).

[32] Merino, Danila, Gutiérrez, Tomy J, and Alvarez, Vera A, "Structural and thermal properties of agricultural mulch films based on native and oxidized corn starch nanocomposites," *Starch-Stärke,* vol. 71, no. 7-8, p. 1800341, (2019).

[33] Tesfaye, Tamrat, Sithole, Bishop B, and Ramjugernath, Deresh, "Preparation, characterization and application of keratin based green biofilms from waste chicken feathers," (2018).

[34] Murat, ATES and Pinar, KUZ, "Starch-based bioplastic materials for packaging industry," *Journal of Sustainable Construction Materials and Technologies,* vol. 5, no. 1, pp. 399-406, (2020).

[35] Raghavan, D, "Characterization of biodegradable plastics," *Polymer-Plastics Technology and Engineering,* vol. 34, no. 1, pp. 41-63, (1995).

[36] Din, Muhammad Imran, Siddique, Nida, Hussain, Zaib, and Khalid, Rida, "Facile synthesis of biodegradable corn starch-based plastic composite film reinforced with zinc oxide nanoparticles for packaging applications," *Inorganic and Nano-Metal Chemistry,* pp. 1-10, (2022).

[37] Jaya, Jaka Darma, Elma, Muthia, Sunardi, Sunardi, and Nugroho, Agung, "Physical and Mechanical Properties of Biodegradable Pot Derived from Oil Palm Empty Fruit Bunch and Sodium Alginate," *Brazilian Archives of Biology and Technology,* vol. 65, (2022).

[38] Hidayah, Nur, Elma, Muthia, Darsono, Putri Vidiasari, and Syauqiah, Isna, "Oil Palm Empty Bunches Adsorber Membrane for CO2 Reduction of Biogas TPA Gunung Kupang in South Borneo," in *Materials Science Forum*, 2022, vol. 1056: Trans Tech Publ, pp. 159-164.

[39] Elma, Muthia *et al.*, "Fabrication of interlayer-free silica-based membranes–effect of low calcination temperature using an organo-catalyst," *Membrane Technology,* vol. 2019, no. 2, pp. 6-10, (2019).

[40] Elma, Muthia *et al.*, "Combination of Coagulation, Adsorption, and Ultrafiltration Processes for Organic Matter Removal from Peat Water," *Sustainability*, vol. 14, no. 1, p. 370, (2021), https://doi.org/10.3390/su14010370.

[41] Elma, Muthia, Rahma, Aulia, Pratiwi, Amalia E, and Rampun, Erdina LA, "Coagulation as pretreatment for membrane-based wetland saline water desalination," *Asia-Pacific Journal of Chemical Engineering,* vol. 15, no. 4, p. e2461, (2020), https://doi.org/10.1002/apj.2461.

[42] Elma, Muthia and Saputro, Gesit Satriaji, "Performance of Cobalt-Silica Membranes through Pervaporation Process with Different Feed Solution Concentrations," in *Materials Science Forum*, 2020, vol. 981: Trans Tech Publ, pp. 342-348, https://doi.org/10.4028/www.scientific.net/MSF.981.342.

[43] Mat Nawi, Normi Izati *et al.*, "A Rotary Spacer System for Energy-Efficient Membrane Fouling Control in Oil/Water Emulsion Filtration," *Membranes,* vol. 12, no. 6, p. 554, (2022).

[44] Lestari, APTAR E, Elma, MUTHIA, Rabiah, SADIDAN, Rampun, Erdina Lulu Atika, Rahma, AULIA, and Pratiwi, Amalia Enggar, "Performance of mesoporous organo silica membrane for desalination," in *Materials Science Forum*, 2020, vol. 1000: Trans Tech Publ, pp. 285-292.

[45] Motuzas, Julius, Darmawan, Adi, Elma, Muthia, and Wang, David K, "Microporous Silica Membrane: Structure, Preparation, Characterization, and Applications," in *Current Trends and Future Developments on (Bio-) Membranes*: Elsevier, 2019, pp. 77-99.

[46] Rahman, Sazila Karina, Rahma, Aulia, Syauqiah, Isna, and Elma, Muthia, "Functionalization of hybrid organosilica based membranes for water desalination–Preparation using Ethyl Silicate 40 and P123," *Materials Today: Proceedings,* vol. 31, pp. 60-64, (2020).

[47] Rampun, Erdina LA, Elma, Muthia, Rahma, Aulia, and Pratiwi, Amalia E, "Interlayer-free silica–pectin membrane for sea-water desalination," *Membrane Technology,* vol. 2019, no. 12, pp. 5-9, (2019).

[48] Wong, Dannee, Andriyana, Andri, Ang, Bee Chin, Lee, Jacky Jia Li, Verron, Erwan, and Elma, Muthia, "Poisson's ratio and volume change accompanying deformation of randomly oriented electrospun nanofibrous membranes," *Plastics, Rubber and Composites,* vol. 48, no. 10, pp. 456-465, (2019).

[49] Waqas, Sharjeel *et al.*, "Effect of membrane properties in a membrane rotating biological contactor for wastewater treatment," *Journal of Environmental Chemical Engineering,* vol. 9, no. 1, p. 104869, (2021).

[50] Sihombing, N, Elma, M, Thala'ah, RN, Simatupang, FA, Pradana, EA, and Rahma, A, "Garlic essential oil as an edible film antibacterial agent derived from Nagara sweet potato starch applied for packaging of Indonesian Traditional Food-Dodol," in *IOP Conference Series: Earth and Environmental Science*, 2022, vol. 999, no. 1: IOP Publishing, p. 012026.

[51] Elma, Muthia *et al.*, "Development of hybrid and templated silica-P123 membranes for brackish water desalination," *Polymers,* vol. 12, no. 11, p. 2644, (2020).

[52] Elma, Muthia, Setyawan, Heru, Rahma, Aulia, Pratiwi, AE, and Rampun, Erdina Lulu A, "Fabrication of Interlayer-free P123 Caronised Template Silica Membranes for Water Desalination: Conventional Versus Rapid Thermal Processing (CTP vs RTP) Techniques," in *IOP Conference Series: Materials Science and Engineering*, 2019, vol. 543, no. 1: IOP Publishing, p. 012076.

[53] Elma, Muthia, "Functionalization of Si-C Using TEOS (Tetra Ethyl Ortho Silica) as Precursor and Organic Catalyst," (2020), http://dx.doi.org/10.1051/e3sconf/202014807008.

[54] Elma, Muthia, "Performance of interlayer-free pectin template silica membranes for brackish water desalination," (2020), https://doi.org/10.1016/S0958-2118(20)30108-7.

[55] Pratiwi, Amalia Enggar, Elma, Muthia, Rahma, Aulia, Rampun, Erdina LA, and Saputro, Gesit Satriaji, "Deconvolution of pectin carbonised template silica thin-film: synthesis and characterisation," *Membrane Technology,* vol. 2019, no. 9, pp. 5-8, (2019), https://doi.org/10.1016/S0958-2118(19)30167-3.

[56] Elma, Muthia and Setyawan, Heru, "Synthesis of silica xerogels obtained in organic catalyst via sol gel route," in *IOP Conference Series: Earth and Environmental Science*, 2018, vol. 175, no. 1: IOP Publishing, p. 012008, https://doi.org/10.1088/1755-1315/175/1/012008.

[57] Susanto, Anto, Elma, Muthia, and Putra, Meilana Dharma, "Palm kernel shell activated carbon using phosphoric acid for crude palm oil clarification: properties and application," *European Chemical Bulletin,* vol. 11, no. 4, pp. 1-1, (2022).

[58] Elma, Muthia, Nata, Iryanti F, Maulida, Noor Aisya, Fitriah, Siti Humaeroh, Rampun, Erdina Lulu Atika, and Rahma, Aulia, "Organosilica Multichannel Membranes Prepared by Inner Coating Method Applied for Brackish Water Desalination," in *Materials Science Forum*, 2022, vol. 1057: Trans Tech Publ, pp. 136-143.

Measuring The Use of Detention Basin as Flood Control System at Campus I Lambung Mangkurat University Banjarmasin, Indonesia

Maya Amalia*[1], Bani Noor Muchamad[2], Maulidya Putri Sutanti[1], Elma Sofia[1], Noordiah Helda[1]

[1]Civil Engineering, Faculty of Engineering, Lambung Mangkurat University, Indonesia
[2]Architecture, Faculty of Engineering, Lambung Mangkurat University, Indonesia

E-mail: *m.amalia@ulm.ac.id (corresponding author)

Abstract. South Kalimantan Province has experienced floods during the rainy seasons due to high intensity of precipitation. In mid-January 2021, flood inundated the Banjarmasin Campus I area of Lambung Mangkurat University (ULM Campus I). In mid-December 2021, other flood case inundated Campus I ULM. The research aims to measure the effectiveness of using detention basin as flood control in Banjarmasin campus area of ULM. Data collection is based on hydrological analysis of rain data to obtain a designed flood discharge with a return period of 20 years using the Der Weduwen method. The hydraulic analysis of the transverse and longitudinal sections on the river section of Campus I ULM was analyzed using the HEC-RAS. Based on the value of the flood volume, an analysis of the volume of the detention basin is carried out, where the volume of the detention basin must be greater than the volume of the flood. Spatial data is used to obtain contours using ArcGIS, to determine the optimal location of the detention basin. Based on the simulation, cumulative planned flood volume was 36,010 m^3 with a discharge value of 89.30 m^3/s. It was designed two detention basin. First detention basin dimension is 160 m x 130 m x 1.5 m located closed to General Building. The second one, located at Master of Law Building with dimension is 120 m x 35 m x 1.5 m First detention basin dimension is 160 m x 130 m x 1.5 m located closed to General Building. The second one, located at Master of Law Building with dimension is 120 m x 35 m x 1.5 m.

Keywords: Flood, Campus I ULM, HEC-RAS, Der Werduwen Method, Detention Basin.

1. Introduction

The city of Banjarmasin has a relatively flat topographical plain (often inundated by water) with an average height of 0.16 meters below sea level for the entire area. Located close to the mouth of the Barito River and divided by the Martapura River. It has a soil slope of 0.13% with a geological composition dominated by clay, fine sand, and alluvial deposits and the soil surface is covered by gravel, sand, and clay that settles in rivers and swamps [1]. Based on this, the city of Banjarmasin is easily inundated with water either due to rain or overflowing river water due to tides.

Rain with moderate to high intensity caused flooding in mid-January 2021 in South Kalimantan Province. This wide-scale flood, triggered by high rainfall since January 9, 2021, also inundated 11 regencies/cities in South Kalimantan. The Banjarmasin Campus I area of Lambung Mangkurat University (ULM Campus I) was inundated to a height of 50 cm. Floods caused by high tides coupled with several days of rain inundated most of the campus area which hindered access to learning and administration [2]. Furthermore, the inability of the drainage channel to drain rainwater that falls into the river has exacerbated the overflow of water that occurs, especially to the Campus I ULM road [3].

Flood control methods applied both structural and non-structural. Non-structural flood control method is the design of a flood inundation area mapping as an informative tool in disaster mitigation [4]. Mapping is carried out to determine inundation height and flood volume with a certain return period. Further, it can function as preventive measures, which leads to permanent solutions that can be carried out in overcoming flooding [5]. Another alternative to flood control system is based on structural method, namely the design of detention basins. Detention basin is considered as one of the alternatives that is quite effective in flood control in urban areas which functions as a detention basin and water catchment. In the condition of the study area, efforts need to be made to overcome the problem of flooding that occurs as an effort to control water and puddles caused by overflowing of river water and rainwater [6]. This study aims to analyze the volume of the flood, the dimensions of the detention basin, and the optimal location of the detention basin.

2. Methodology

2.1 Research Location

The research location is in the Banjarmasin Campus I area of Lambung Mangkurat University which is located on Brigadier General H. Hasan Basri Kayu Tangi road, North Banjarmasin District, Banjarmasin City, South Kalimantan Province. The picture of the research location can be seen in Figure 1.

Figure 1. Lambung Mangkurat University Banjarmasin Campus
Source: [7]

2.2 Data Analysis and Method

Rainfall data obtained from the BMKG Syamsuddin Noor station in the form of maximum annual rainfall data for the last 20 years. Hydrological statistical parameters were calculated, and the results were corresponding with the requirements for each type of distribution, namely Normal, Log Normal, Gumbel and Log Pearson III. Then the distribution suitability test was carried out with the Chi Square and Smirnov Kolmogorov Test and continued with the calculation of the designed rainfall. The design rainfall is calculated from maximum rainfall analysis then calculates the designed flood discharge using the Der Weduwen method. The cross-sectional analysis includes the layout of Campus I Banjarmasin ULM has obtained using Google Earth data meanwhile topographic data taken using field survey investigation. Inundation conditions at ULM Campus I area obtained through field survey measurements.

The results of flood volume will be used as a reference for designed the detention basin. The dimensions volume of detention basin will be designed greater than designed flood volume. According to the dimensions of detention basin as a result the flooding can be controlled optimally. Spatial data and the layout of Campus I Banjarmasin ULM will be taken into consideration in determining the location for the detention basin.

2.3 Hydrological Analysis

Hydrology is the study of the distribution of water above and below the earth's surface, involving the reaction of the physical and chemical properties of water on earth and the environment. The most influential hydrological factor is rainfall (precipitation). Rainfall volume in an area is one of the factors that determine the amount of flood discharge that occurs in the area that receives it [8].

2.4 Parameters of Statistics and Logarithms

Equations for calculating statistical parameters and logarithms [9] are:

a. Mean ($\bar{X}$) and Standard Deviation (S)

$$\bar{X} = \frac{1}{n}\sum_{i=1}^{n} Xi$$

$$\bar{X} = \frac{1}{n}\sum_{i=1}^{n} \log X_i$$

$$S = \sqrt{\left[\frac{1}{n-1}\sum_{i=1}^{n}(Xi - \bar{X})^2\right]} \quad (1)$$

Where:

X = rainfall (mm)
$\bar{X}$ = mean rainfall (mm)
n = amount of data
S = standard deviation (mm)

b. Coefficient of Skewness (Cs), Coefficient of Variation (Cv) and Coefficient of Kurtosis (Ck)

$$Cs = \frac{n\sum_{i=1}^{n}(Xi-\bar{X})^3}{(n-1)(n-2)s^3}$$

$$Cv = \frac{s}{\bar{X}}$$

$$Ck = \frac{n\sum_{i=1}^{n}(Xi-\bar{X})^4}{(n-1)(n-2)(n-3)s^4} \quad (2)$$

Where:

X = rainfall (mm)
$\bar{X}$ = mean rainfall(mm)
n = amount of data
S = standard deviation (mm)
Cs = coefficient skewness
Cv = Coefficient of variation
Ck = coefficient of kurtosis

Table 1. Determination of Distribution Type

Distribution	Requirements	Distribution	Requirements
Normal	Cs = 0 and Ck = 3	Gumbel	Cs = 1.14 and Ck = 5.4
Log Normal	$Cs \approx Cv^3 + 3Cv$	Log Pearson III	Beside above values
	$Ck \approx Cv^8 + 6\,Cv^6 + 15Cv^4 + 16Cv^2 + 3$		

Source: [10]

2.5 Designed Rainfall Analysis

a. Normal Distribution

The Normal distribution method equation [9] is:

$$X_T = \bar{X} + K_T.S \tag{3}$$

Where:

X_T = planned period T year
$\bar{X}$ = mean rainfall (mm)
K_T = probability factor (value of Gauss reduction variable)
S = standard deviation

b. Log Normal Distribution

The Log Normal distribution method equation [9] is:

$$Log\,X_T = Log\,\bar{X} + K_T.S \tag{4}$$

Where:

X_T = planned period T year
Log $\bar{X}$ = logarithm value of mean rainfall (mm)
K_T = probability factor (value of Gauss reduction variable)
S = standard deviation

c. Gumbel Distribution

The Gumbel distribution method equation [9] is:

$$X_T = \bar{X} + \frac{S}{Sn}(Yt - Yn) \tag{5}$$

Where:

X_T = planned period T year
$\bar{X}$ = mean rainfall (mm)
S = standard deviation
Sn = standard deviation of total data (n)
Yt = the reduction value of the variable period T years
Yn = variable reduction value from data (n)

d. Log Pearson III Distribution

The Log Pearson Type III distribution method equation [9] is:

$$Log\,X_T = Log\,\bar{X} + K.S \tag{6}$$

Where:

Log X_T = logarithm value of planned period T year
Log $\bar{X}$ = logarithm value of mean rainfall (mm)
K = probability factor
S = standard deviation

This method is intended to determine the suitability of the distribution equation that is selected and can represent the sample of the analyzed data. There are two types of Test, the Chi Square and Smirnov Kolmogorov tests. In these tests, the expected values of the calculation results are usually observed [11].

2.6 Designed Flood Discharge of Der Weduwen

One of flood discharge method is Der Weduwen for a catchment area less than 100 km^2 as follows [12]:

$$Q_n = \alpha.\beta.q_n.A \tag{7}$$

$$\alpha = 1 - \frac{4.1}{\beta.q+7} \tag{8}$$

$$\beta = \frac{120+\frac{t+1}{t+9}A}{120+A} \tag{9}$$

$$q_n = \frac{R_n}{240} \cdot \frac{67,65}{t+1,45} \tag{10}$$
$$t = 0,25. L. Q^{-0,125}. I^{-0,25} \tag{11}$$

Where:
Q_n= flood discharge (m^3/s) with probability of not being fulfilled n%
R_n= maximum daily rainfall (mm/day) with probability of not being met n%
α = run off coefficient
β = area reduction coefficient for watershed rainfall
q_n = rainfall ($m^3/s.km^2$)
A = flow area (km^2)
t = duration of rainfall (hours)
L = river length (km)
I = gradient of river or terrain

2.7 Flood Inundation Mapping

Flood inundation mapping process requires data from hydraulic and hydrologic analysis. The equation of river slope [13] is:

$$river\ slope = \frac{upstream\ elev. - downstream\ elev.}{length}\ x\ 100\% \tag{12}$$

where:
downstream elevation = elevation at the observed bottom point
Upstream elevation = elevation at the observed top point
Length = distance between 2 observed points

2.8 Detention Basin

Detention basin is a pond that is used to accommodate excess rainwater, especially during the rainy season. The type of detention basin can be grouped into 4 conditions, namely based on the purpose of its construction, its use, the flow of water, and the material forming it [14]. Detention basins is a highly effective hydraulic construction to reduce the frequency of erosive floods. At least the first 0.5 inch of runoff from the design stormwater discharge. This kind of structure is recommended to larger, more effective developed on areas of over 20 acres [15]. Practically, in Japan conventional policy is to build embankments to separate the locations of hydraulic infrastructure, such as river channels and detention basins from floodplains [16].

2.9 Flood Simulation Using HEC-RAS

HEC-RAS get known and used for river simulation by hydraulic engineers and different researchers because of its capabilities and abilities to simulate and identifies flood-plain areas where allows the researcher to visualize the flood extent along a river route [17]. The simulation results provide information on the potential elevation of the flood water level in each cross section of the river based on the elevation of the bottom cross section of the river and riverbank so that the maximum flood water level is obtained based on the return flood discharge and flood volume.

3. Result and discussion

3.1 Hydrological Analysis

Used rain data for the last 20 years from 2002-2021. Based on the analysis of rainfall data in 2002 and 2003, it was declared invalid and could not be used because the data had the exact same value. Meanwhile, data in 2010 and 2013 were omitted because the data were not recorded. The rainfall at Syamsuddin Noor Station used in the analysis process can be seen in Table 2.

Table 2. Maximum Annual Rainfall at Syamsuddin Noor Station

Year	Rainfall (mm)	Year	Rainfall (mm)
2004	136.2	2014	118.2
2005	62.2	2015	117.5
2006	80	2016	98.9
2007	109.3	2017	111.8

2008	145	2018	112.1
2009	105.2	2019	70.5
2011	113	2020	122.1
2012	92.4	2021	249

Source:[18]

3.2 *Statistic and Logarithmic Parameters*

The results of the calculation of statistical and logarithmic parameters can be seen in Table 3.

Table 3. Statistic and Logarithmic Parameters

Parameter	Statistic	Logaritmic
Mean (mm)	115,206	2,040
Standard Deviasi (S)	41,858	0,136
Coefficien Variation (Cv)	0,363	0,067
Coefficien Skewness (Cs)	0,138	0,001
Coefficien Kurtosis (Ck)	0,040	0,023

Table 4. Type of Distribution Based Statistic Parameter, Chi Square Test and Smirnov Kolmogorov Test

Distribution	Statistic Parameter	Chi Square Test	Smirnov Kolmogorov
Normal	Unsuitable	Unsuitable	Suitable
Log Normal	Unsuitable	Suitable	Suitable
Gumbel	Unsuitable	Unsuitable	Suitable
Log Pearson III	Suitable	Suitable	Unsuitable

Based on Table 4, the type of Distribution that is considered suitable is Log Pearson III. There are two test to carry out the distribution suitability test, the Chi Square Test and the Smirnov Kolmogorov Test. Based on the results of the Chi Square Test, the type of Distribution that is considered suitable and can be used in further analysis is the Log Pearson III Distribution. The results show, the Smirnov Kolmogorov test is unsuitable. On the different method, Log Normal, is not suitable according to distribution analysis. Log Normal have two suitable condition due to Chi Square Test and Smirnov Kolmogorov Test. Table 5 shows the value for Log Normal and Log Pearson III return period, the difference between two distributions for 20 years return period is less than 1.5%. This condition determined the Log Normal method had chosen.

Designed rainfall results from the Log Pearson III and Log Normal with return periods of 2, 5, 10, and 20 years is presented in Table 5.

Table 5. Designed rainfall of Log Pearson III and Log Normal with various return periods.

Return Periods T (year)	Log Pearson III (mm)	Log Normal (mm)	Difference (%)
2	109.62	109.62	0.000
5	142.80	142.71	0.063
10	163.96	163.85	0.067
20	180.90	183.46	1.415

3.3 *Designed Flood Discharge of Der Weduwen*

The designed flood discharge with Der Weduwen method can be used at catchment area less than 100 km^2, based on the results of the analysis using Google Earth, it was found that the study area in this research is 3.98 km^2. Then this method is considered eligible and can be used. According to field measurements survey, the upper elevation (P_1) in the middle of the river is 5.88 meters above sea level and the lowest elevation (P_{13}) in the middle of the river is 5.62 meters above sea level with a river length of 536.6 m. The slope of the river is 0.0005 or 0.05%. The results of designed flood presented in Table 6. The flood discharge value from the 20-year return period is 89.30 m^3/s, the value used as a boundary condition for HEC-RAS simulation model.

Table 6. Designed Flood Discharge of Der Weduwen

Return Periods (year)	2	5	10	20
Q (m^3/s)	48.68	66.63	78.35	89.30

3.4 Hydraulics Analysis with HEC-RAS

Hydraulic analysis using the HEC-RAS application downloaded on the official website, www.hec.usace.army.mil and with the help of DEMNAS data obtained from the official website, Tanahair.indonesia.go.id according to the study area, namely Campus I ULM with DEMNAS_1712-51 data. The results of longitudinal cross-section analysis of the inundation water level are shown in Figure 2 It can be seen that water level exceeds the riverbank. Final results of the water level of the Long Section River Campus I ULM are shown in Figure 5.

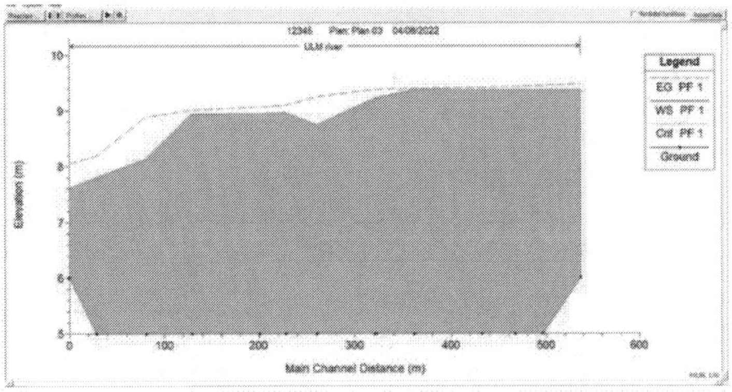

Figure 2. Water level condition of 20 years discharge value as Q_{20} = 89.30 m^3/s.

Analysis of cross-sectional view, height of water level in the upstream, middle, and downstream parts, has different heights due to the cross-sectional dimensions, different base elevations, river slopes, and the width of the river cross-section. In the downstream part, the inundation tends to be lower, while in the upstream part the inundation tends to be higher.

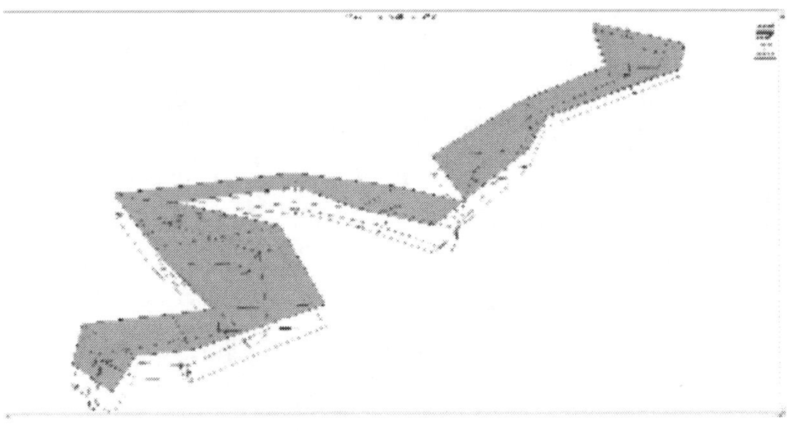

Figure 3. Simulation results of 3D inundation height as Q_{20} = 89.30 m^3/s.

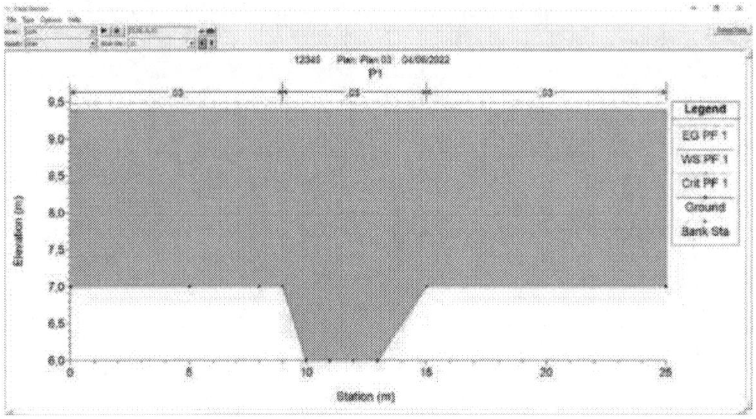

Figure 4. Upstream part water level condition (P1)

It can be seen in Figure 4 that water level on the Upstream side of ULM Campus I River at +9.38 m, and water level reached 2.38 m from the side elevation of the river which at +7 m. Figure 5 shows that water level in the middle parts of the Campus I ULM River +8.74 m, water level reached level of 1.74 m on the left and 2.74 m on the right side of the riverbank. Which is at +7 m on the left side and +6 m on the right side.

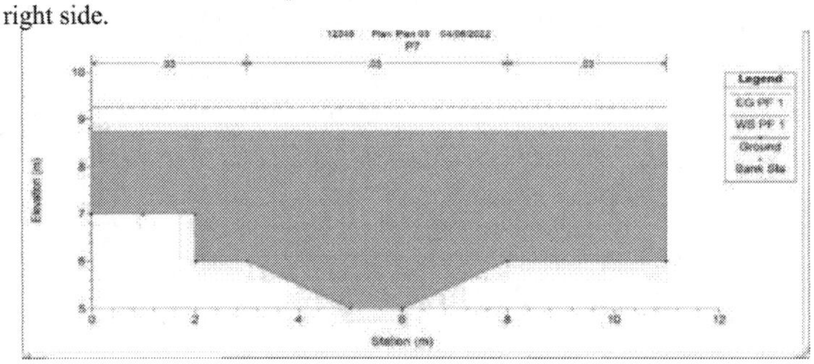

Figure 5. Flood Water Level Middle (P7) Cross Section River Campus I ULM

Figure 6 shows water level in the Lower reach of river at station P13 was at +7.61 m, water level reached 0.61 m from the side elevation of the river which is at +7 m.

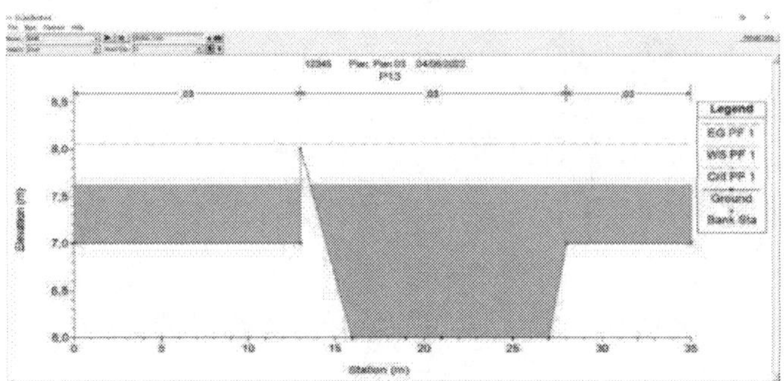

Figure 6. Water Level Condition at Downstream part (P13)

The calculation of the total volume of the designed flood from HEC-RAS with a return period of

20 years and a discharge $Q_{20} = 89.30\ m^3/s$, as follows:

$$V_{Total} = V_{left} + V_{channel} + V_{right}$$
$$V_{Total} = (8.01 + 18.22 + 9.78) \times 1000\ m^3$$
$$V_{Total} = 36.010\ m^3$$

The obtained Total Volume was 36,010 m^3 base on Q_{20} simulation. A detention basin volume must be larger, to cover the designed flood volume. The detention basins designed must be prevent the overflow of the river.

3.5 Detention basin Analysis

Free space of Campus I is an area located around the General Building with 47,307.48 m^2. At the location of detention basin 1, the actual conditions in the field indicate that the area is dominated by illegally managed rice fields, shrubs, and swamps. Based on the actual conditions, it was concluded that the location could be used as the location of detention basin 1 and green open space in the Banjarmasin Campus I area.

It is designed for detention basin 1 with the following dimensions:

$$Area_{DB\,1} = 160\,.\,130 = 20.800\ m^2$$

If the detention basin is designed with a depth of 1.5 m, then the volume of the detention basin obtained is:

$$V = l.w.d = 160\,.\,130\,.\,1{,}5 = 31.200\ m^3$$

The dimensions of detention basin 1 were 160 x 130 x 1.5 with a detention basin 1 accommodate volume of 31.200 m^3.

Figure 7. Design and Location of Detention Basin 1 Nearby to the General Building

Based on Figure 8, it can be seen that there is an open area opposite the Master of Law Building with an area of 8,215.54 m^2. At the location of detention basin 2, the actual conditions in the field indicate that the area is dominated by shrubs and swamps and has a lower elevation than the surrounding area. So the key is that this location can be used as the location of detention basin 2 and green open space in the Banjarmasin Campus I area.

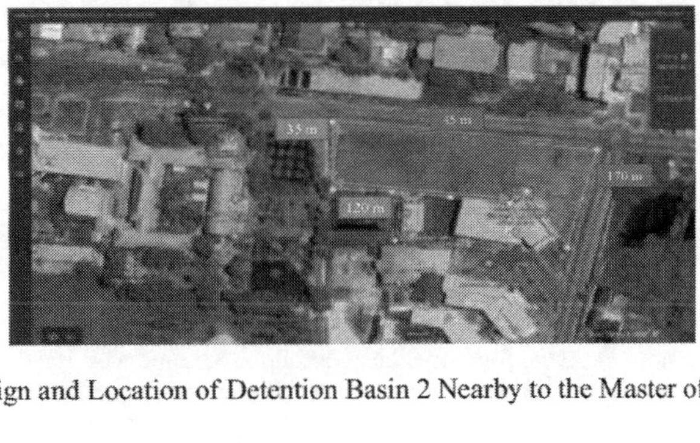

Figure 8. Design and Location of Detention Basin 2 Nearby to the Master of Law Building

It is planned for detention basin 2 with the following dimensions:

$Area_{DB\,2} = 120\,.\,35 = 4.200\ m^2$

If the detention basin is designed with a depth of 1.5 m, then the volume of the detention basin obtained is:

$V = l.w.d = 120\,.\,35\,.\,1{,}5 = 6.300\ m^3$

The dimensions of detention basin 2 were 120 x 35 x 1.5 with a detention basin 1 accommodate volume of 6.300 m^3.

Then the cumulative volume of the two detention basins, were:

$V_{cum} = V_1 + V_2$

$V_{cum} = 31.200 + 6.300$

$V_{cum} = 37.500\ m^3$

Based on the results of the design flood analysis using HEC-RAS for a 20-year return period with a discharge of 89.30 m^3/s and a flood volume of 36,100 m^3, the cumulative volume of the designed detention basin, which is 37,500 m^3, is concluded to be applicable. Shallow-designed detention basins are considered appropriate to prevent excessive costs for excavation and foundation planning, as well as preventing damage due to excessive water pressure caused by excessive depth.

Based on the results of the ULM Campus I analysis obtained from the application of ArcGIS and spatial data obtained from Google Earth, the contours were obtained which were used as comparisons to determine the location of the detention basin according to the height of the land. The results of contour analysis with ArcGIS can be seen in Figure 9.

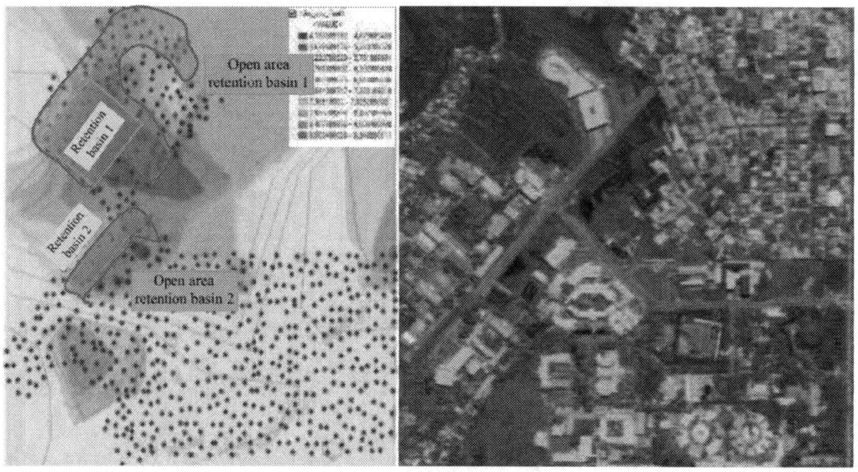

Figure 9. Contour Analysis Results with ArcGIS in the Banjarmasin Campus I Area Lambung Mangkurat University

Based on Figure 9, it can be concluded that the area with the lowest elevation is around the General Building and the Master of Law Building with an elevation between + 4.7 masl to +6.3 masl, 0.5 meters to 2.2 meters lower than the area surrounding areas which are at an elevation of +6.8 masl to +8.5 masl.

Final result in this research compare to the previous research in India, Mumbai city. The detention pond has the capacity to mitigate inundation, which would inundate past drainage system, located to upstream of the catchment [19]. The location of detention basin in this research located at two different location, first located in Master Program of Law is South side and the North side is located at General Building. Detention basin located at downstream area, different from the previous research because of the free space availability at study location.

4. Conclusion

According to the simulation the cumulative designed flood volume was 36,010 m^3 with a discharge value of 89.30 m^3/s. According to Google Earth imaging, the open area of Campus I ULM was located

IOP Conf. Series: Earth and Environmental Science 1184 (2023) 012013 doi:10.1088/1755-1315/1184/1/012013

at 2 points, at the open area around the General Building, it was designed the detention basin 1 with dimensions of 160 m x 130 m x 1.5 m with a storage capacity of 31,200 m^3 and an open area in front of the Master of Law Building was designed detention basin 2 with dimensions of 120 m x 35 m x 1.5 m with a storage capacity of 6,300 m^3 and the cumulative detention basin volume was 37,500 m^3. Analysis using ArcGIS shows that these 2 locations are optimal as detention basin locations because they have a lower elevation than the surrounding area.

References

[1] BPK Banjarmasin, "Profil Kota Banjarmasin," 2022. [Online]. Available: https://kalsel.bpk.go.id/kota-banjarmasin/#.

[2] Kanal Kalimantan, "Banjir Rob Rendam Kampus ULM, Mahasiswa Terpaksa Pakai Sandal," 2021. [Online]. Available: https://www.kanalkalimantan.com/banjir-rob-rendam-kampus-ulm-mahasiswa-terpaksa-pakai-sandal/.

[3] Apa Habar, "Diterjang Hujan Deras dan Air Pasang, ULM Banjarmasin Kebanjiran," 2021. [Online]. Available: https://apahabar.com/2021/12/diterjang-hujan-deras-dan-air-pasang-ulm-banjarmasin-kebanjiran/.

[4] A. Nisa and C. R. H, Perencanaan Detail Embung UNDIP Sebagai Pengendali Banjir Pada Banjir Kanal Timur, Semarang: Fakultas Teknik Universitas Diponegoro, 2008.

[5] M. Belladona, "Pemetaan Daerah Genangan Dan Kajian Genangan Banjir Pada Kawasan Komersial DI Kelurahan Rawa Makmur," *Majalah Teknis Simes,* pp. 40-44, 2017.

[6] D. Ferdian, A. Saggaff and Sarino, "Efektifitas Pengendalian Banjir Dengan Embung: Studi Kasus Taman Firdaus Universitas Sriwijaya," *Cantilever Jurnal Penelitian dan Kajian Bidang Teknik Sipil,* pp. 57-62, 2020.

[7] Google Earth Web, "Universitas Lambung Mangkurat," 2022. [Online]. Available: https://earth.google.com/web/search/ulm+banjarmasin/@-3.29574941,114.58529252,5.28496146a,1228.93963926d,35y,-131.86653625h,45.00110863t,-0r/.

[8] P. E. Black, "Watershed Functions 1," *JAWRA Journal of the American Water Resources Association, 33(1),* pp. 1-11, 1997.

[9] Suripin, Sistem Drainase Perkotaan yang Berkelanjutan, Yogyakarta: ANDI Offset, 2004.

[10] B. Triatmodjo, Hidrologi Terapan, Yogyakarta: Beta Offset, 2008.

[11] Soewarno, Hidrologi Aplikasi Metode Statistik Untuk Analisa Data, Bandung: Nova, 1995.

[12] J. Loebis, Banjir Rencana Untuk Bangunan Air, Jakarta: Departemen Pekerjaan Umum, 1987.

[13] D. Rosgen, Applied River Morphology, widland Hydrology, Colardo: Pagosa Springs, 1996.

[14] I. Soedibyo, Teknik Bendungan, Jakarta: PT Sentra Sarana Abadi, 2003.

[15] https://oa.mo.gov, Detention Ponds and Basins, Access January 2023.

[16] S. Harada and T.Ishikawa, "Evaluation of the Effect of Hamao Detention Pond on Excess Runoff from the Abukuma River in 2019 and Simple Remodeling of the Pond to Increase Its Flood Control Function," *Applied Science Journal*, 2021

[17] H. Tamiru and M. O. Dinka, "Application of ANN and HEC-RAS model for flood inundation mapping in lower Baro Ako bo River Basin, Ethiopia," *Journal of Hydrology: Regional Studies 36,* 2021.

[18] BMKG.go.id, "Data Iklim," 2022. [Online]. Available: https://dataonline.bmkg.go.id/data_iklim.

[19] Yazdi, Jafar. "Optimal Operation of Urban Storm Detention Ponds for Flood Management," Water Resources Management, 33, 2019.10.1007/s11269-019-02228-5., 2019

IOP Conf. Series: Earth and Environmental Science 1184 (2023) 012014 doi:10.1088/1755-1315/1184/1/012014

Rainwater Harvesting Tank Sizing: A Case in Urban Catchments in Metro Cebu

A F A Oraya [1,2], G J G Cuba [1], L L M Varquez [1] and J C R Hermosa [1]

[1]Department of Civil Engineering, Cebu Technological University, M.J Cuenco Ave., Corner R. Palma St. Cebu City 6000, Philippines

[2]aureflo.oraya@ctu.edu.ph

Abstract. Rainwater is an abundant resource in the Philippines and when properly collected and managed, can help address flood problems. Rainwater harvesting has been practiced for years and has been viewed as one of the better options for flood mitigation in urban areas compared to other storm water management practices. It does not require large land area and cost per area is high in these locations. However, there is a challenge in selecting the suitable tank size. This paper investigates if the increase in stormwater runoff brought about by the changes in landuse in urbanized catchments can be represented by a single coefficient similar to what is being presented in the Philippine Green Building Code and in other local ordinances in the country. Simulations through HEC-HMS of three urban catchments in Metro Cebu showed that the increase in volume is highly varying such that the coefficient produced for each catchment was so distinct that no single average coefficient can be drawn. Thus, analysis should be catchment-based to determine the suitable tank size in rainwater harvesting systems as an option for flood mitigation.

1. Introduction

When the amount of precipitation is greater than the infiltration rate of the ground surface, runoff occurs. Much water in river system comes directly from surface runoff, defined as runoff from the land surface. When rain falls on the saturated or impervious ground, flooding occurs [1].

Rainwater harvesting (RWH) is one of the methods employed and followed to support water conservation. According to the Philippine Green Building Code (PGBC), rainwater is considered one of the cleanest sources of water and can be used for non-potable purposes. Good quality rainwater can also be used for domestic use. Scarcity of water in some areas is of concern and RWH has been one of the identified solutions. Aside from its significance for its usage, it can also lessen flooding in impervious areas [2]. In Cebu City, there is Ordinance 1711 which requires providing rainwater tanks or containers to conserve potable groundwater and to lessen flood incidences in the city.

Implementation of the RWH system in urban catchment areas has a significant impact on urban water management. In particular, RWH system significantly reduces floods and avoids potential drainage system disruptions during storms [3].

Similar to water supply systems, RWH system has three major components – a catchment system, conveyance and storage facility. The collection system is the roof area, the delivery system are the pipes and drainage ditches to transport rainwater to storage tanks. However, tank sizing should be given great attention in a RWH system since this is considered the most expensive part. With a given amount of conditions, the suitable size can be determined [4]. The sizing of storage tanks in cubic meters according to PGBC is calculated by dividing the building footprint area by 75.

This study aims to determine the appropriate coefficient that can be used in sizing rainwater harvesting tanks in urban catchments. This is derived from the changes in land use in the area and the corresponding increase in runoff volume due to these changes.

2. Materials and Methods

This study focuses on urban catchments in Metro Cebu as it is referred to as the principal urban settlement of Region 7 (Central Visayas) [5]. Specifically, the areas within the Cities of Mandaue and Cebu are highly urbanized [6] and are susceptible to flooding due to heavy precipitation and occasional typhoons [7]. In addition, about 1,864 mm (73.4 inches) of rain falls in Cebu each year within an average of 190 rainy days [8].

This study aims to obtain a coefficient that can help in sizing the rainwater tanks efficiently in urbanized catchments. This study can benefit the policy-makers as the recommended coefficient of the PGBC may not fully mitigate flooding in urban catchments. This can also benefit the residents in highly urbanized cities wherein they can use tanks that are appropriate in size and save costs in non-potable water.

2.1. Study Area

This study covers three urban catchments in the Cities of Mandaue and Cebu. It is observed that runoff increases as land cover changes in a catchment [9] [10]. In Butuanon River Catchment, land development increases with time with the largest increase of 8.062% happening from 2007-2011 to 2014-2017 [9]. The properties of the catchment are presented in the table below.

Table 1. Catchment Properties

Watershed	Catchment Area	Outlet	Elevation	Soil Type	Land Use
Butuanon River	62,893,266 m^2	10° 20' 25.99" N and 123° 57' 39.47" E	3.36 meters to 726.93 meters	Clay Loam	Upstream portion is mainly covered with trees, agricultural lands and barren lands; Downstream portion mainly covered with built-up areas
Guadalupe River	18,076,712 m^2	10° 17' 21.04" N and 123° 53' 47.16" E	1.2 meters to 695 meters	Clay Loam	Upstream portion filled with grasslands and trees; Downstream portion has built-up lands and developed areas

Mahiga Creek	17,920,324 m^2	10° 18' 41.10" N and 123° 55' 30.45" E	1 meter to 245 meters	Clay Loam	Mainly impervious areas wherein commercial and residential areas are prominent

2.2. Terrain Preparation and Catchment Delineation
In describing the terrain of the watershed of interest, digital elevation model (DEM) was used. DEM are acquired through techniques that rely on aerial photography [11]. In the Philippines, high resolution elevation datasets using the light detection and ranging technology were initially acquired through the Disaster Risk and Exposure Assessment for Mitigation Program [12].

A 1-meter resolution digital elevation model was used for both Guadalupe River and Mahiga Creek. Due to the size of the catchment, the elevation model was resampled to 10 meters for Butuanon River. Prior to catchment delineation, the elevation model was pre-processed to identify the direction of the flow and stream networks. Catchment delineation was done using Geographic Information System software.

2.3. Direct Runoff Calculation
The direct runoff in the three catchments was modelled using the Soil Conservation Service – Curve Number (SCS-CN) method. It calculates the runoff as affected by land use practices in small agricultural catchments and then adapted to urban catchments. This method makes use of a curve number (CN) which is a function of the soil group, cover complex and antecedent soil moisture conditions [13].

The CN values were calculated with the aid of Google Earth Pro. The process of individually and manually identifying the current land use classification per catchment led to the creation of Figure 1.

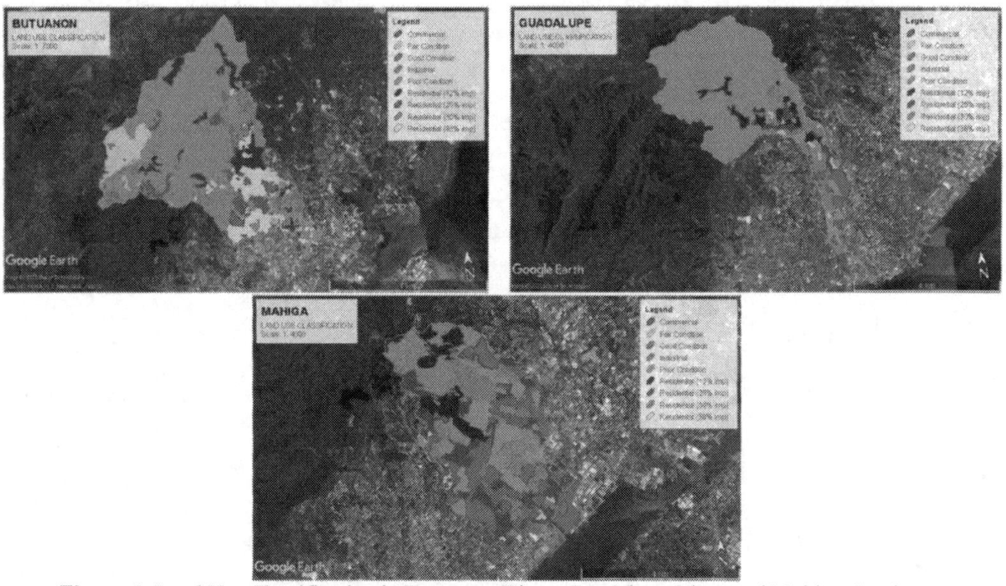

Figure 1. Land Use Classification in Butuanon River, Guadalupe River and Mahiga Creek

The SCS unit hydrograph requires the calculation of the time of concentration which was estimated using Kirpich's formula

$$T_c = 0.0195L^{0.77}S^{-0.385}$$

The rainfall data used has duration of 12 hours. As this study involves rainwater harvesting, the return period suggested are 2 or 5 years. In this study, rainfall-runoff analysis will also be shown via time versus direct runoff graph and provide observations from the said relationship.

2.4. *Rainfall-Runoff Simulation*

There are various software that can be used for hydrologic simulations but the Hydrologic Engineering Center – Hydrologic Modeling System (HEC-HMS) is the most widely used modeling programs in rainfall-runoff simulation [11]. Primary input parameters in simulation are rainfall data and elevation data.

Rainfall data or design storms with duration of one hour are prevalently used in hydrologic modeling applications because it simplifies the modelling process [14]. The design storms are usually derived from established intensity-duration-frequency curves. The alternating block method which is used in the United States is the most extensively and internationally used among other models. When compared with other methods, this design storm also has the same features as that of Safilda of Czechoslovakia, Linear/Exponential Method of Canada and Average Variability Method of Australia [14].

2.5. *Storage Size Calculation*

After the rainfall-runoff simulation, storage size calculation of RWH tanks was done. The sizing was done with the assumption that the increase in runoff caused by the changes in land use in urban catchments should be reversed by collecting rainwater through tanks. Thus, the volume of the tank is calculated as the increase in runoff due to development changes in the catchment. According to Hazarika et al., building footprint represents the total area of a building and it also provides the outline of a building drawn along the exterior walls [15]. The building footprint area is used in the calculation of the size of the RWH tank in the PGBC.

3. Results and Discussions

Urban catchments have been chosen for this study to model the RWH tank sizing coefficient. The catchments are located in Metro Cebu where the annual growth rate of urban sprawl is at 0.7108 $km^2/year$ in the years 1994-2019 [16]. Three hydrologic models each corresponding to a river basin was generated using HEC-HMS. Simulation duration covers 36 hours. The time series data used was derived from intensity-duration-frequency curve of a station in Mactan, Cebu with 37 years of recorded data.

Various scenarios representing the changes in developed area were also created. This was done by changing pervious areas to impervious areas in the catchment. The progressive developments for impervious areas were +10%, +20%, +50%, +75% and +100%.

IOP Conf. Series: Earth and Environmental Science 1184 (2023) 012014 doi:10.1088/1755-1315/1184/1/012014

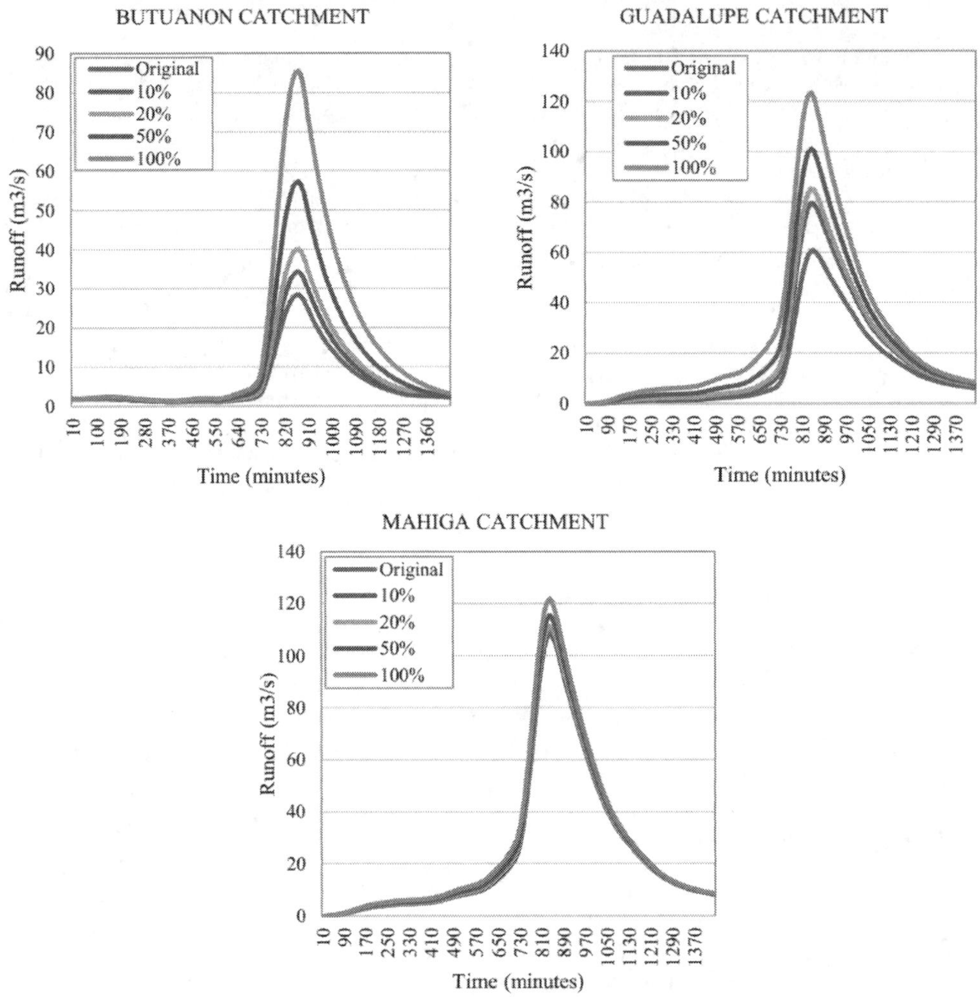

Figure 2. Simulated Runoff Hydrographs of Various Percent Impervious Area Scenarios

The runoff hydrographs for the three different catchments are presented in Figure 2. The peak discharge increases as the impervious areas increased in the catchment. However, the peak discharge varies depending on the ratio of the impervious and pervious areas within the catchment.

Table 2. Coefficient Calculation in the Three Urban

		Change in Impervious Area				
		10%	20%	50%	75%	100%
Butuanon River	Runoff Volume (m^3)	708,600	818,460	1,010,749	1,133,678	1,480,470
	Increase in Volume (m^3)	112,200	222,060	414,349	537,278	884,070
	Increase in Area Developed (m^2)	4,231,811	8,463,621	21,159,054	31,738,580	42,318,107
	Coefficient	37.72	38.11	51.07	59.07	47.87
Guadalupe	Runoff Volume (m^3)	1,594,140	1,713,540	2,050,740	2,306,319	2,539,860

River	Increase in Volume (m^3)	340,680	460,080	797,280	1,052,859	1,286,400
	Increase in Area Developed (m^2)	1,413,857	2,827,716	7,069,290	10,603,934	14,138,579
	Coefficient	4.15	6.15	8.87	10.07	10.99
Mahiga Creek	Runoff Volume (m^3)	2,257,440	2,258,762	2,377,782	2,419,177	2,520,416
	Increase in Volume (m^3)	32,880	34,202	153,222	194,617	295,856
	Increase in Area Developed (m^2)	439,466	878,933	2,197,332	3,295,998	4,394,664
	Coefficient	13.37	25.70	14.34	16.94	14.85

The calculated coefficient for RWH tank sizing is presented in Figure 3. As revealed in the plot, the points are too varied to be able to conclude a single or average coefficient. The areas between each catchment vary greatly from its size resulting to an irregular pattern of coefficients.

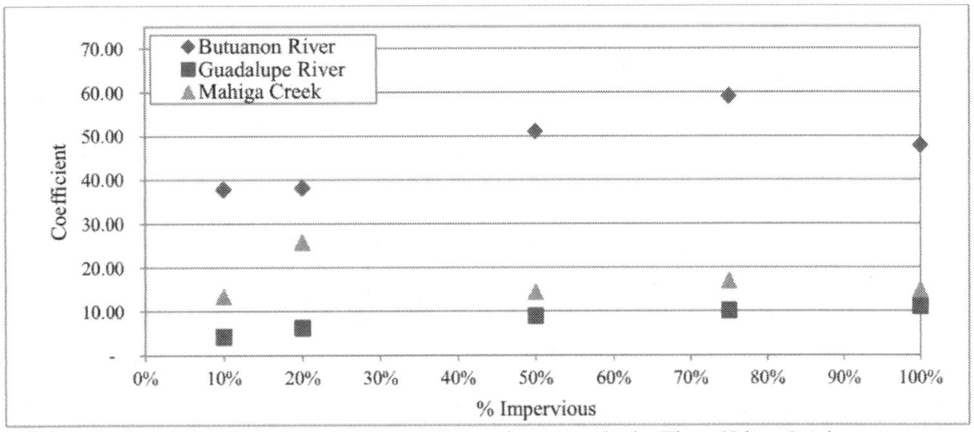

Figure 3. Plot of Coefficient versus Percent Impervious Area in the Three Urban Catchments

The average coefficient for RWH tank sizing for Butuanon River is 46.77 with a standard deviation of 9.05. For Guadalupe River the average coefficient is 8.05 with a standard deviation of 2.84. For Mahiga Creek, the average coefficient is 17.04 with a standard deviation of 5.01. It can be noted that the obtained coefficients are less than the recommended coefficient of 75 by the PGBC. Calculating the tank sizes using the derived coefficients in this study will yield larger tank sizes compared to those obtained using the recommended value of the PGBC.

4. Conclusion and Recommendation

The three urban catchments analysed are Butuanon River, Guadalupe River and Mahiga Creek. The rainfall-runoff model was developed for each catchment using the SCS-CN method in HEC-HMS. Several components were taken into consideration and the results have shown direct relationship between the developed areas and the runoff volume.

In RWH tank sizing, the PGBC utilizes a coefficient of 75. However, this study suggested modifying the coefficient as areas are constantly developing which leads to increase in runoff. By simulating various scenarios in HEC-HMS, increase in runoff volume were identified for each catchment. Volume changes were observed as the progressive developments for impervious areas changes which were +10%, +20%, +50%, +75% and +100% were inputted in the model. Coefficients were calculated by dividing the increase in runoff volume to the increase in developed areas.

The calculated coefficients were analysed and it was then concluded that no single average coefficient can be drawn from the simulated scenarios in the three urban catchments. Thus, a catchment-based analysis should be used to determine the appropriate tank size for reducing runoff in urban catchments.

References

[1] U.S. Geological Survey Water Science School. (2019). Surface Runoff and the Water Cycle

[2] Lumbera, M P, Macuha, R N and Castro, P P M. (2013). Integrating Storage Sizing and Water Treatment for Rainwater Harvesting in the Philippines *Iconda CIB Library*

[3] Freni, G and Liuzzo, L. (2019). Effectiveness of Rainwater Harvesting Systems for Flood Reduction in Residential Urban Areas *Multidisciplinary Digital Publishing Institute*

[4] Ward, S, Memon, F A and Butler, D. (2010). Rainwater harvesting: Model-based design evaluation *Water Science & Technology*. 61 (1): 85-96

[5] Mercado, R. (1998). Metropolitan Cebu: The Challenge of Definition and Management *PIDS Discussion Paper Series, No. 1998-1, Philippine Institute for Development Studies*

[6] Philippine Statistics Authority. (2002). Cebu: The Scond Most Populated Province in the Philippines

[7] Organization for Economic Cooperation and Development. (2017). OECD Green Growth Studies, Building Resilient Cities

[8] Climate Data Org. Weather By Month // Weather Averages Cebu City (2019). *https://en.climate-251 data.org/asia/philippines/cebu/cebu-city-3183/*

[9] Maglana, J M R, Parmes, H O O, Yu, A C C, Fornis, R L and Oraya, A F A. (2020). An attempt to classify landcover of the Butuanon river catchment using Landsat images *AIP Conference Proceedings* **2278**, 020044

[10] Santillan, J R, Amora, A M, Makinano-Santillan, M, Gingo, A L and Marqueso, J T. (2019). Analyzing the Impacts of Land Cover Change to the Hydrologic and Hydraulic Behaviours of the Philippines' Third Largest River Basin *ISPRS Annals of the Photogrammetry, Remote Sensing and Spatial Information Sciences*. IV-3

[11] Bedient, P B, Huber, W C and Vieux, B E. (2012). Hydrology and Floodplain Analysis 5th Edition Pearson

[12] Andaya, K J, Alviar, J, Mars, P, Cruz, C, Sarmiento, C J, Balicanta, L and Paringit, E. (2015). Airborne LIDAR surveying in the Philippines: Data Acquisition of the Nationwide Disaster Risk and Exposure Assessment for Mitigation (DREAM) program in 18 major river basins *Proceedings of the 36th Asian Conference on Remote Sensing*

[13] United States Soil Conservation Service. (1954). SCS National Engineering Handbook *University of Michigan Library*

[14] Balbastre-Soldevila, R, Garcia-Bartual, R and Andres-Domenech, I. (2019). A Comparison of Design Storms for Urban Drainage System Applications *Water*. **11** 757

[15] Hazarika, M K, Nashrrullah, S, Alam, M, Chen, L, van Westen, C, Dwijananto, A and Roxas, M L. (2021). Building footprint maps. CDEMA

[16] Patiño, C, Campomanes, F P and Flores, M J L. (2019). Estimation of Urban Sprawl and Vegetation Loss of Metro Cebu Through Multi-Temporal Image Analysis *Asian Conference on Remote Sensing*

Roof Runoff Analysis: A Case for Cebu Technological University – Main Campus

A F A Oraya [1,2], B V Alburo [1], M K S Cauban [1] and J B Canene [1]

[1]Department of Civil Engineering, Cebu Technological University, M.J Cuenco Ave., Corner R. Palma St. Cebu City 6000, Philippines

[2]aureflo.oraya@ctu.edu.ph

Abstract. The volume of runoff varies depending on the surface on which it flows. Cebu Technological University – Main Campus, which is situated in Cebu City at the intersection of M.J. Cuenco Avenue and R. Palma Street, has been selected as location for the investigation of the effect of roofing properties on roof runoff. The average runoff coefficient was calculated under various rainfall scenarios. These were also related to the roofing structure, design patterns and materials whether these are sloping roofs or flat roofs. This investigation aims to enhance efficiency in rainwater harvesting systems. The calculated values of runoff coefficient in various roofs in the campus range from 0.78 to 0.92. A comparative analysis was used to differentiate the gap between the calculated runoff coefficient and the established values from literature. The recommended size for the rainwater tanks were calculated by taking into consideration the amount of volume that can be collected in the roofs in the campus.

1. Introduction

On Earth, water makes up the majority of the planet's surface. It is necessary for the survival of all forms of life as well as the proper operation of the world's ecosystems and the continued preservation of the natural world [1]. However, the most significant problem with water worldwide is a lack of availability. Not only are locations with a limited water supply more likely to experience water scarcity, but regions with adequate water supply systems are also at risk [2][3]. As the most common substance, water is a primary infection source. Infections are spread by water, and 3.1% of fatalities are caused by unclean water; in 2020, 75% of the world's population has safe water, and 25% do not [4].

Rainwater harvesting and storage systems are one method that can help reduce one's consumption of potable water. Rainwater harvesting collects and stores rainwater for human use from rooftops, land surfaces, or rock catchments using simple and engineered procedures. It looks to be the most viable alternative for supplying fresh and high-quality water to substantial area that lack access to water supply in the face of increasing water scarcity [5][6].

IOP Conf. Series: Earth and Environmental Science 1184 (2023) 012015 doi:10.1088/1755-1315/1184/1/012015

Meanwhile, most rainwater harvesting systems, on the other hand, are centered on tanks and collect rainfall from roofs. The quality of roof runoff is affected by the type of roof as well as environmental factors [7]. Several studies were conducted to analyze the quality of rainwater collected from different roofing materials. Nicholson et al. investigated rainfall from galvanized metal, cedar shake, asphalt shingle, treated wood and green roofs [8]. Green, galvanized, and asphalt shingles had the highest pH level. Treated wood and galvanized metal produced the most copper and zinc. Moreover, according to Yaziz et al., as indicated in their water samples, the longer the dry period between rainstorm events, the more pollutants are deposited on roof surfaces [9]. Studies on copper (Cu) roofing materials [10], vegetated roofs [11], zinc runoff from a variety of galvanized materials [12], concrete and asphalt paving [13], lead and other metals incorporated into the paint itself [14], among others, have been conducted.

The rational method will be utilized to examine roof runoff. Runoff is estimated using area, rainfall intensity, and a runoff coefficient [15]. This strategy is most accurate for impermeable areas. Moreover, the runoff coefficient is defined as the fraction of total rainfall appearing at the conveyance as runoff.

The runoff coefficient is a variable used to calculate runoff from rainfall data; it is constant yet variable between events [16]. In addition, the runoff coefficient is generally taken from a table based on type of cover and infiltration capacity of the soil. Roof runoff coefficients for regular storms are estimated to be 0.70 to 0.95 [7].

Further study by Scharer et al. shows that the lab measurements gave different runoff coefficients depending on the types of roofs and the amount of added block rain [17]. Nevertheless, the roof systems with an extruded clay layer had a much lower runoff coefficient than all the other types. It shows that rainwater needs to be held in place by a porous flow-based layer. Furthermore, Monteiro et al. demonstrated that coefficients change during the year; therefore, annual averages should not be used to size tanks and other rainwater harvesting system components in a Mediterranean climate [18]. Also, the study indicates that runoff coefficient determination relies on the green roof (GR) type. With this, a study on the analysis of runoff and the effect of the various roof conditions was conducted within the Main Campus of Cebu Technological University

2. Main Discussion

The main campus of Cebu Technological University, which can be found in Cebu City at the intersection of M.J. Cuenco Avenue and R. Palma Street, has been selected as the location for the investigation into the effect that roofing materials have on roof runoff. The campus features scorching temperatures, high humidity, and two wet and dry seasons. Strong storms or continuous rainfall may cause campus flooding. The likelihood of this happening increases as the severity of the weather event increases. The campus is covered with various roofing materials, including galvanized iron sheets, cement finishes, ceramic tile finishes, and homogenous porcelain tile. Moreover, Cebu Technological University was founded in 1911; therefore, the roof has significantly altered over time, making this study necessary for runoff analysis. As a response, a rainwater harvesting system would be advantageous.

2.1. Rainfall Data

Rainfall data is gathered from the Meteostat database which is an approved provider of open-source meteorological and climate data. The data collected is at the closest weather station to the research location which is the Mactan Weather Station with corresponding time and date.

2.2. Accumulation Area

The roof area of the four chosen buildings is computed based on the provided structural plan of the following buildings which include the 4-story mechatronics building (ME), 3-story technology building 1 & 2 (T), 9-story administrative building (AD), and 7-story multi-purpose school building (MP).

2.3. *Calculation of Actual Runoff*
For the actual discharge, it was calculated by collecting actual rainfall from each building using a container. Rainfall was collected from a downspout for each selected building and the number of total downspouts for each building was divided from its respective accumulation area. The discharge equates to the collected volume of water per unit of time.

2.4. *Calculation of Runoff Coefficient*
The actual quantities of runoff coefficient for each building are obtained using the modified rational formula:

$$C = \frac{Q}{iA}$$

where: C is the runoff coefficient, Q is the peak discharge, i is the rainfall intensity and A is the catchment area

For areas with two types of roofs, the weighted average of the two materials was utilized for the calculation of C. In the study area, this is true for the Mechatronics Building.

2.5. *Calculation of Theoretical Runoff*
The theoretical runoff is calculated to differentiate the value of theoretical runoff coefficient from the accumulated runoff coefficient using actual data. Roof structure whether it's a sloped roof or flat roof is classified and is used in identifying the suitable coefficient to employ. The coefficient to be used is gathered from previous study by Farreny et al.[7].

2.6. *Roof Conditions*
The researchers identified the roof conditions using drone footage from the ICT department. The assessment of roofing condition is done by inspecting roof symptoms classified from the study of Roof Deterioration and Impact: A questionnaire survey [19]. After data was collected, the rooftops' damage was categorized as small or large. The outside of a building's roofing sheet materials for a rainwater harvesting system was inspected for debris, drainage, physical damage, coating condition, granular loss, punctures, cracks and blisters, ponding, roof material, surface condition, deformed edges, shingles, metal and fasteners. The roof condition assessment will be evaluated based on the number of "No" recorded and is classified as follows; 0-1 (Critical), 2-4 (Poor), 5-7 (Good), 8-9 (Very Good).

2.7. *Rainwater Harvesting System Tank Size*
The Philippine Green Building Code suggests a set of guidelines for rainwater harvesting systems. The calculation for the tank states that the building footprint should be divided by 75 to determine the minimum amount of storage tank in cubic meters. This was used as basis in recommending tank sizes in the campus.

3. Results and Discussions
Rainfall data from three different rainfall events were collected using the Metostat website. The events happened last August 1, 4 and 12, 2022 with rainfall intensities are 3.2 mm/hr, 4.2 mm/hr and 2.4 mm/hr, respectively.

Roof areas for the selected buildings were also calculated. The actual discharges obtained in these buildings were also tabulated. Furthermore, the runoff coefficient derived from the Modified Rational Formula was calculated using the obtained rainfall intensity, actual discharge and roof area. These are presented in detail in the table below.

IOP Conf. Series: Earth and Environmental Science 1184 (2023) 012015 doi:10.1088/1755-1315/1184/1/012015

Table 1. Calculated Runoff Coefficient in Various Buildings

Building	Roof Runoff (L/s)	Roof Area (m^2)	Rainfall Intensity (mm/hr)	Calculated Runoff Coefficient
Admin Building (AD)	0.59	942	2.40	0.93
	0.78	942	3.20	0.93
	1.02	942	4.20	0.92
Mechatronics Building (ME)	0.24	452	2.40	0.79
	0.32	452	3.20	0.79
	0.40	452	4.20	0.75
Multi-Purpose Building (MP)	0.30	480	2.40	0.94
	0.40	480	3.20	0.92
	0.51	480	4.20	0.91
College of Technology Building (T)	0.97	1,685	2.40	0.85
	1.25	1,685	3.20	0.83
	1.60	1,685	4.20	0.81

In addition to the calculation of the actual discharge and actual runoff coefficient, the theoretical discharge was also calculated. This is done by utilizing the published values of runoff coefficient, the same rainfall intensity and roof area. These values of theoretical discharge were compared with the actual discharge collected and presented in the table below. As can be seen in the table, the actual discharge values are lower than the theoretical discharge for all buildings with difference ranging from 0.01 to 0.07 liter per second.

Table 2. Comparison of the Actual and Theoretical Roof Runoff

Building	Rainfall Intensity (mm/hr)	Roof Runoff (L/s)	Theoretical Roof Runoff (L/s)	Difference (L/s)
Admin Building (AD)	2.40	0.59	0.60	0.01
	3.20	0.78	0.80	0.02
	4.20	1.02	1.05	0.03
Mechatronics Building (ME)	2.40	0.24	0.26	0.02
	3.20	0.32	0.35	0.03
	4.20	0.40	0.46	0.06
Multi-Purpose Building (MP)	2.40	0.30	0.31	0.01
	3.20	0.40	0.41	0.01
	4.20	0.51	0.54	0.03
College of Technology Building (T)	2.40	0.97	0.95	(0.02)
	3.20	1.25	1.27	0.02
	4.20	1.60	1.67	0.07

The conditions of the roof in various buildings were evaluated based on the prepared assessment described in Section 2.6. The answers in the assessment were converted to the corresponding roof assessments presented in the table below. The discrepancies in values of discharge have been attributed to the percentage of abstraction of the roof materials during a rainfall which then affects the runoff. It can be noted that the percentage abstraction increases with increasing rainfall intensity. As the roof condition improves for AD, ME and MP, the percentage abstraction decreases. However, this is not true for T. The roof conditions and the calculated abstraction percentages are presented in the table below.

Table 3. Comparison of Roof Assessment, Rainfall Intensity and Percentage Abstraction

Building	Roof Assessment	Rainfall Intensity (mm/hr)	Difference	Percentage
Admin Building (AD)	Very Good	2.40	0.01	2%
		3.20	0.02	3%
		4.20	0.03	3%
Mechatronics Building (ME)	Poor	2.40	0.02	8%
		3.20	0.03	9%
		4.20	0.06	13%
Multi-Purpose Building (MP)	Good	2.40	0.01	3%
		3.20	0.01	2%
		4.20	0.03	6%
College of Technology Building (T)	Poor	2.40	(0.02)	-2%
		3.20	0.02	2%
		4.20	0.07	4%

Tank sizes to collect rainwater have been calculated following the provisions of the Philippine Green Building Code. The volume for Admin Building is 12.6 cubic meters, for Mechatronics Building is 6.0 cubic meters, for Multi-Purpose Building is 6.4 cubic meters and for the College of Technology Building is 22.5 cubic meters.

4. Conclusion and Recommendation

The development of an appropriate design approach and operational evaluation of RWH systems is essential to improving system performance. In this analysis of the influence of roof conditions to runoff, four roof catchments were investigated with an average runoff coefficient of 0.92, 0.78, 0.92 and 0.83. The percentage abstraction in roofs increases with increasing rainfall intensity. Moreover, it was observed in three buildings that as the condition of the roofs improved, the percentage abstraction reduces It is recommended that the same framework be applied to other buildings with different roofing materials and conditions to further improve the derived correlation. Various tank sizes derived from the provisions of the Philippine Green Building Code has also been recommended for the four buildings.

References

[1] Adeboye, O B, & AlAtiSe, O M. Surface water potential of the river Osun at apoje sub-basin Nigeria (2008). *Soil and Water Research* 3(2), 74

[2] Jury, W A, & Vaux, H. The role of science in solving the world's emerging water problems (2005). *Proceedings of the National Academy of Sciences* 102(44), 15715-15720

[3] Mun, J S & Han, M Y. Design and operational parameters of a rooftop rainwater harvesting system: definition, sensitivity and verification (2012). *Journal of Environmental Management*

93(1), 147-153

[4] Health, P. World Health Organization (WHO) Drinking Water Standards Table - Public Health. Public Health (2020) *www.publichealth.com.ng*

[5] Zhu, K, Zhang, L, Hart, W, Liu, M, & Chen, H. Quality issues in harvested rainwater in arid and semi-arid Loess Plateau of northern China (2004). *Journal of Arid Environments* 57(4), 487-505

[6] Baek, C W, & Coles, N A. Defining reliability for rainwater harvesting systems (2011). *19th International Congress on Modelling and Simulation* pp. 12-16

[7] Farreny, R, Morales-Pinzón, T, Guisasola, A, Tayà, C, Rieradevall, J, & Gabarrell, X. Roof selection for rainwater harvesting: Quantity and quality assessments in Spain (2011). *Water Research* 45(10), 3245-3254

[8] Nicholson, N, Clark, S E, Long, B V, Spicher, J, & Steele, K A. Rainwater harvesting for non-potable use in gardens: a comparison of runoff water quality from green vs. traditional roofs (2009). *World Environmental and Water Resources Congress 2009: Great Rivers* pp. 1-10

[9] Yaziz, M I, Gunting, H, Sapari, N, & Ghazali, A W. Variations in rainwater quality from roof catchments (1989). *Water Research* 23(6), 761-765

[10] Pennington, S L, & Webster-Brown, J G. Stormwater runoff quality from copper roofing, Auckland, New Zealand (2008). *New Zealand Journal of Marine and Freshwater Research* 42(1), 99-108

[11] Berndtsson, J C, Bengtsson, L, & Jinno, K. Runoff water quality from intensive and extensive vegetated roofs (2009). *Ecological Engineering* 35(3), 369-380

[12] Sullivan, J H, & Worsley, D A. Zinc runoff from galvanised steel materials exposed in industrial/marine environment (2002). *British Corrosion Journal* 37(4), 282-288

[13] Clark, S E, Steele, K A, Spicher, J, Siu, C Y, Lalor, M M, Pitt, R, & Kirby, J T. Roofing materials' contributions to storm-water runoff pollution (2008). *Journal of Irrigation and Drainage Engineering* 134(5), 638-645

[14] Davis, A P, & Burns, M. Evaluation of lead concentration in runoff from painted structures (1999). *https://doi.org/10.1016/S0043-1354(98)00509-0*

[15] Schärer, L A. Comparing experimentally measured runoff coefficients with field observations for detention-based roofs (2018). *Master's thesis, NTNU*

[16] Şen, Z, & Altunkaynak, A. A comparative fuzzy logic approach to runoff coefficient and runoff estimation (2006). *Hydrological Processes: An International Journal* 20(9), 1993-2009

[17] Schärer, L A, Busklein, J O, Sivertsen, E, & Muthanna, T M. Limitations in using runoff coefficients for green and gray roof design (2020). *Hydrology Research* 51(2), 339-350

[18] Monteiro, C M, Calheiros, C S, Pimentel-Rodrigues, C, Silva-Afonso, A, & Castro, P M. Contributions to the design of rainwater harvesting systems in buildings with green roofs in a Mediterranean climate (2016). *Water Science and Technology* 73(8), 1842-1847.

[19] Ahluwalia, S S & Hegazy, T. Roof deterioration and impact: A questionnaire survey (2010). *Journal of Retail & Leisure Property* 9(4), 337-348

STUDY OF POTENTIAL DEVELOPMENT OF RAINFED LAND IN MALUKA BAULIN SWAMP IRRIGATION AREA

Abdul Holis Rahmadi[1], Novitasari Novitasari[1*], Ulfa Fitriati[1]

[1]Civil Engineering, Faculty of Engineering, Lambung Mangkurat University, Banjarbaru, South Kalimantan, Indonesia

*novitasari@ulm.ac.id

Abstract. Agricultural land in South Borneo has high rice productivity. The supporting factors for rice productivity are irrigation areas, one of which is the Maluka Baulin Swamp Irrigation Area (DIR). It is located in Kurau, Tanah Laut Regency. It has an area of 421 ha (Decree of the Minister PUPR of the Republic Indonesia) now to 286 ha (land function and administrative other villages) with use as agricultural land. It so far has a low average rice production and cropping index of one crop in one year. The purpose of this study was to analyze the availability and demand of water well as the provision of irrigation water based on the water balance for the development of cropping patterns and schedules. This study collects primary and secondary data. Preliminary data are location survey, instantaneous discharge, and questionnaire. Secondary data are climatological data, maps of land types, and data on the discharge Tabanio-Bajuin River. The water requirement is obtained from the calculation of land preparation and consumptive use of water for plants. Water availability is obtained by analyzing rainfall, calculating evapotranspiration, calibrating F.J. Mock Model discharge, and calculating mainstay discharge. The temporary result obtained the water balance is land area is obtained with cropping pattern and schedule that increases to paddy-paddy-palawija, which was previously only paddy-paddy. Then, agricultural production, especially rice and secondary crops, will directly grow.

1. Introduction

Paddy fields in lebak (non-tidal) swamps and tidal swamps that have the potential to be optimized and developed are in the province of South Kalimantan. According to [1], rice productivity in South Kalimantan is the highest compared to 4 other provinces on the island of Kalimantan, which is 4,19 tons/ha. Rice productivity in South Kalimantan is still much lower at 79.42% of the average national rice productivity of 5,34 tons/ha. Agricultural Survey Data Land in 2015 showed that the paddy fields planted with rice were 450,024 ha, consisting of 47,877 ha of irrigated paddy fields, 141,033 ha of rainfed land, 166,317 ha of tidal swamps, and 94,797 ha of lebak swamps. The total paddy field planted with rice once a year is 408,870 ha (90.86%), planted with rice twice a year is 41,014 ha (9.11%), and planted with rice three times a year is 140 ha (0.03%).

The total paddy field, which is a supporting factor for rice productivity in the land, is the irrigation area. According to [2], South Kalimantan Province has 535 Swamp Irrigation Areas (DIR) under the authority of district governments. One of them is the DIR in Tanah Laut Regency, and the name is DIR Maluka Baulin which has an area of 421 ha (Permen-PUPR No. 14/2015). DIR Maluka Baulin is located in Kurau District, Tanah Laut Regency.

The problem that is often faced by the farmers of DIR Maluka Baulin is the lack of water, especially during the dry season, causing several problems, such as reduced yields, vulnerability to pest attacks, and uneven distribution of water which causes competition for water to irrigate paddy fields. According to secondary data obtained from the Agricultural Extension Center (BPP) of Kurau District in 2021, the average rice production in Kurau District is around 3.4 tons/ha, while the average cropping index (IP 100) or one crop per hectare one year. Based on research conducted by [3], rice production in Kurau District can be said to be low, classified below the median or average. One of the causes of low rice production is the uneven water balance in the Maluku Baulin DIR area. Water balance is a balance between water availability and water demand.

Water balance is closely related to water balance. The water balance is a quantitative relationship model between the amount of water available above and in the soil and the amount of rainfall that falls on an area within a certain time [4]. Therefore, it is necessary to use irrigation water optimally by streamlining the distribution and use of irrigation water so that existing agricultural land can be irrigated optimally. In order to make efficient use of irrigation water, it is necessary to conduct research on adjusting the amount of water demand for plants and setting cropping patterns according to the availability of water in the DIR Maluka Baulin.

2. Materials and Methods

2.1 Collection Data

Primary Data

Data collection was obtained directly in the field, namely DIR Maluka Baulin. The primary data in this study are as follows: (i) Research site survey; (ii) Filling out questionnaires (land type, cropping index, pattern, and planting schedule); (iii) Measuring the discharge of DIR Maluka Baulin.

Secondary Data

Data collection was obtained from the relevant agencies. Requests for secondary data for this study include the Department of Agriculture, Horticulture and Plantation, Tanah Laut Regency, Agricultural Extension Center (BPP) in Kurau District, and Class 1 Banjarbaru Climatology Station. The secondary data required are as follows: (i) The geospatial data of Kurau District consists of: Map of land and irrigation network, a map of land area; (ii) Rainfall data consists of Banjarbaru Climatology Data for 20 years (2002-2021); (iii) Climatological data from BMKG Banjarbaru Station.

2.2 Water Availability Analysis

Calculating Planned

Rainfall Planned rainfall is calculated on a monthly basis; a cumulative value of 15 days will be taken. The planned rainfall will get the mainstay of rainfall. The mainstay rainfall obtained is the mainstay rainfall (R_{80}) and (R_{50}). R_{80} means that out of 10 planned rainfall events, eight times will be exceeded. At the same time, R_{50} means that out of 10 planned rainfall events, five times will be exceeded.

Calculating Evapotranspiration

Evapotranspiration is calculated based on the FJ Mock Method. Before calculating the evapotranspiration value, climatological data is needed first. The climatological data will be used to calculate the value of potential evapotranspiration that occurs at the study site; daily evapotranspiration is calculated based on the Penman-Monteith, Thornthwaite, and Hargreaves formulas.

Calculating the Discharge Calibration of the FJ Mock Model

Method *Mock* was developed by Dr. FJ Mock. Method *Mock* is used to estimating the amount of discharge in a watershed based on the concept of water balance. Rainwater (precipitation) that falls will experience evapotranspiration according to the vegetation that covers the catchment area. Evapotranspiration in the *Mock* is evapotranspiration which is influenced by the type of vegetation, soil surface, and the number of rainy days [5].

Discharge data of the Tabanio-Bajuin River is used for the calibration process to obtain the parameters used in the simulation or experiment stage. The parameters obtained include the infiltration coefficient, initial soil moisture, soil moisture capacity, and initial groundwater storage so that the simulation process can be calculated with the reference year 2001-2020; this simulation discharge data will be tested with measured discharge data from field observations at DIR Maluka Baulin.

Calculating Mainstay Discharge

The analysis is continued by calculating the mainstay discharge using the Weibull method to meet water demand at the research site. The mainstay discharge analysis is calculated based on the results of water availability by taking the probabilities of 80% and 50% for the function of irrigation water demand.

IOP Conf. Series: Earth and Environmental Science 1184 (2023) 012016 doi:10.1088/1755-1315/1184/1/012016

2.3 Water Demand Analysis

Calculating Land Preparation

The method developed by [6] to calculate irrigation needs during land preparation was used. The method is based on a constant water rate in lt/s/ha during the land preparation period. According to [7] from several studies, it was found that the Penman-Monteith Model provides accurate estimates, so FAO recommends its use for estimating standard evapotranspiration in estimating water demand for plants.

Calculating Plant Water

Use Consumptive use is the amount of water used by plants for the photosynthesis process of these plants. The calculation of plant water demand is assisted by the CROPWAT 8.0 Application. The main functions of *CROPWAT,* according to [8], are: (i) To calculate evapotranspiration references; (ii) To calculate crop water requirements; (iii) To calculate the need for irrigation water; (iv) to prepare irrigation schedules; (v) to make a pattern of water availability; (vi) to evaluate rainfall; and (vii) to evaluate the efficiency of irrigation practices.

2.4 Maluka Baulin DIR Water Balance Optimizing the Maluka Baulin

DIR is done after getting the water balance between water availability and water demand at the research site. There are two conditions resulting from the water balance; namely, the availability of water is greater than water demand ($Q_{available} > Q_{requirement}$), or water availability is smaller than water demand ($Q_{available} < Q_{requirement}$). Optimization is carried out in 3 ways, namely optimizing land area, optimizing cropping patterns, and optimizing planting schedules.

3. Results and Discussions

3.1 Water Demand Analysis

Explained by [9] that the cropping pattern is the order of the types of plants cultivated on a plot of agricultural land in one year, including the tillage period and the rest period. The cropping pattern in this study uses paddy-paddy-palawija. Calculation of water needs is contained in the calculating irrigation water needs. The calculation of irrigation water needs is based [10]. The results of these calculations are shown in **Table 1**.

Table 1. Water Demand

Month		Water Demand lt/sec/ha	Type Plant	Month		Water Demand lt/sec/ha	Type Plant
Jan	I	0,06	Palawija	Jul	I	5,56	Paddy
	II	0,15			II	0,95	
Feb	I	0,44		Agt	I	1,04	
	II	5,39			II	1,04	
Mar	I	5,26	Paddy	Sep	I	1,09	
	II	5,53			II	0,83	
Apr	I	0,57		Oct	I	7,93	
	II	0,61			II	7,84	
May	I	0,72		Nop	I	6,84	Palawija
	II	0,87			II	0,03	
Jun	I	0,59		Dec	I	0,10	
	II	4,78			II	0,11	

The start planting in **Table 1.** was carried out on the second of March. The maximum water demand in the first October was 7.93 lt/sec/ha, and the minimum water demand in the second November was 0.03 lt/sec/ha.

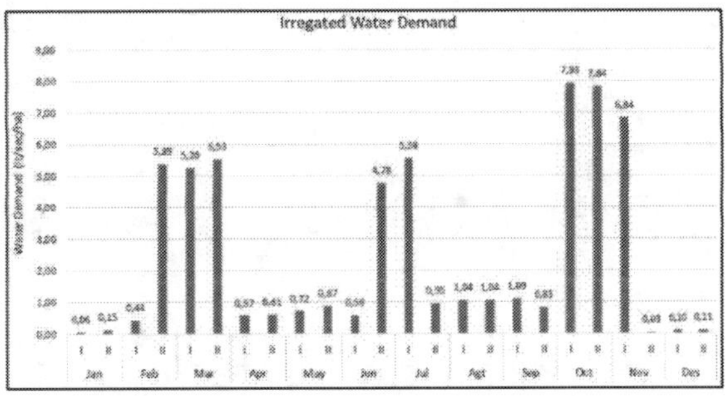

Figure 1. Barchart of Irrigated Water Demand

3.2 Water Availability Analysis

Calibration F.J. Mock Method

The availability of water at DIR Maluka Baulin can be calculated using the FJ Mock Method. The data needed to calculate using the F.J. Mock Method include the following:

1. Potential evapotranspiration
2. Rainfall
3. Measurable discharge data

The obstacle faced in general when calculating water availability is that the available data is very minimal. The discharge data used in this study is the discharge date of the Tabanio-Bajuin River due to the unavailability of discharge data in the rivers closest to the study site. The steps for calculating water availability using the FJ Mock Method are described in detail as follows:

Table 2. Tabanio-Bajuin Semi-Monthly Measured Discharge Data

Month		Average	Month		Average
Jan	I	16,19	Jul	I	4,35
	II	23,27		II	6,63
Feb	I	22,05	Agt	I	3,98
	II	8,87		II	6,60
Mar	I	7,36	Sep	I	5,89
	II	16,23		II	4,43
Apr	I	16,43	Oct	I	4,57
	II	5,74		II	6,57
May	I	5,48	Nop	I	8,95
	II	4,65		II	13,49
Jun	I	14,55	Dec	I	27,28
	II	6,68		II	25,04

The calibration results obtained a *volume error* of 0.048 and a correlation of 0.714, which means that it has a slight error rate and a strong relationship so that the calculation can proceed to the simulation stage. The graph of the calibration results is shown in **Figure 2**.

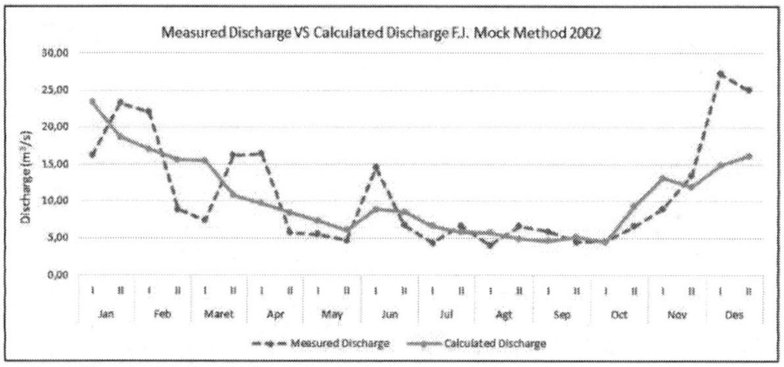

Figure 2. Calibration F.J. Mock Method of Measured Discharge VS Calculated Discharge

It can be concluded that the value of the calibration parameters that will be used in the next stage, namely the simulation stage, is as follows, which is attached in the form of **Table 3**.

Table 3. Parameter of Calibration Result and Specification

Parameter	Calibration Result	Specification
DIC	0,80	$0{,}0001 \le DIC \le 0{,}9999$
WIC	0,65	$0{,}0001 \le WIC \le 0{,}9999$
ISM	200	$ISM \ge 0{,}0001$
SMC	188	$50 \le SMC \le 250$
IGWS	1000	$IGWS \ge 0{,}0001$
k	0,84	$0{,}0001 \le k \le 0{,}9999$

The calibration result in the next step of simulation F.J. Mock Method shows in **Figure 3**.

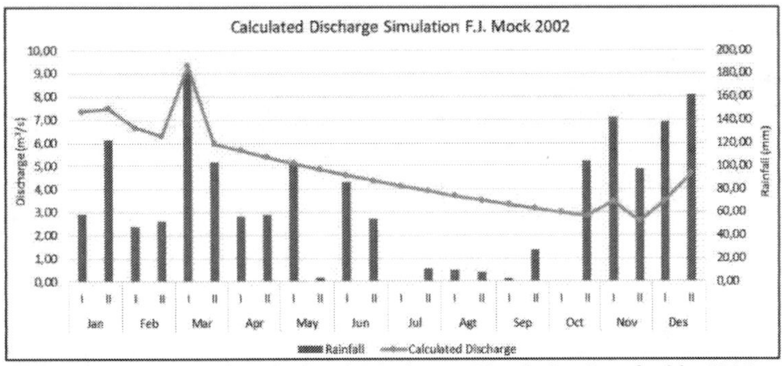

Figure 3. F.J. Mock Calculated Discharge Simulation Resulted in 2002

The amount of the debit value is based on calculations in 2002; the maximum debit value occurred in the first March with a debit value of 9.33 m^3/s, and the minimum debit occurred in the second November with a debit value of 2.62 m^3/s.

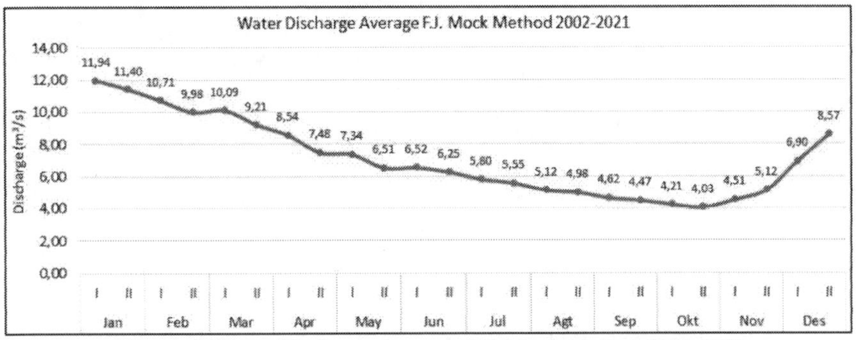

Figure 4. Water Discharge Average F.J. Mock Method 2002-2021

The average monthly discharge value is obtained by averaging the debit data each year in the same month so that the maximum debit value occurs in the first January with a debit value of 11.94 m^3/s and the minimum debit value occurs in the second October with a value discharge of 4.03 m^3/s.

Analyze Water Availability

Water Availability with the Weibull method is shown chart shown in **Figure 5**.

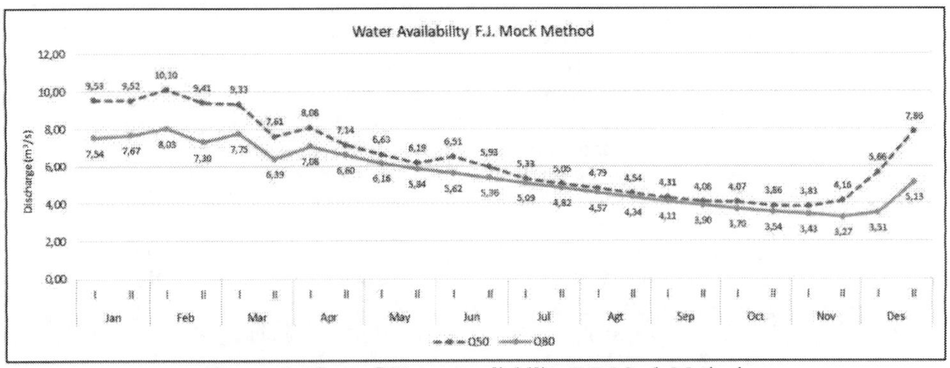

Figure 5. Chart of Water Availability F.J. Mock Method

Discharge of irrigation water needs using the F.J. Mock method; the maximum for secondary crops (Q_{50}) occurred in the first February with a discharge value of 10.10 m^3/s, and the minimum occurred in the first November with a discharge value of 3.83 m^3/s. The maximum irrigation water demand discharge for rice (Q_{80}) occurred on the first of February with a discharge value of 8.03 m^3/s, and the minimum occurred on the second of November with a discharge value of 3.27 m^3/s. The availability of water in the Maluka Baulin DIR will be used for the service area of the irrigation area, especially for groundwater storage, crop water needs, and water needs in Kurau District, Tanah Laut Regency. The availability of water at DIR Maluka Baulin is expected to be used at the end of each rainy season to be used during the dry season.

3.3 Water Balance

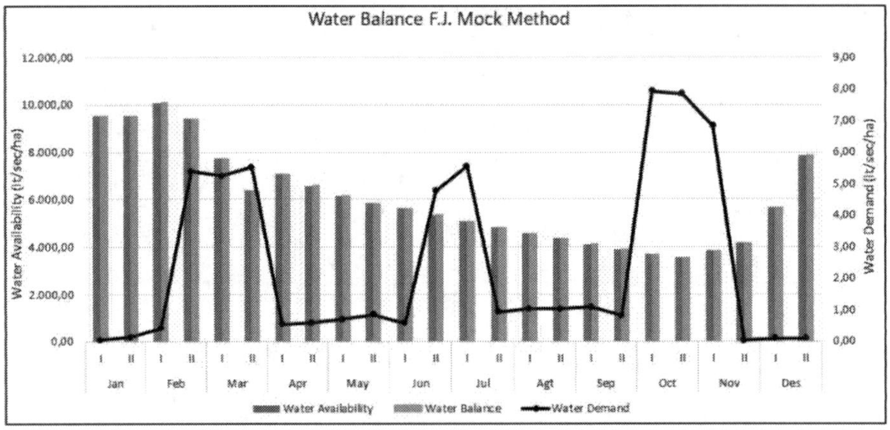

Figure 6. Water Balance F.J. Mock Method

Availability of water in **Figure 6**, the maximum in the first February was 10,100 lt/sec/ha, and the minimum in the second October was 3,540 lt/sec/ha. The maximum water demand on the first October was 7.93 lt/sec/ha, and the minimum water demand in the second November was 0.03 lt/sec/ha. The maximum water balance in the first February was 10,099.56 lt/sec/ha, and the minimum in the second October was 3,532.16 lt/sec/ha during the land preparation period.

4. Conclusions

The maximum water demand in the first October was 7.93 lt/sec/ha, and the minimum water demand in the second November was 0.03 lt/sec/ha. The calibration results obtained a *volume error* of 0.048 and a correlation of 0.714, which means that it has a slight error rate and a strong relationship so that the calculation can proceed to the simulation stage. And then, the water availability maximum in the first February was 10,100 lt/sec/ha, and the minimum in the second October was 3,540 lt/sec/ha. The maximum water requirement on the first October was 7.93 lt/sec/ha, and the minimum water requirement in the second November was 0.03 lt/sec/ha. The maximum water balance in the first February was 10,099.56 lt/sec/ha, and the minimum in the second October was 3,532.16 lt/sec/ha during the land preparation period.

This research needs to be strengthened by rainfall that is close to the location under study, for example, satellite data on rainfall or rain-captive equipment at the research location, so that the results of water availability and water demand are more accurate.

References

[1] Fathurrahman. (2016). Program Peningkatan Produksi Tanaman Pangan dan Hortikultura Mendukung Pertanian Berkelanjutan di Kalimantan Selatan. Prosiding Seminar Nasional Inovasi Teknologi Pertanian, page 40-47.
[2] Regulation of the Minister of Public Works and Public Housing of the Republic of Indonesia Number 14/PRT/M/2015.
[3] Setyowati, D. (2018). Klasifikasi data produksi padi Pulau Jawa menggunakan algoritmeclassification version 4.5 (c4. 5). Universitas Sebelas Maret. Solo.

[4] Tufaila, M., Mpia, L., & Karim, J. (2017). Analisis neraca air lahan terhadap jenis tanah yang berkembang pada daerah karts di Kecamatan Parigi Kabupaten Muna Sulawesi Tenggara. *Agritech: Jurnal Fakultas Teknologi Pertanian UGM*, *37*(2), 215-219.

[5] Indra, Z., Jasin, M. I., Binilang, A., & Mamoto, J. D. (2012). Analisis debit sungai Munte dengan metode Mock dan metode Nreca untuk kebutuhan pembangkit listrik tenaga air. *Jurnal Sipil Statik*, *1*(1).

[6] Van de Goor and Zijlsha (1968); Hermanto, K., & Utami, SF (2019). Forecasting Water Demand for Land Preparation Using Cyclical Method (Case Study of Batu Bulan Dam Irrigation Area, Moyo Hulu District). *Unisda Journal of Mathematics and Computer Science (UJMC)*, *5*(01), 25-34.

[7] Manik, T. K., Rosadi, R. B., & Karyanto, A. (2012). Evaluasi metode Penman-Monteith dalam menduga laju evapotranspirasi standar (ET0) di dataran rendah Propinsi Lampung, Indonesia. *Jurnal Keteknikan Pertanian*, *26*(2).

[8] Shalsabillah, H., Amri, K., & Gunawan, G. (2018). Analisis Kebutuhan Air Irigasi Menggunakan Metode Cropwat Version 8.0. *Inersia: Jurnal Teknik Sipil*, *10*(2), 61-68.

[9] Septyana, D. 2016. Optimasi Pola Tanam Untuk Meningkatkan Keuntungan Hasil Pertanian dengan Program Linear (Studi Kasus Daerah Irigasi Rambut Kabupaten Tegal Provinsi Jawa Tengah). Tesis. Magister Pengelolaan Sumber Daya Air Institut Teknologi Bandung. Bandung.

[10] Department of Public Work. (1986). Irrigation Planning Standards. *Irrigation Network Planning Criteria (KP-01)*.

Assessment of physical and chemical characteristics of bio-organic compost from seafood sludge

Nguyen Thi Phuong[1] **and Ho Nguyen**[1, 2*]

[1] Department of Land management, Dong Thap University, 783 Pham Huu Lau Street, 870000 Cao Lanh City, Dong Thap, Viet Nam
[2] Institute of Landscape Ecology, University of Münster, Heisenbergstr. 2, 48149 Münster, Germany

*Corresponding author: Ho Nguyen, Email: nguyenho@dthu.edu.vn

Abstract. Seafood sludge (SS) was one of the solid wastes generated during the wastewater treatment processes of seafood processing factories, and it increasingly discharged to the environment. This study aimed to reuse sludge from seafood wastewater treatment plants for producing microbial-organic fertilizer to handle environmental pollution. SS samples were collected at seafood factories in Dong Thap province. Other materials, including sugarcane filter cake (SC) and bagasse (B) were collected in Hau Giang province. Five treatments with three replications for each consisted of a mixture ratio of SS, SC and B at (1) 10-60-30, respectively; (2) 10 + 70 + 20; (3) 20 + 60 + 20; (4) 10 + 70 + 20 and (5) 20 + 80 + 0. All composting mixtures were analyzed for temperature, moisture content, pH, electrical conductivity, volume reduction, organic carbon (OC), total nitrogen (TN), total phosphorus (TP) and total potassium (TK) at stages of 0 and 49 days after incubation. Results after 49 incubation days showed that the pH value of both kinds of sludge was slightly acid to neutral (6-7). Electrical conductivity values were suitable from 0.74 to 1.57 mS/cm. The weight loss of incubated samples was 35.87-77.56% after a 49-day incubation. Organic carbon contents were at a high level (over 30% C). After 49 incubation days, the TN, TP, and TK content were high at 2.01-3.03%N; 6.71-9.95%P_2O_5; and 2.62-3.15%K_2O, respectively. The C/N ratio ranged from 12.15 to 19.1. The optimal ratio was mixed with 20% seafood sludge and 80% sugarcane filter cake. The period of 49 days was seen as the suitable incubation time for composting maturity and stability.

Key words: Seafood sludge, microbial-organic, reuse, environmental treatment.

1. Introduction

In Vietnam, there were more than 313 thousand tons of solid seafood sludge (SS) produced annually from wastewater treatment plants of seafood factories [1]. Since 1998, sludge from the wastewater treatment process have been banned from being discharged into the sea. These sludges were treated as incineration or landfill by many factories. Whole of SS were released into the soil environment, which had a harmful impact on community activities due to soil, water, and air pollution [2–4]. Only a small amount was reused for agricultural purposes due to the cost of sludge treatment, which accounted for about 40-50% of the total costs of the wastewater treatment process in factories [5,6]. In Vietnam, landfill accounted for about 50-75%, about 25-35% was reused in agricultural activities [1].

Moreover, seafood sludge had high nutrient contents, consisting of nitrogen, phosphorus, and other carbonhydrate contents. Whenever seafood sludge was reused as organic compost, those could supply multiple nutrients for plant growth. Sugarcane filter cake (SC) and bagasse (B) residues were the by-products of the sugar production process. The adding of SC and B aimed to increase carbon content and the initial C/N ratio during organic composting. Therefore, it is possible to combine reuse of SS with SC and B to produce a microbial-organic fertilizer for agricultural land.

IOP Conf. Series: Earth and Environmental Science 1184 (2023) 012017 doi:10.1088/1755-1315/1184/1/012017

The aim of this study was not only to reuse seafood sludge discharged into the environment but also was to provide a new bio-organic fertilizer in agricultural cultivation.

2. Materials and methods

2.1. Materials

All the material samples were collected at the Vietnamese Mekong Delta. The SS collected at seafood factories in Dong Thap province. Most of SS was dewatered to give a sludge cake of around 70% (w/w) solid to allow composting. Samples of SC and B were selected from Hau Giang Sugarcane Factory.

2.2. Composting process

The composting process was carried out from October to November 2020. Five composting treatments with three replications for each were handled (Table 1). The total dry weight (in kg) for each mixture ratio was 100 kg. Each treatment could be supplemented with the probiotic *Trichoderma*-DHCT (in Vietnam), with a content of 100g per cubic meter. The initial C/N ratio and moisture content ranged around 30 and 65% (w/w), respectively. Each treatment was turned once each week until the incubation period reached 28 days. All composting mixtures were analyzed for temperature, moisture content, pH, electrical conductivity, volume reduction, organic carbon (OC), total nitrogen (TN), total phosphorus (TP), total potassium (TK)

Table 1. Composition of composting piles, mixture ratio (dry weight-d.w.) and C/N ratio

Treatments (T)	Seafood sludge	Sugarcane filter cake	Bagasse	C/N ratio
1	10	60	30	30
2	10	70	20	29
3	20	60	20	29
4	10	70	20	29
5	20	80	0	28

Chemical indicators included pH, electrical conductivity (EC), organic matter (OM), total nitrogen (TN), phosphorus (TP) and potassium (TK) contents. pH and EC values were measured by using a pH digital meter and an EC meter. OM was determined in a muffle furnace at 830^{0}C for 2 hours. The TN was measured by the Kjeldahl method. TP was determined colorimetrically by flow injection analysis. The TK was using a flame photometric method.

The physical parameters consisted of composting temperature and moisture. Moisture content was determined by drying the samples at 105^{0}C for 24 hours.

2.3. Statistical analysis

All of analysed values were indicated by mean value with three replicates using IBM SPSS Statistics for Windows [version 23.0]. Comparison of differences between all treatments was performed by using the Duncan Multiple Range Test (p-values ≤ 0.05).

3. Results

3.1. Temperature and moisture profile

In the initial stage, the temperature fluctuated between 40 and 50^{0}C, an essential condition for the destruction of weed seeds and pathogenic microorganisms in the compost profile (Fig. 1) [7–9].

Between 14 and 21 days of composting, even though the whole of the treatments reached the maximum temperature, namely 55^{0}C, there were not significant differences among all mixture ratios. The temperature fluctuated down from 21 days afterward until the end of the incubation stage. At the end of the composting period, the temperatures of all treatments were fluctuated average 32^{0}C as same as the ambient temperatures.

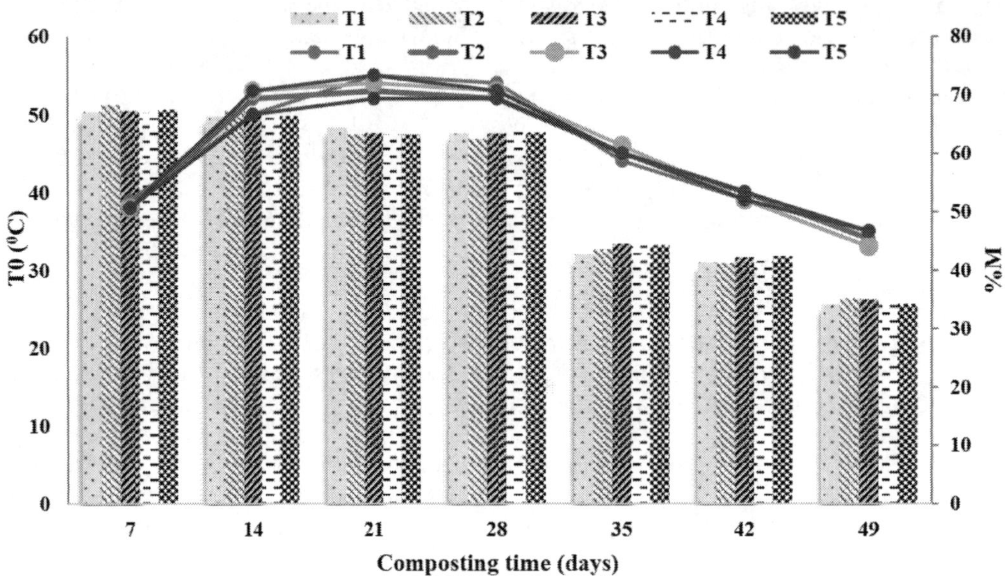

Figure 1. Changes in treatment temperature and moisture during the composting process.
**T: treatment. Points are indicated mean of three replications, n=3, Line: show temperature value (T0), and column show moisture value (M).*

Over the whole incubation period, T5 was characterized by a statistically significantly lower moisture level ($p < 0.05$) than the others, which ranged from 67.5% at initial stage to 34.1% at the end of the incubation stage. Our findings are consistent with the results of Paredes et al. [10], who have suggested that moisture content can reach 65-67% when organic fraction is mixed with sludge. Similar to Dương et al.'s report [11], the results indicated that the moisture during the composting mixture between pig manure and sugarcane cake sludge reached so high, ranging from 60-70%.

All mixture ratios showed high moisture contents at the end of composting (around 40%), higher than the recommendation of Shammas and Wang [7], and Füleky and Benedek [12], who advocated a lower moisture content at the end of composting. Furthermore, it was also higher than the maximum allowable limit of 35% by Vietnamese standards (10TCN 526:2002). Therefore, all composts should be dried in the air.

3.2. pH

The whole mixture treatments remained at a pH of almost 7.17 during the incubation period in the starting time. The pH values of all treatments decreased slowly from starting until 49-day composting (pH between 6,13-7,25) but T3 and T5 increased slightly to pH 7,14 and 7,25, respectively, after 49 days (fig.2). Comparing the different composting treatments, the pH of compost in T5 was the highest (7.25) followed by the pH of the compost in T3 (7.25), T4 (6.52), T2 (6.47), and T1 (6.13).

In Viet Nam, according to the regulations of the Ministry of Agriculture and Rural Development, a compost to be considered suitable for use in agriculture must have an optimal pH value of 6-8 for matured compost. Our results show that a mixture of SS and SC with ratio 20:80 and those of SS and SC and B with a ratio of 20:60:20 are optimal mixtures for composting.

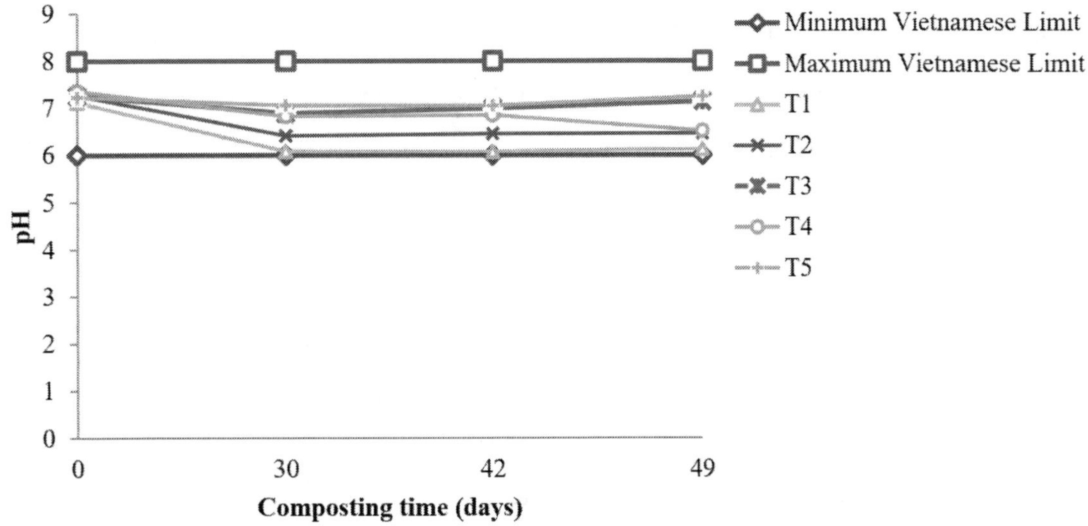

Figure 2. Changes in pH during the composting process.
**T: treatment. Points are indicated mean of three replications, n=3.*

3.3. Electrical conductivity (EC-mS/cm)

In all the mixture ratios, EC values were low at the 49-day stage of composting, ranging between 0.74 and 1.57 mS/cm (Fig. 3). These results are lower than those reported by Karak et al. [8] and Shammas and Wang [7] where EC = 4 mS/cm and EC = 3,03dS/m, respectively.

Among these mixtures, T5 with a ratio of SS and SC (20:80 mixture ratios) gave the lowest EC value during decomposing time. According to the Ministry of Agriculture and Rural Development (2010) optimum EC value of matured compost should be below 4 dS/m because most agricultural crops can survive and grow healthy whenever electricity conduction indicated below <4dS/m. The EC value of T1 was higher than that of T5 because of the higher amount of bagasses in treatment T1. In addition, according to Việt et al. [13], bagasse contained a lot of dissolved salts, that leaded to the EC value of T1 was quite high.

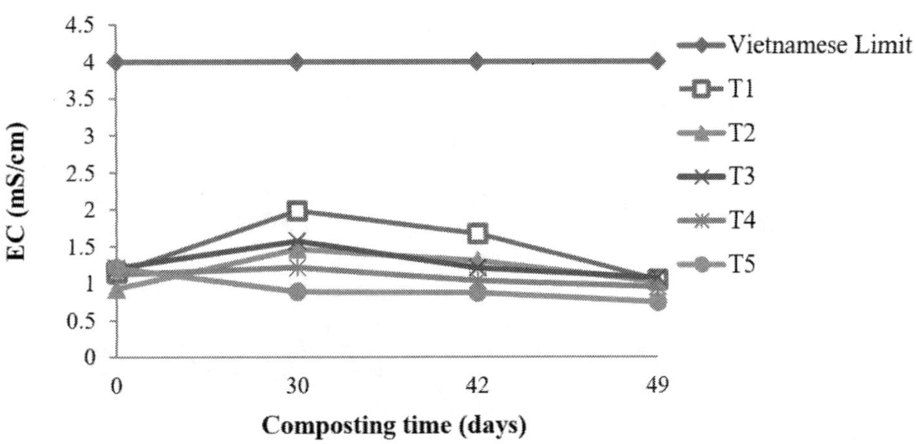

Figure 3. Changes in EC during the composting process.
**T: treatment. Points are indicated mean of three replications, n=3.*

3.4. Volume reduction

The dynamic fluctuation in compost volume during the bioorganic incubation process was shown in Figure 4. The result highlighted that the highest volume reduction was announced by T5, with a weighting dropout of around 47.56% at the end of the incubation time.

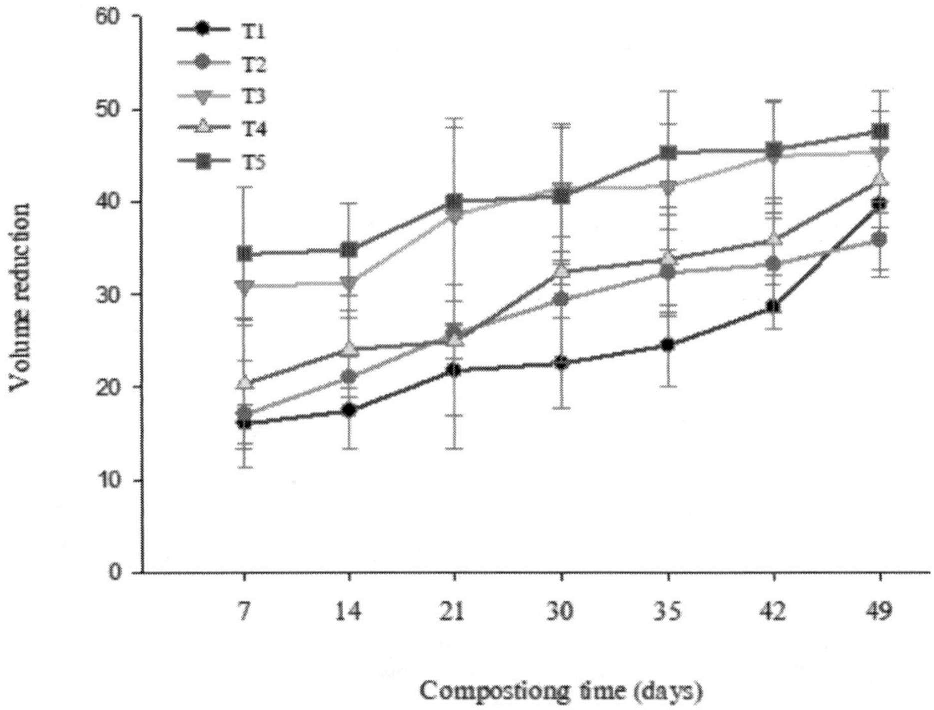

Figure 4. Volume reduction during the composting process.
**T: treatment. Points are indicated mean of three replications, n=3.*

For other composting mixture ratios, volume loss may range from 35.87% to 45.29% during the last stage of composting. The volume reduction in the 49th composting day trended dramatically: T5 > T3 > T4 > T1> T2. The case of T2 showed the lowest reduction because of undecomposed compounds such as cellulose and lignin composition in the initial materials [3,14].

3.5. Organic carbon, total nitrogen and C/N ratio

The results of organic carbon compositions, total nitrogen, total phosphorus, total potassium contents were recorded in Table 2. As shown in Table 2, the organic matter (OM) content of all treatments tended to drop off in the end stage of the composting process, compared with the first stage of the incubation process. OM losses of 22.45, 22.16, 17.4, 14.3, and 11.05% for T3, T1, T5, T2 and T4, correspondingly were recorded after 49 days of the composting. Generally, the OM compounds were ranging from 31.25–36.62% at the end of the composting stage, which is in line with Vietnamese standards for bio-organic fertilizers (OM ≥ 22%).

In addition, total N achieved the maximum content and a non-significant difference among T3, T4 and T5 with 2.96, 3.03 and 2.92%, respectively (Table 2) ($p < 0.05$). The total N concentration of compost in T5 was significantly higher than that of T1 and T2 after 49 days of composting because T5 contains a higher amount of seafood sludge.

Table 2. Organic carbon, total Nitrogen, Phosphorus and potassium and C/N ratio among treatments

Treatment	Chemical properties of organic fertilizer after incubating								
	C		N		P		K		C/N
	0	49	0	49	0	49	0	49	49
1	40.17b	31.27c	1.98a	2.01c	3.54b	6.71c	2.96a	3.15a	17.57b
2	48.54b	41.61a	1.92a	2.31b	2.01c	7.14b	1.85b	2.62cd	19.1a
3	41.07b	31.85c	2.01a	2.96a	4.46b	7.47b	1.99b	2.85bc	12.15d
4	41.17b	36.62b	1.88a	3.03a	4.01b	7.38b	2.02b	2.98ab	13.69c
5	44.13a	36.45b	1.97a	2.92a	6.20a	9.95a	1.82b	2.56d	17.71b

[*] The same letters within column indicate no significant difference at 5% level.

Carbon to nitrogen ratio (C/N) was one of the important parameters to assess the decomposed degree of organic materials. In general, initial C/N ratios between 28 and 30 were considered ideal for the composting process [11,15] (Table 1). The C/N ratio of all treatments changed down continuously up to 49 days. At the 49 composting days, C/N ratios were found between 12.15 and 19.10. This range of C/N ratio was in the acceptable range for mature compost along the Vietnamese standard (C/N < 20).

3.6. Total Phosphorus (%P_2O_5) and total potassium (%K_2O)

The total P content of organic fertilizer at the end of composting jumped significantly with the addition of seafood sludge (Table 2). Highest total P concentrations were recorded for compost in T5 (9.95 %P_2O_5), followed by compost in T2 (7.14 %P_2O_5), T3 (7.47 %P_2O_5) and T4 (7.38 %P_2O_5) with the increment of 37.5%, 51.3%, 30.1% and 33.7% of initial P, respectively.

Total K in different composting was shown in Table 2. Total K concentration increased significantly ($p < 0.05$) and gained a high amount in all composting ratios, ranging between 2.56 and 3.15 %K_2O at 49 days after composting.

4. Conclusion

The results of this study pointed toward composting using seafood sludge along with agricultural wastes such as sugarcane cake and bagasse. The qualities of organic compost showed high nutrients as well as NPK contents. The treatment mixture of seafood sludge and sugarcane cake with ratios of 20 and 80, respectively, was the optimal mixture ratio for composting. Moreover, our results indicated that the optimal composting stage could be 49 days due to the maturity and stability of compost. For commercial production, further research should be carried out with a larger incubation volume.

References

[1] Võ Phú Đức 2013 *Producing microbial organic composting of seafood sludge from wastewater treatment plants of seafood factories* (Dong Thap, Viet Nam: Department of Science and Technology of Dong Thap province)

[2] Feng Y, Wang X, Logan B E and Lee H 2008 Brewery wastewater treatment using air-cathode microbial fuel cells *Appl Microbiol Biotechnol* **78** 873–80

[3] Negro M 1999 Composting of sweet sorghum bagasse with other wastes *Bioresource Technology* **67** 89–92

[4] Olajire A A 2020 The brewing industry and environmental challenges *Journal of Cleaner Production* **256** 102817

[5] Parawira W, Kudita I, Nyandoroh M G and Zvauya R 2005 A study of industrial anaerobic treatment of opaque beer brewery wastewater in a tropical climate using a full-scale UASB reactor seeded with activated sludge *Process Biochemistry* **40** 593–9

[6] Külcü R and Yaldiz O 2014 The composting of agricultural wastes and the new parameter for the assessment of the process *Ecological Engineering* **69** 220–5

[7] Shammas N K and Wang L K 2009 Biosolids Composting *Biological Treatment Processes* ed L K Wang, N C Pereira and Y-T Hung (Totowa, NJ: Humana Press) pp 669–714
[8] Karak T, Sonar I, Nath J R, Paul R K, Das S, Boruah R K, Dutta A K and Das K 2015 Struvite for composting of agricultural wastes with termite mound: Utilizing the unutilized *Bioresource Technology* **187** 49–59
[9] Nakhshiniev B, Biddinika M K, Gonzales H B, Sumida H and Yoshikawa K 2014 Evaluation of hydrothermal treatment in enhancing rice straw compost stability and maturity *Bioresource Technology* **151** 306–13
[10] Paredes C., M.P.Bernal, A.Roig, and J. Cegarra 2001 Effects of Olive Mill Wastewater Addition in Composting of Agro Industrial and Urban Wastes *Biodegradation* **12** 225–34
[11] Dương Minh Viễn, Trần Kim Tính, and Võ Thị Gương 2011 *Microbial composting and its effectiveness in crop growth and yield and soil quality* (Ha Noi, Viet Nam: Agricultural Publishing House)
[12] Füleky G and Benedek S 2010 Composting to Recycle Biowaste *Sociology, Organic Farming, Climate Change and Soil Science* Sustainable Agriculture Reviews vol 3, ed E Lichtfouse (Dordrecht: Springer Netherlands) pp 319–46
[13] Việt L H, Ngân N V C, Uyên T T and Đức N V 2017 Kết hợp keo tụ hóa học với tuyển nổi điện hóa xử lý sơ cấp nước thải sản xuất mía đường *CTU Journal* **50** 66 (in Vietnamese)
[14] Wang W, Lai D Y F, Wang C, Pan T and Zeng C 2015 Effects of rice straw incorporation on active soil organic carbon pools in a subtropical paddy field *Soil and Tillage Research* **152** 8–16
[15] Morales A B, Bustamante M A, Marhuenda-Egea F C, Moral R, Ros M and Pascual J A 2016 Agri-food sludge management using different co-composting strategies: study of the added value of the composts obtained *Journal of Cleaner Production* **121** 186–97

IOP Conf. Series: Earth and Environmental Science 1184 (2023) 012018 doi:10.1088/1755-1315/1184/1/012018

Effect of Nutrition on Microalgal Growth and Pollutant Decrease in Contaminated River

C Nisa[1], A P Ramadani[2], D K Dewangi[2], A Nugroho[1], N A Fauzana[1], P Mardina[2] and M D Putra[1,2,*]

[1]Department of Management of Natural Resources and Environment, Lambung Mangkurat University, South Kalimantan, Indonesia

[2]Department of Chemical Engineering, Lambung Mangkurat University, South Kalimantan, Indonesia

E-mail: mdputra@ulm.ac.id

Abstract. Microalgae have the ability to remove nutrients, heavy metals, and pollutants in the form of organic substances. Thus, it is expected that the use of microalgae can reduce pollutant levels in contaminated rivers. This study aims to reduce the concentration of pollutants in the Martapura river contaminated with industrial waste by utilizing microalgae. The maximum microalgal growth was obtained on day 4 with a nutritional variation of 5.25 mL as the concentration was 0.0115 g/L. The decrease in the level of pollutants occurred during the treatment with a decrease in sulfate levels from 18.9 mg/L to 11 mg/L. The amount of cadmium decreased from 0.0014 mg/L to an undetectable point. On the other hand, the iron content dropped quite drastically from 1.8 mg /L to 0.8 mg/L.

Keywords: Pollutant, microalgae, nutrition, growth

Introduction

Population increase and industrialization are the main pollutant contributors to water bodies. This waste is very dangerous for humans and ecosystems and requires comprehensive and effective handling before being discharged into water bodies [1]. Textile industry wastewater contains organic dyes, heavy metals, phosphates, nitrates and sulfates, and high COD concentrations, which degrade the environment in several ways [2]. This pollutant can cause negative impacts if it is directly discharged into the waters without being processed first. Some of them can cause carcinogenic, and eutrophication, and are dangerous for plants and living things in the waters and dangerous for the health of humans who use these waters. Several technological developments both physicochemical and biological have been developed for the remediation of textile waste processing [3,4] however they are not yet efficient and not cost-effective. Recently, the biological treatment of wastewater by microalgae has received wide attention [5].

The ability of microalgae to grow, assimilate, and toxin-resistant conditions in wastewater makes it a versatile organism for treating various types of wastewater originating from various industries and agriculture [6]. Microalgae isolated from peat swamps showed excellent performance with a faster growth time of 10 days and higher biomass [7]. The use of microalgae has a high value of photosynthetic conversion efficiency compared to other photosynthetic organisms [8]. Microalgae also have the ability to thrive in diverse ecosystems [9]. Due to the capacity of microalgae for nutrition and removal of organic matter in wastewater treatment, microalgae are often used [10].

This study aims to reduce the concentration of pollutants in the Martapura river contaminated with industrial waste by utilizing microalgae. The growth of microalgae cultured in river water polluted by textile waste was also evaluated. The efficiency of reducing pollutant parameters in Martapura river water using microalgae was also observed.

Materials and methods

The commercial microalgae Chlorella sp. was pre-cultured for 7 days using F-2 Guilard nutrition. The different ratios of microalga substrate to waste of 5%, 7.5 and 10% were conducted in cultivation. One milliliter sample was periodically taken and analyzed. The sample was put in a pre-weighed Eppendorf collection tube. The sample was centrifuged to separate the microalgae and solution. The microalgae was dried at 80°C overnight until the weight was constant. The concentration of the cell mass was determined using the dry weight method by subtracting the initial weight of the empty tube from that with dried cell mass. The pollutant content in the solution was then analyzed to evaluate sulfate, sulphur, chlorine, cadmium, and iron.

Results and discussion

Table 1 shows the variations of microalgae grown in river wastewater with varying amounts of Guilard nutrients. The CO2 flow rate concentration was 15 ml/minute with a harvest volume of 15 mL for further analysis. Figure 2 shows the growth of microalgae on a variety of nutrients. Variation 1 with the lowest nutrition showed the best microalgal growth. The higher amount of nutrients indicates a decrease in microalgal growth, this is possible because it is sufficiently saturated with the addition of nutrients, resulting in the phenomenon of growth inhibition due to nutrients [11]. On the other hand, variations 2 and 3 also experienced an increase in the number of microalgae, resulting in self-inhibition by microalgae. It was in accordance with the literature result regarding the self-inhibition [12].

Table 1. The result of swelling test.

River waste	Microalgae	Nutrition	Flow rate concentration CO_2	Harvest volume
2850 mL	150 mL	5.25 (mL) (F-2 Guilard)	15 mL/min	15 mL
2700 mL	300 mL	10.50 (mL) (F-2 Guilard)	15 mL/min	15 mL
2550 mL	450 mL	15.5 (mL) (F-2 Guilard)	15 mL/min	15 mL

Figure 1 shows the growth of microalgae on a variety of nutrients and the number of microalgal cultures mixed with river wastewater. From the figure, it can be seen that in variation 1, microalgal growth begins in the lag phase for 1 day which is then continued with the logarithmic phase for 3 days. The duration of lag and logarithmic phase depended on the substrate media and microorganism [13]. On day 4, there was a stationary phase with peak growth conditions at 0.0115 g/L. After that, there was a death phase (death) as indicated by a decrease in the number of microalgae for 4 days. This also indicates that the optimum time for growth is on day 5 which is marked at peak conditions. The significant drop in cell mass observed was due to the inhibition phenomenon [14]. Unlike the case with microalgal growth conditions in variations 2 and 3, microalgae did not experience a significant increase in growth as previously described. In the photobioreactor, it is clear that the microalgae have died thus they settle on the bottom of the photobioreactor.

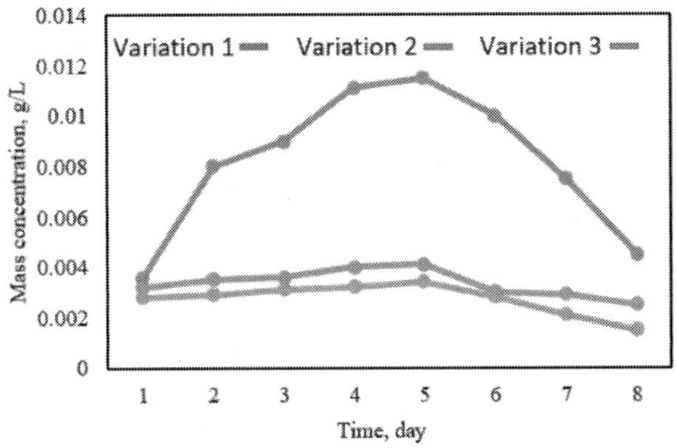

Figure 1. Microalgal growth on various nutrients

Table 2 shows the river water waste's content, including sulfate, sulfur, chlorine, cadmium and iron. It can be seen that after processing using microalgal growth, there was a decrease in sulfate levels from 18.9 to 11 mg/L. Chlorine levels slightly increased from 0.02 to 0.058 mg/L. This makes it possible to release the microalgal content which is experiencing a death phase. The death phase observed was due to the extreme osmotic state imposed on the microorganism [15]. The amount of cadmium decreased from 0.0014 mg/L to undetectable. And iron levels fell quite drastically from 1.8 to 0.8 mg/L. This indicates that the process of cultivating microalgae in river water waste, in addition to increasing the growth of microalgae [16,17], also has an impact on reducing the level of waste in the river water.

Table 2. Pollutant reduction levels in river wastewater.

Content of River Waste Water	Before treatment	After treatment
Sulphate (SO_4) (mg/L)	18.855	11.048
Sulphur as H_2S (mg/L)	<0,001	<0,001
Free chlorine (Cl_2) (mg/L)	0.0219	0.058
Cadmium (Cd) (mg/L)	0.0014	ND
Iron (Fe) (mg/L)	1.8281	0.8116

Conclusion

The use of microalgae in the cultivation process of the contaminated river is successfully conducted. The maximum microalgal growth was obtained on day 4 with a nutritional variation of 5.25 mL as the concentration was 0.0115 g/L. The decrease in the level of pollutants occurred during the treatment with a decrease in sulfate levels from 18.9 mg/L to 11 mg/L. The amount of cadmium decreased from 0.0014 mg/L to an undetectable point. On the other hand, the iron content dropped quite drastically from 1.8 mg /L to 0.8 mg/L.

Acknowledgments

The authors thank the Ministry of Education, Culture, Research and Technology, Republic of Indonesia that provided a Research Grant with contract No. 036/E5/PG.02.00/2022 for generous support and funding of this work.

References

1. Mishra, N. and N. Mishra, Utilization of microalgae for integrated biomass production and phycoremediation of wastewater. Journal of Algal Biomass Utilization, 2017. 8: p. 95-105.

2. Fazal, T., et al., Bioremediation of textile wastewater and successive biodiesel production using microalgae. Renewable and Sustainable Energy Reviews, 2018. 82: p. 3107-3126.

3. Shen, L., et al., The critical utilization of active heterotrophic microalgae for bioremoval of Cr(VI) in organics co-contaminated wastewater. Chemosphere, 2019. 228: p. 536-544.

4. Fazal, T., et al., Integrating adsorption and photocatalysis: A cost effective strategy for textile wastewater treatment using hybrid biochar-TiO_2 composite. Journal of Hazardous Material, 2020. 390: p. 121623.

5. Lage, S., A. Toffolo, and F.G. Gentili, Microalgal growth, nitrogen uptake and storage, and dissolved oxygen production in a polyculture based-open pond fed with municipal wastewater in northern Sweden. Chemosphere, 2021. 276: p. 130122.

6. Al-Jabri, H., et al., Treatment of Wastewaters by Microalgae and the Potential Applications of the Produced Biomass—A Review. Water, 2020. 13: p. 27.

7. Padil, et al., Prospective peat swamp water as growth medium for microalgal cultivation and kinetic study. Alexandria Engineering Journal, 2021. 61: p. 2552-2562.

8. Papry, R.I., et al., Integrated effects of important environmental factors on arsenic biotransformation and photosynthetic efficiency by marine microalgae. Ecotoxicology and Environmental Safety, 2020. 201: p. 110797.

9. Rastogi, R.P., et al., Resilience and self-regulation processes of microalgae under UV radiation stress. Journal of Photochemistry and Photobiology C: Photochemistry Reviews, 2020. 43: p. 100322.

10. Hom-Diaz, A., et al., Performance of a microalgal photobioreactor treating toilet wastewater: Pharmaceutically active compound removal and biomass harvesting. Science of The Total Environment, 2017. 592: p. 1-11.

11. Kim, T.K. et al., Quality characteristics of pork loin cured with green nitrite source and some organic acids. Meat Science, 2019. 152: p. 141-145.

12. Thomas, P.K. et al., Tolerance to allelopathic inhibition by free fatty acids in five biofuel candidate microalgae strains. Bioresource Technology Reports, 2023. 21: p. 101321.

13. Putra, M.D. et al. A green process for simultaneous production of fructose and ethanol via selective fermentation. Journal of Cleaner Production, 2017. 162: p. 420-426.

14. Sulieman, A.K, et al. Kinetic modeling of the simultaneous production of ethanol and fructose by Saccharomyces cerevisiae. Electronic Journal of Biotechnology, 2018. 34: p. 1-8.

15. Putra, M.D. et al. Production of fructose from highly concentrated date extracts using Saccharomyces cerevisiae. Biotechnology Letters, 2014. 36: p. 531–536.

16. Olsson, J. et al. Co-digestion of cultivated microalgae and sewage sludge from municipal waste water treatment. Bioresource Technology, 2014. 171: p. 203-2010.

17. Afifah, A.S., et al. Microalgae production using photo-bioreactor with intermittent aeration for municipal wastewater substrate and nutrient removal. Communications in Science and Technology, 2020. 5: p. 107-111.

RELATIONSHIP BETWEEN NORMALIZED DIFFERENCE VEGETATION INDEX (NDVI) AND RICE GROWTH PHASES IN DANDA JAYA SWAMP IRRIGATION AREA REGENCY BARITO KUALA

Naufal Adlie Nafarin[1,*], **Novitasari Novitasari** [1]
[1]*Lambung Mangkurat University, Jl. A. Yani KM 36, Banjarbaru, Kalimantan Selatan, Indonesia*

*Corresponding author: novitasari@ulm.ac.id

Abstract. One of the factors that influence the success of rice productivity is rice growth, from land preparation to harvesting. With remote sensing technology, the phase of rice growth can be known more quickly. The purpose of this study was to analyze the NDVI value as a representation of vegetation density characteristics and analyze cluster data grouping based on the results of the identification of the rice growth phase using Sentinel-2 Imagery in the 2019 planting season cycle in Danda Jaya Swamp Irrigation Area. The method used in this study is a quantitative descriptive method with the interpretation of Sentinel-2 Image. Image data were collected in the form of time series in 2019 and downloaded via the Google Earth Engine (GEE) platform. Rice productivity was calculated using NDVI to identify plant vegetation density. Then calculate the statistical zone using a GIS application to get the average NDVI value per rice field plot. NDVI values were analyzed using the K-means Cluster method for class determination. These results support the hypothesis that the time series NDVI value in the sentinel-2 image with a high spatial resolution (10 m) computed from the Google Earth Engine produces a strong correlation with the rice growth phase. The next research development is to analyze the estimation model of rice productivity on Sentinel-2 Image by utilizing the relationship between rice productivity and NDVI.

Keywords: NDVI, Rice Productivity, K-Means, Danda Jaya

IOP Conf. Series: Earth and Environmental Science 1184 (2023) 012019 doi:10.1088/1755-1315/1184/1/012019

1. Introduction

The Danda Jaya Swamp Irrigation Area was developed in 1969 by the Department of Ministry of Public Works, located in Rantau Badauh District, Barito Kuala Regency, South Kalimantan Province. The potential of 2,400 ha of land for tidal swamp farming allows farmers to innovate technology that is suitable for the location to produce high-quality products that meet market demand and are produced efficiently [1]. Therefore, to increase productivity, farmers need access to the right technology tools [2]. One method of monitoring rice or other seasonal crops is monitoring through remote sensing satellite data, which has high temporal resolution and adequate spatial resolution (1-10 m). The use of remote sensing imagery can explain more detailed information for monitoring paddy fields where each land has a different growth phase. One of the remote sensing satellite data that has an adequate spatial and temporal resolution is Sentinel-2 Image [3]. Sentinel-2 is a remote sensing satellite with passive sensors made in Europe. Sentinel-2 is multispectral and has 13 channels (bands), four channels spatial resolution of 10 m, six channels spatial resolution of 20 m, and three channels with a spatial resolution of 60 m [4]. Analysis of the growth phase of rice plants was carried out after the paddy fields were successfully classified for their growth age using the vegetation index algorithm. Image multispectral that shows the aspect of vegetation density so that it can be used to perform plant growth phase analysis. The algorithm used in this study is the Normalized Difference Vegetation Index (NDVI). NDVI is the greenness index of vegetation which is determined from the photosynthetic activity of vegetation. Kurniawan (2020) had previously conducted research on planting phase analysis using the NDVI algorithm on Sentinel-2 Image. However, the data processing process in this study requires high data processing device capabilities and requires large data storage space, so it takes longer to process the data [4]. This study will use a cloud, namely Google Earth Engine (GEE) [5]. Cloud-based computing is a combination of the use of computer technology with internet-based development. GEE allows users to create and run custom algorithms and fast computations that enable easy global scale analysis. Google has archived many datasets and integrated them into cloud computing for open source, one of which is the medium-resolution satellite image Sentinel-2 Image [4]. This study intends to relate the NDVI value to the rice growth phase of DIR Danda Jaya, Barito Kuala Regency. This analysis aims to analyze the NDVI value as a representation of vegetation density characteristics and analyze cluster data grouping based on the results of the identification of the rice growth phase using Sentinel-2 Imagery in the 2019 planting season cycle.

2. Materials and Methods

2.1. Study Area

The research location is in the Danda Jaya Swamp Irrigation Area, Rantau Badauh District, South Kalimantan Province. Danda Jaya is geographically located between 3°4'50"S - 3°8'10"S and 114°41'50"E - 114°38'10"E. The study area location can be seen in Figure 1. The topography of the area ranges between -3 and 4 m [6]. Based on the analysis of hydro-topographic maps, the Danda Besar area consists of three hydro-topographic classes, namely the areas of class B and C are relatively the same, namely 820,78 ha and 815,17 ha, while class A is only 800,20 ha. The actual conditions in the field are mostly used for agricultural crops (rice), and a small portion is used for annual crops, especially in the yard [7]. Scheme of the Danda Besar Unit water management network, which has primary canals, secondary canals, and 54 Tertiary canals, which are divided into 27 Left Tertiary canals and 27 Right Tertiary canals as well as a tidal pool that is a no more extended function [8]. Tropical monsoon is the type of climate in Barito Kuala; climatological conditions throughout the year are hot and humid with a dry season of about 2-3 months with rainfall depth of less than 100 mm/month. The wet season around 5-6 months, has a rainfall depth of more than 200 mm/month [9]. From the analysis result of

precipitation data for Mandastana rain station from 2000-2020, the result of the regression analysis model of rainfall from 2000-2020 shows an increasing tendency pattern of 18,944 mm/year.
There was a shift in the dry season in period II. 8 months of the rainy season in the period I start from October to May, nine months in period II start from October to June, and the dry season from July to September. Analysis of annual average rainfall in the rainy season in periods I and II decreased by -2,663 mm/year and -11,542 mm/year, respectively. In the dry season, the value of annual average rainfall increased in the period I by 4,886 mm/year and period II by 3,083 mm/year [10]. The dry season in Indonesia occurs as an effect of the southeast winds pushing dry Australian continental air masses toward the archipelago, while the wet season occurs when the monsoon winds blow from a northwest direction [11]. The climate of South Kalimantan, especially in Barito Kuala, is strongly influenced by the monsoon. The Monsoon type of rain pattern is U-shaped. The pattern thus has a wet season around October to March; this pattern was also strongly influenced by the movement of winds from the northwest towards the southeast starting in October and ending in March. The dry season starts from April to September as an effect of the wind from the southeast to the northwest, which is from April to September. In addition, this pattern is also strongly influenced by extremely wet situations (La-Nina) and extremely dry situations [12]. The monthly average temperature in South Kalimantan is 26,22°C; the minimum temperature mainly occurs in January and December. The maximum temperature occurs in May [13].

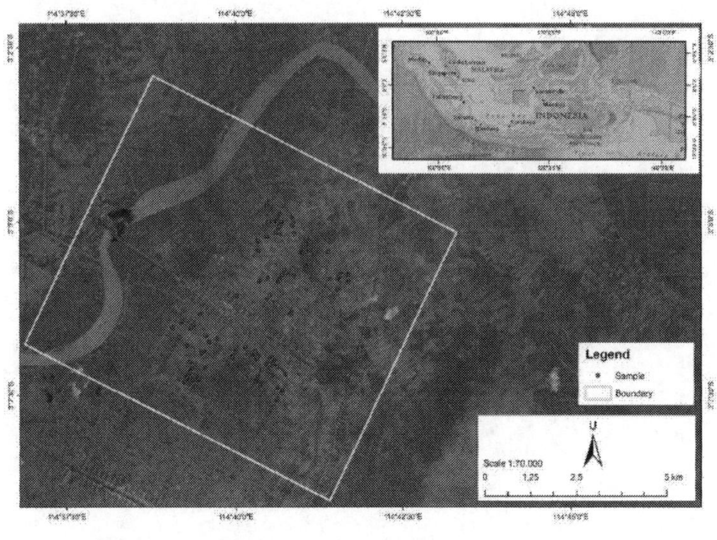

Figure 1. Map of Study Area

2.2. Rice Cultivation Practices

Rice fields in Danda Jaya have two cropping seasons. Crops during the wet season (November to March) are called the main season or are locally known as *Musim Hujan*. The rice planted is superior rice varieties such as *Inbrida* (*Inpari 42 & Inpara 02*) and *Hibrida* (*Mapan 05, Suppadi, and Sembada*) with a planting period of 3 months. Crops planted during the dry season (April to September) are called the offseason or are locally known as *Musim Kemarau*. The rice is planted in local rice varieties such as *Siang Mayang* and *Karang Dukuh*, with a planting period of 5-6 months. Based on the cropping pattern Danda Jaya has three cropping patterns; the 1-time local rice cropping pattern carried out by farmers encounters a shortage of water or a deficit in the middle of the first month of July. Planting patterns using superior rice can be done up to 2 times a year, but the recommended planting period is the beginning of the first planting period, November-February, and the second planting period, March-June. However, in the third scenario, namely two local rice plantings, water shortages were encountered for several planting periods. Increasing the cropping index can be done if the planting period and the type of rice used can be adjusted, namely superior rice with a shorter planting period [14].

2.3. Satellite Imagery Datasets

The search for sentinel imagery was only carried out in 2019 (January to December) due to the difficulty in finding satellite imagery with less than 50% cloud cover at the research location and research focused on the dry season. Table 1 presents the characteristics of the Sentinel-2 data used in this study.

Table 1. Product of the Satellite Data Used In This Study

Satellite	Product ID	Date	Optic Sensor
Sentinel-2	S2B_MSIL2A_20190220T022729_N0211_R046_T49MHS	February 20	MSI
Sentinel-2	S2B_MSIL2A_20190322T022549_N0211_R046_T49MHS	March 22	MSI
Sentinel-2	S2A_MSIL2A_20190506T022551_N0212_R046_T49MHS	May 6	MSI
Sentinel-2	S2A_MSIL2A_20190605T022551_N0212_R046_T49MHS	June 5	MSI
Sentinel-2	S2A_MSIL2A_20190705T022551_N0212_R046_T49MHS	July 5	MSI
Sentinel-2	S2A_MSIL2A_20190725T022551_N0213_R046_T49MHS	July 25	MSI
Sentinel-2	S2A_MSIL2A_20190814T022551_N0213_R046_T49MHS	August 14	MSI
Sentinel-2	S2A_MSIL2A_20190913T022551_N0213_R046_T49MHS	September 13	MSI
Sentinel-2	S2A_MSIL2A_20191023T022741_N0213_R046_T50MKB	October 23	MSI

The image data collected is in the form of time series in 2019 and downloaded via the Google Earth Engine (GEE) platform. The image used has been geometrically corrected and radiometrically corrected. Band 4 (Red) and band 8 (NIR) from Sentinel-2 multispectral image (MSI) data were used to calculate NDVI. The spatial resolution of the images was 10 m.

2.4. Methodology

Figure 2 presents a flowchart of the methodology for this study. Each step is described in the following sections.

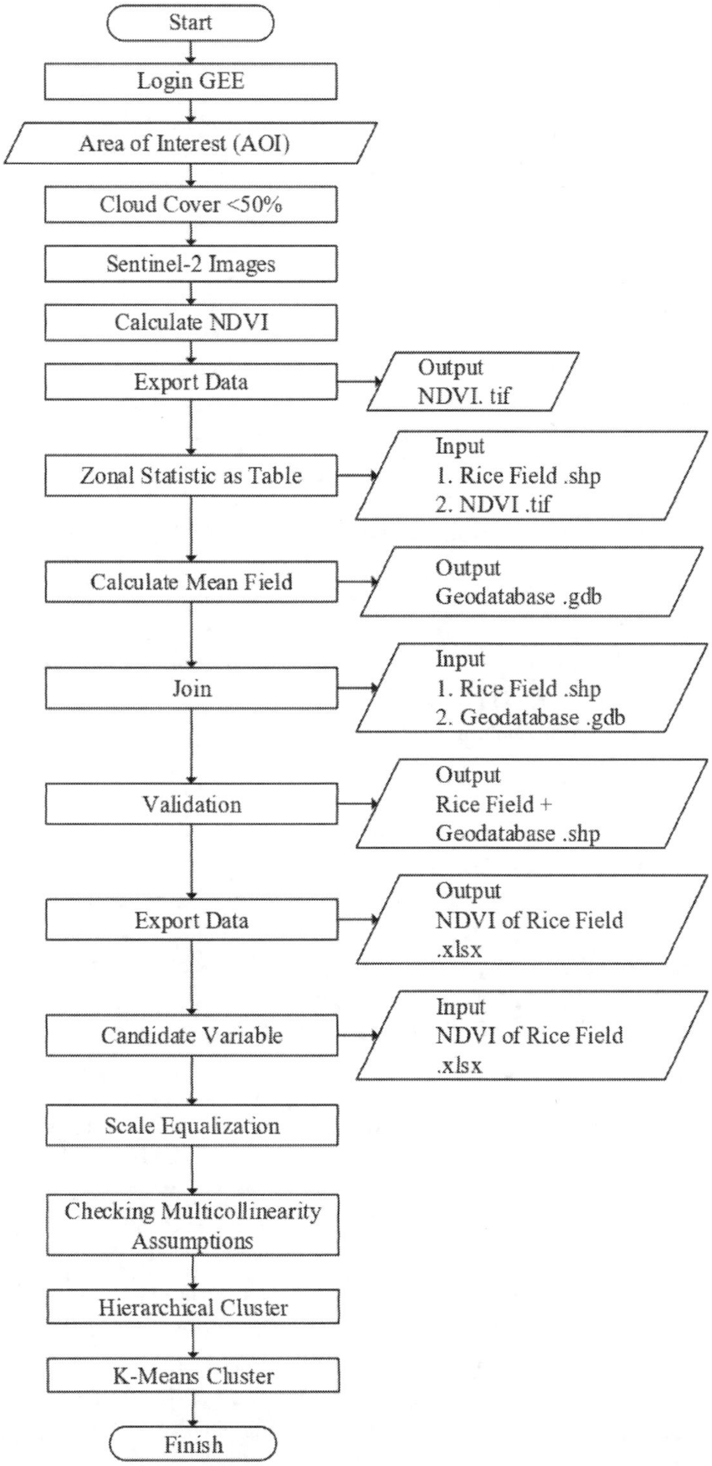

Figure 2. Research Flow Chart

Based on the flow chart presented in Figure 2, if the user does not have a GEE account, the user can register on the website (https://signup.earthe-ngine.google.com/) and can use the account e-mail which has existed. It usually takes 1-2 days to get the notification e-mail agreement. Then the next step is to enter location data with the format shape file. On the Code Editor menu, open Assets → New → Table Upload → Shape Files, and select the format (.shp; .shx; .dbf; .prj; and .zip). When the upload is complete, the user can see the uploaded file on the Asset tab. The next step is inserting script code on the Scripts tab; the first is location data, the second is cloud cover, the third is a collection of Sentinel-2 imagery based on date and <50% cloud cover, and the fourth is calculating normalized index values such as NDVI. Calculated the NDVI in the red (band 4) and near-infrared (band 8) as measured by Sentinel-2 as follows.

$$\mathrm{NDVI} = \frac{(\mathrm{NIR} - \mathrm{RED})}{(\mathrm{NIR} + \mathrm{RED})} \quad (1)$$

The next step is the Zonal Statistic As Table analysis; the input data includes rice field polygon data and NDVI data. Then the value of each NDVI pixel that is in the polygon area of the rice field plots will be averaged to get an average NDVI value that is in the polygon area so as to produce information in the type geodatabase. The next step is the rice field polygon data, as well as the information in the type geodatabase, which will be combined using analysis join. Then on to the analysis join, the two will be validated to match the data with each other. The polygon data of the paddy field area, which contains the file geodatabase then it will be exported to Microsoft Excel for cluster analysis. The next step is to prepare candidate variables, which consist of information columns for paddy field polygon data and rows for the date of collection of NDVI data. Then the next step is to do scale equalization with the method z-score. Then the next step is checking multicollinearity assumptions, where variables have a strong correlation with each other, then proceed with hierarchical cluster analysis, which assumes that each existing data is a cluster at the beginning of the process. If the number of data is n, and the number of clusters is k, then the magnitude of n = k. Then the distance between clusters is calculated using the Euclidean distance based on the average distance between objects. Furthermore, from the results of the distance calculation, the minimum distance is selected and combined so that the value is n = n -1. When two clusters are merged, the distance between the two clusters that are joined to the other cluster is updated. Cluster merging will continue and will stop if it satisfies the condition for the sum of k = 1. At the end of the hierarchical clustering stage, a dendrogram is obtained, which shows the order in which each member is grouped in the cluster. The next step is K-means cluster analysis. The first step is to determine how many k-clusters you want to form from the hierarchical cluster results. Then calculate the distance of each input data to each centroid using the distance formula (Euclidean Distance) until the closest distance from each data to the centroid is found. The next step is to classify each data based on its proximity to the centroid (smallest distance). Then update the centroid value. Score centroid newly obtained from the cluster average. Sample training using the slovin formula with a confidence level of 5% [16]. Then calculate the cluster with k-means [17].

3. Results

3.1. Cloud Cover Analysis

Based on climate observations, the percentage of clouds is influenced by the Asian Monsoon Wind pattern. These winds carry a lot of water vapor which results in greater potential for cloud formation. The clouds that form block the radiation or solar heat that should be released into the atmosphere. The temperature from solar radiation that is absorbed by the earth and reflected into the atmosphere is prevented from being released because there is still a lot of cloud cover. Finally, the sun's heat returns to the earth so that the earth's temperature rises [7].

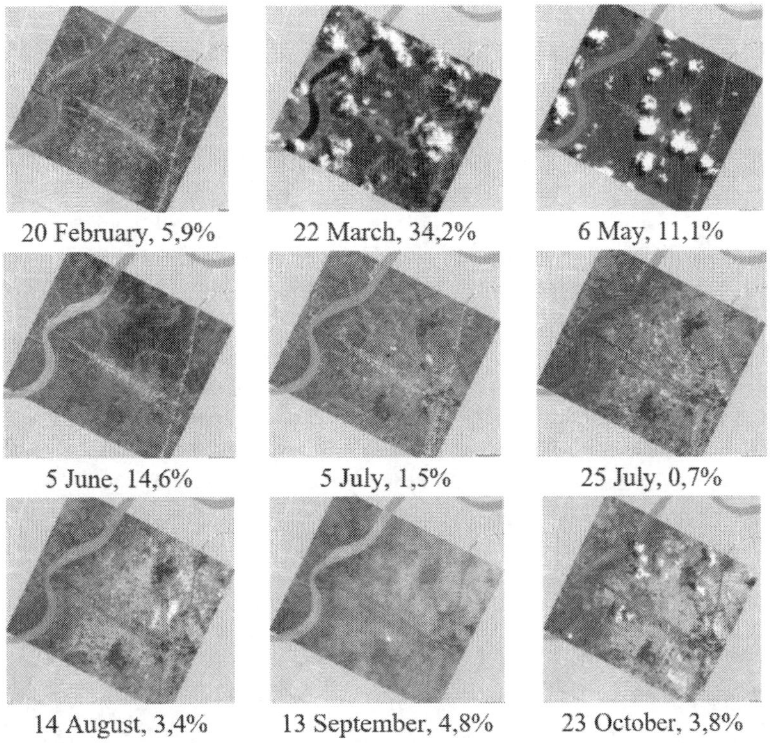

20 February, 5,9% 22 March, 34,2% 6 May, 11,1%

5 June, 14,6% 5 July, 1,5% 25 July, 0,7%

14 August, 3,4% 13 September, 4,8% 23 October, 3,8%

Figure 3. Sentinel-2 Image Collection

Image search results on Google Earth Engine (

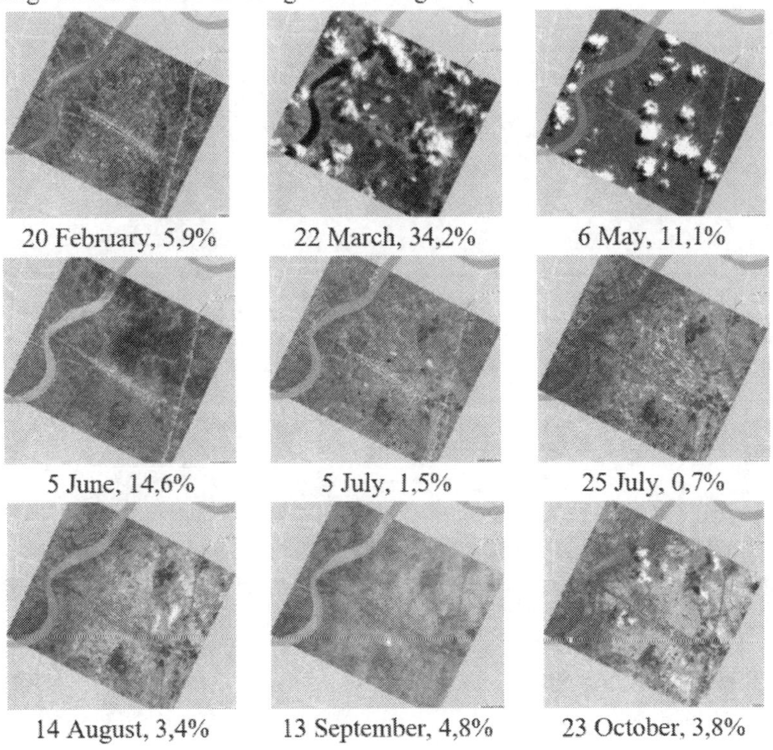

20 February, 5,9% 22 March, 34,2% 6 May, 11,1%

5 June, 14,6% 5 July, 1,5% 25 July, 0,7%

14 August, 3,4% 13 September, 4,8% 23 October, 3,8%

Figure 3) show that nine images of Sentinel-2 were selected for analysis from a total of 61 images between February - October 2019. Images with >50% cloud cover were not used for analysis because the near-infrared sensor will only absorb the color of the cover—lands such as plant vegetation, soil, and water. Therefore, thick clouds in paddy fields are considered to interfere with the reflectance value of land cover so that it can affect the NDVI value. The highest cloud cover occurred on March 22 at 34,2%; this is influenced by the Southwest Monsoon, which blows from October to April. This wind pattern is wet and produces a lot of rain. The lowest cloud cover occurred on July 25 at 0,7%; this is influenced by the East Monsoon, which blows from April to October. This wind pattern is dry because it brings a period of dry air; the impact occurs in the dry season.

3.2. NDVI Analysis

NDVI is an index that describes the level of greenery of plants. The value of the greenness of the plant ranges from -1 to +1. The relationship between NDVI and rice plants lies in the change in the color of the rice leaves in each growth phase. The color of the leaves will continue to be greener, in line with the growth of leaves that grow tall and large. The green color of the leaves will be absorbed more by near-infrared sensors and red light.

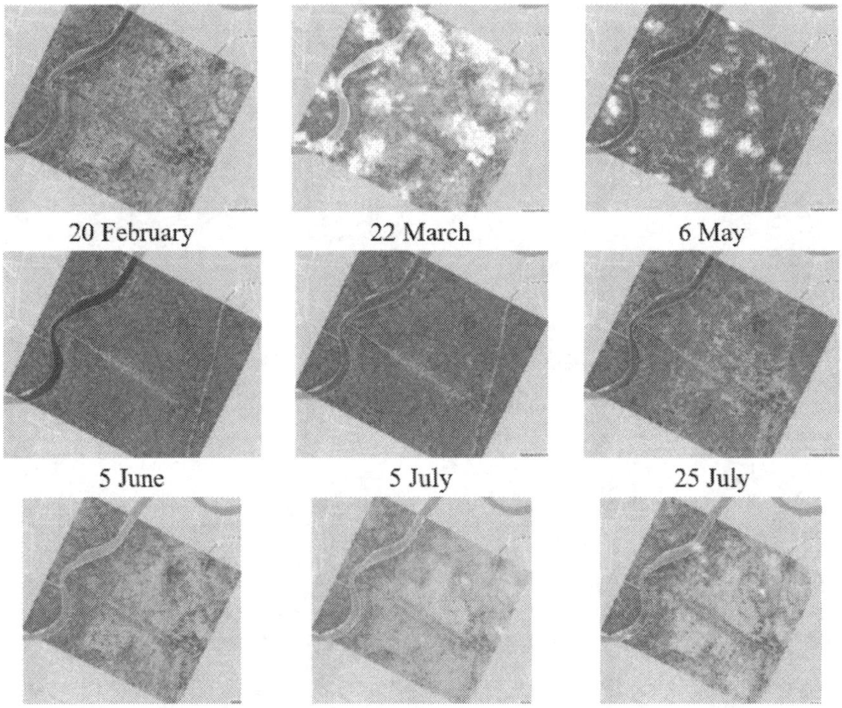

Figure 4. Calculate NDVI

Figure 4 presents the highest greenery level of rice plants that occurs on June 5; rice leaves grow tall and wide and receive full sun or degrees of heat in June or the term called photoperiodic sensors so that near-infrared sensors absorb more rice leaves. While the lowest greenery level of rice plants occurred on February 20, rice was still in the seedling/nursery phase, so the infrared sensor absorbed very few rice leaves and absorbed more soil and water.

3.3. Cluster Analysis

The number of paddy fields that are not covered by clouds is 135 plots, then using sample training with a confidence level of 5%, so that the paddy fields used for the sample are 100 plots. Cluster analysis aims to group data based on similarity. Cluster analysis using SPSS. The analysis procedure is to prepare candidate variables or the average value of the NDVI time series in quantitative form. Then determine the type of cluster analysis with hierarchical (hierarchical cluster) and non-hierarchical (k-means). Hierarchical cluster analysis aims to present all possible clusters optimally, while the k-means cluster analysis presents cluster results for a predetermined k value. Based on hierarchical cluster analysis, the optimal number for grouping is four clusters or k = 4. Next, the k-means cluster is performed to calculate the number of variables in each cluster. The results of the k-means analysis are shown in Table 2.

Table 2. K-means Final Cluster Analysis

Date	Planting Age (days)	Cluster			
		1	2	3	4
February 20	-53	0,28	0,32	0,31	0,33
March 21	-24	0,21	0,25	0,22	0,40
April 14	0	0,33	0,47	0,38	0,56
May 6	22	0,37	0,71	0,56	0,74
June 5	52	0,75	0,79	0,79	0,78
July 5	82	0,75	0,76	0,77	0,72
July 25	102	0,64	0,61	0,70	0,40
August 14	122	0,42	0,32	0,49	0,34
September 13	152	0,29	0,28	0,27	0,29
October 23	192	0,24	0,26	0,26	0,25

Based on the analysis, the maximum cluster occurred on June 5 in cluster 3, which was 0,75, and the minimum final cluster occurred on March 21 in cluster 1, which was 0,21. The cluster value in April is calculated using forecasts because, during discussions with Balai Penyuluh Pertanian (BPP), local rice varieties are planted during the dry season between Aprils – September. The amount of data used as a sample is 100 data, and each data has a lot of similarity data.

1. Cluster 1

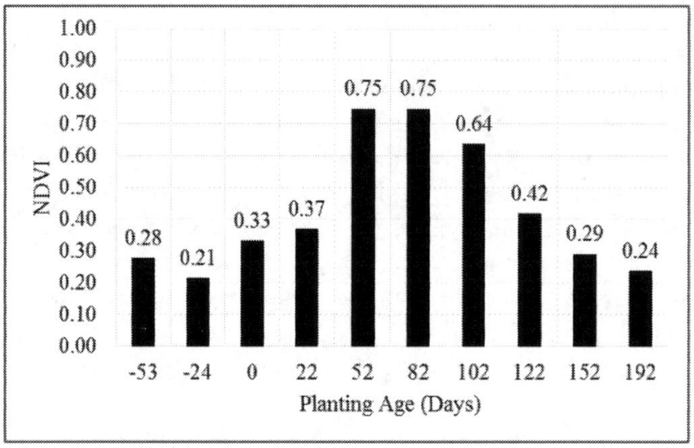

Figure 5. Analysis of Cluster 1

Based on the graph, the NDVI value for planting age (–53) – (-24) days decreased because the land was still in the preparation stage. The plowed land is left until the soil is saturated with water; at the planting age of -24 – 0 days, the harrowing stage is carried out to destroy and muddy the soil. The NDVI value for planting age 0-22 days, there was an insignificant increase due to rice in the seedling/nursery phase until rice entered the vegetative state at 52 days of planting. Entering the generative phase of the planting age of 82 days, the NDVI value did not change but entered the reproductive phase until the planting age of 102 days; the value decreased because rice was in the flowering phase. At the planting age of 122 days, the NDVI value decreased significantly because it was already in the ripening phase until, at the planting age of 152-192 days, the rice was ready to be harvested.

2. Cluster 2

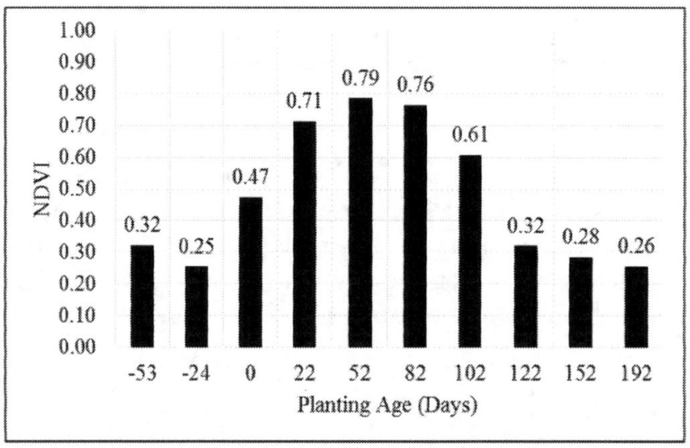

Figure 6. Analysis of Cluster 2

Based on the graph, the NDVI value for planting age (–53) – (-24) days decreased because the land was still in the preparation stage. The NDVI value of planting age -24 – 0 days there was a significant increase; this means that rice began to be sown at the planting age of -15 days until the rice entered vegetative at the planting age of 52 days. Entering the generative phase of the planting age of 82 days, the NDVI value did not change but entered the reproductive phase until the planting age of 102 days; the value decreased because rice was in the flowering phase. At the planting age of 122 days, the NDVI value experienced a significant decrease which caused the rice to be ready for harvest.

3. Cluster 3

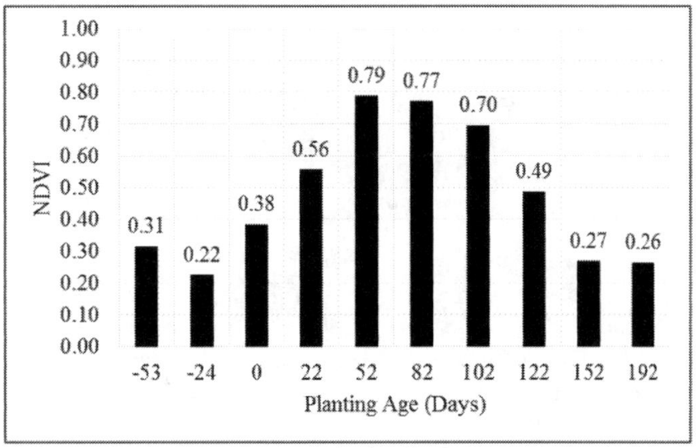

Figure 7. Analysis of Cluster 3

Based on the graph, the NDVI value for planting age (–53) – (-24) days decreased because the land was still in the preparation stage. The plowed land is left until the soil is saturated with water; at the planting age of -24 – 0 days, the harrowing stage is carried out to destroy and muddy the soil. The NDVI value of planting age 0-22 days increased in the phase of rice stalks growing large and tall until rice entered vegetative at 52 days of planting. The NDVI value at the planting age of 82-102 days did not decrease significantly but entered the reproductive phase until at the age of 122 days; the NDVI value decreased; this phase of rice was in the flowering phase. The NDVI value at the planting age of 152-192 days experienced a significant decrease which caused the rice to be ready for harvest; this means that the flowering phase of rice occurred at the planting age of 134 days.

4. Cluster 4

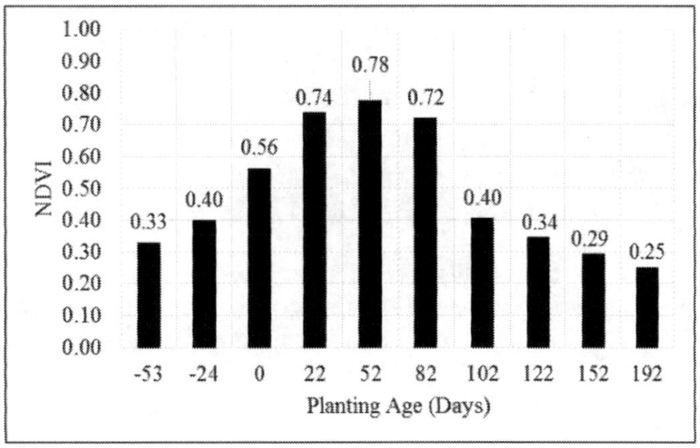

Figure 8. Analysis of Cluster 4

Based on the graph, the NDVI value of planting age (–53) – (-24) days, there was a difference in the increase from grades 1 – 3; rice had been sown first until the rice entered vegetative at the planting age of 52 days. Entering the generative phase of the planting age of 82 days, the NDVI value did not change but entered the reproductive phase until the planting age of 102 days; the value decreased because rice was in the flowering phase. At the planting age of 122 days, the NDVI value decreased because it was already in the ripening phase until, at the planting age of 152-192 days, the rice was ready to be harvested.

Table 3. Comparison of Cropping Calendars from Clusters

Rice Cluster	Planting Age (Days)									
	-53	-24	0	22	52	82	102	122	152	192

1	LP	V1	V2	R	M	H
2	LP	V1	V2	R	H	
3	LP	V1	V2	M	H	
4	V1	V2	M	H		

Description:
LP = Land Preparation, V1 = Vegetative- 1, V2 = Vegetative-2, R = Reproductive, M = Maturity, H = Harvest

From the crop calendar data, the cluster 1 planting calendar is the normal planting age for rice plants in the study area and is in accordance with discussions with them. In contrast to the cluster 1 planting calendar, in the cluster 2 calendar, the vegetative phase 1 is shorter, and the vegetative phase 2 is longer. It is suspected that the application of excess fertilizer makes the plants grow tall and large so that the harvest period is faster. The cluster 3 planting calendar has similarities with the cluster 1 planting calendar, but there is a difference where the vegetative phase 2 is longer. In the planting calendar, cluster 4 is a different cluster from the others. The vegetative phase 1 is longer, and it is suspected that land preparation will be carried out in January.

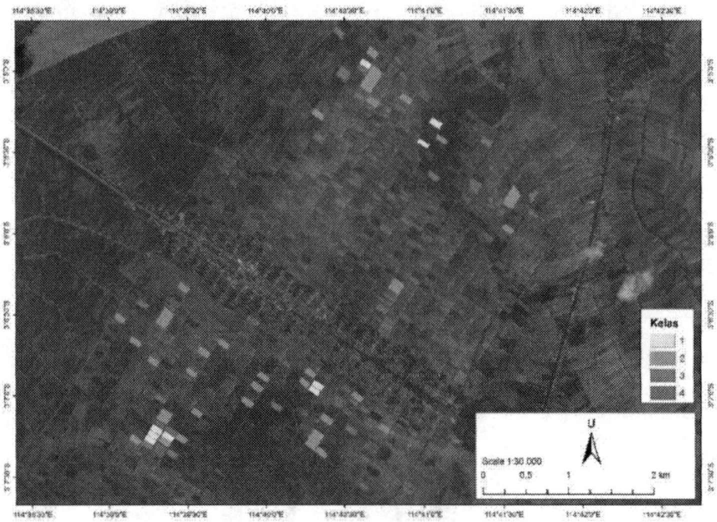

Figure 9. Cluster Map

Figure 9 shows the distribution of the cluster map. The results of the k-means analysis show that cluster 1 has 9 data, cluster 2 has 43 data, cluster 3 has 18 data, and cluster 4 has 30 data. Cluster 1-4 is the rice growth phase for local rice varieties. The various growth phases are influenced by several factors, namely the duration of plowing of paddy fields, fertilizer application, and the preparation of planting plans by farmers.

3.4. *Rice* Productivity Estimation Model

The formation of an equation model for rice estimation is carried out using rice field plots that have productivity data. The data used is data with the highest NDVI value on June 5 or 52 days after planting. Rice productivity was obtained through a field survey of rice field owners or cultivators from 100 plots of land with an average area of 1 hectare. The value obtained from this equation has been converted to tons/ha. The average productivity of each class can be seen in Table 4 below.

Table 4. NDVI Relationship with Productivity

Cluster	NDVI	Productivity (ton/ha)
Cluster 1	0,75	2,77
Cluster 2	0,79	2,95
Cluster 3	0,79	2,94

Cluster	NDVI	Productivity (ton/ha)
Cluster 4	0,78	2,87

Based on the classification results in Table 4, the highest productivity is in class 2 of 2,95 tons/ha with an NDVI value of 0,79, while the lowest is in class 1 of 2,77 tons/ha with an NDVI value of 0,75. To get the regression value, the linear equation model used is linear, which can be seen in Figure 10 as follows.

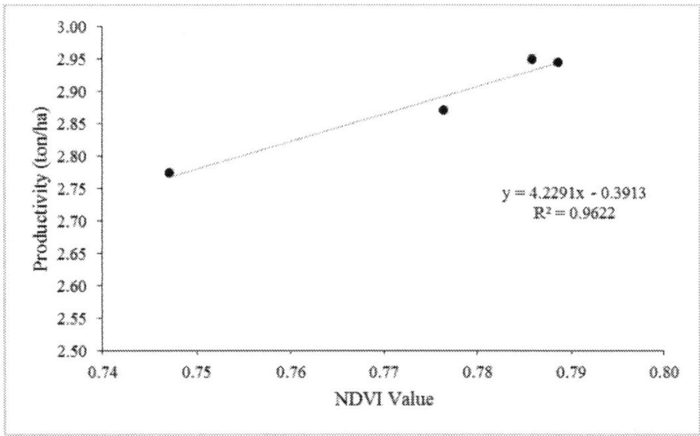

Figure 10. NDVI Coefficient of Determination and Productivity

Based on the classification results in Figure 10, the NDVI value of each class has a significant influence on productivity which has a coefficient of determination (R2) of 0,962 and has an equation model, namely productivity = 4,2291 (NDVI) – 0,3913 with standard error of 0,019, which means that the estimated productivity value can shift up to 0,019 tons/ha. The deviation of this value is small, so the rice productivity estimation model has good precision. According to the results in the field, the average productivity of rice is 3 tons/ha, so this method is close to the results in the field.

4. Conclusions

The use of NDVI on the productivity of rice plants using the main data source, namely Sentinel-2 imagery level 2A from February-October according to the planting phase, has a fairly strong relationship. The sample used was 100 data samples which were then analyzed into four clusters. The productivity of rice plants has four clusters, namely cluster 1 produces 2,77 tons/ha of rice, cluster 2 produces 2,95 tons/ha of rice, cluster 3 produces 2,94 tons/ha of rice, and cluster 4 produces 2,87 tons of rice. /Ha. the NDVI value of each class has a significant effect on productivity which has a coefficient of determination (R2) of 0,962 and has an equation model, namely productivity = 4,2291 (NDVI) – 0,3913 with a standard error of 0,019, which means that the estimated productivity value can shift up to 0,019 tons/ha. The deviation of this value is small, so the rice productivity estimation model has good precision. According to the results in the field, the average productivity of rice is 3 tons/ha, so this method is close to the results in the field.

5. Acknowledgment

The authors thank the data providers, namely the European Space Agency and Google Earth Engine, and the authors thank the Badan Penyuluh Pertanian (BPP) of Rantau Badauh District for taking the time to discuss the development of rice plants in the village of Danda Jaya.

References

[1.] BPP Kecamatan Rantau Badauh. Progama Pertanian Kecamatan Rantau Badauh. Barito Kuala: Kelompok Penyuluh Pertanian; 2021. 46 p.

[2.] Darmawan A. Pembangunan Model Hyperspectral Untuk Estimasi Produktivitas Vegetasi Padi Berdasarkan Metode Derivatif Regresi Linier. 2012;(ITS 2012).
[3.] Rahmi AT, Sukojo BM, Hayati N. Analisis Kebutuhan Air Petak Tersier Berdasarkan Hasil Identifikasi Fase Tanam Menggunakan Citra Sentinel-2 dan Google Earth Engine (Studi Kasus: Daerah Irigasi Sampean Kab.Situbondo) Analysis of Tertiary Field Crop Water Demand Based on Identification of. 2021;17(1):1–20.
[4.] Kawamuna A, Suprayogi A, Wijaya AP. Analisis Kesehatan Hutan Mangrove Berdasarkan Metode Klasifikasi NDVI pada Citra Sentinel-2 (Studi Kasus: Teluk Pangpang Kabupaten Banyuwangi). J Geod Undip. 2017;6(1):277–84.
[5.] Gorelick N, Hancher M, Dixon M, Ilyushchenko S, Thau D, Moore R. Google Earth Engine: Planetary-scale geospatial analysis for everyone. Remote Sens Environ [Internet]. 2017;202(2016):18–27. Available from: http://dx.doi.org/10.1016/j.rse.2017.06.031
[6.] Badan Informasi Geospasial. Seamless Digital Elevation Model (DEM) [Internet]. 2018. Available from: https://tanahair.indonesia.go.id/demnas/#/
[7.] Triadi LB. Studi Hidro-Topografi Persawahan Kalimantan Selatan. 2008;21–3.
[8.] Supriyo A. Gatra Budidaya Padi Dalam Pengembangan Pertanian Lahan Pasang Surut Mendukung Ketahanan Pangan (Studi Kasus Danda Besar, Kabupaten Barito Kuala). J Pertan Agros [Internet]. 2016;18(2):93–9. Available from: http://e-journal.janabadra.ac.id/index.php/JA
[9.] Riza IA. Kesesuaian Lahan Pasang Surut, Potensi, Kendala dan Kesiapan Teknologi Sistem Usaha Tani di Wilayah Kalimantan Selatan dan Kalimantan Tengah. Banjarbaru: Balittra; 1997.
[10.] Amalia M, Sofia E, Munanjar MC. Analysis Of Rainfall Variability In Danda Jaya Swamp Irrigation Area Due. 2021;12(9):52–8.
[11.] Kirono DGC, Butler JRA, McGregor JL, Ripaldi A, Katzfey J, Nguyen K. Historical and future seasonal rainfall variability in Nusa Tenggara Barat Province, Indonesia: Implications for the agriculture and water sectors. Clim Risk Manag [Internet]. 2016;12:45–58. Available from: http://dx.doi.org/10.1016/j.crm.2015.12.002
[12.] Rusmayadi G, R. W, Z R. Adaptation to Temperature and Evapotranspiration Increasing with Reconstruction Tidal Local Rice Line at Monsoon Type in Sei Rangas South Kalimantan. 2016;
[13.] Achyadi MA, Ohgushi K, Morita T. Impacts of Climate Change on Agriculture for Local Paddy Water Requirement Irrigation Barito Kuala, South Kalimantan, Indonesia. J Wetl Environ Manag. 2019;7(2):140.
[14.] Amalia M, Sofia E, Munanjar MC. Analisis Potensi Air Untuk Peningkatan Indeks Pertanaman Pada Lahan Pertanian Dir Danda Jaya Barito Kuala. 2022;7(April).
[15.] Holben BN. Characteristics of maximum-value composite images from temporal AVHRR data. Int J Remote Sens. 1986;7(11):1417–34.
[16.] Kartika T, Parsa IM. Penentuan Training Sampel pada Metode Segmentasi untuk Klasifikasi Sawah dan Non-Sawah Menggunakan Data SPOT-6 (Kabupaten Maros, Sulawesi Selatan). Lapan [Internet]. 2017; Available from: http://repositori.lapan.go.id/id/eprint/1396%0Ahttp://repositori.lapan.go.id/1396/1/Bunga Rampai_Tatik Kartika_Bab 2_ Hal.17-33_2017.pdf
[17.] Orleans B, Putra EP. Clustering Algoritma (K-Means) [Internet]. 2022 [cited 2022 January 1]. Available from: https://sis.binus.ac.id/2022/01/31/clustering-algoritma-k-means/
[18.] Ika. Banyak Awan Sebabkan Suhu Panas di DIY [Internet]. 2019 [cited 2022 Oct 25]. Available from: https://www.ugm.ac.id/id/berita/17870-banyak-awan-sebabkan-suhu-panas-di-diy

IOP Conf. Series: Earth and Environmental Science 1184 (2023) 012020 doi:10.1088/1755-1315/1184/1/012020

RISK ANALYSIS OF COAL HAULING ROAD UPGRADE PROJECT AT PT. BORNEO INDOBARA TO SUPPORT SUSTAINABLE DEVELOPMENT

Koirin[1,2] and Irfan Prasetia[1]

[1]Graduate School of Civil Engineering, Faculty of Engineering, Lambung Mangkurat University

[2]Road Maintenance Dept., PT. Borneo Indobara

Email: iprasetia@ulm.ac.id

Abstract. This study aims to analyze the risks during the construction implementation phase and to propose mitigation strategies. This study was framed by case study design with the project for improving the coal haul road of PT. Borneo Indobara as the chosen case. This project still adheres to support sustainable development concept where the new design can minimize dust pollution and air pollution. Data were collected through interviews, brainstorming, similar previous studies, and questionnaires. All risk variables in the questionnaire were tested for validity and reliability, and then descriptive analysis was carried out. The Godfrey method analysis matrix was used to determine the dominant risks. Identification of valid risk variables found as many as 41 risks divided into 9 sources. Based on the analysis results, 14 dominant risks regarding the impact of time delays and 8 dominant risks regarding the impact of cost losses were found. Based on the results of the study 8 risk mitigation strategies are recommended: 1) Implementation of Management of change; 2) Determination of the implementation method that best suits the needs and conditions in the field; 3) Strengthened supply chain management; 4) Application of construction technology; 5) Strengthened maintenance management of materials and equipment; 6) Tightened supervision and monitoring of implementation activities; 7) Improved coordination between project owners and implementation contractors including stakeholders; 8) Strengthened human resources management and its implementation.

1. Introduction

To meet the target of increasing production in 2022 and to fulfil the Work Plan and Budget (RKAB) approved by the Government, PT. Borneo Indobara is upgrading the infrastructure in the form of a coal haul road. The current capability of the coal hauling road is 12 tons per axle, while the addition of modes of transportation using single trailers and double trailer trucks requires an axle load of 22 tons per axle. The increasing ability of the coal haul road may facilitate the the achievement of annual coal production up to the production peak [1, 3].

In every construction project, risks are always present at every stage of the project work process. These risks arise as a result of an uncertain event or activity. The larger the scale of a construction project, the greater the risks encountered. During the implementation stage or a construction, the coal haul road improvement did not run without risks. The contractor is most responsible party for the implementation of road construction. The success of a construction implementation is determined by its ability to manage various aspects related to construction, including risk aspects, in order to achieve the project objectives that have been set and agreed upon [2]. The project of improving the coal haul road still adheres to the concept of sustainable development towards mining governance. The upgraded road structure design can accommodate the ultimate load, making it more durable.

In addition, the new design can minimize dust pollution and air pollution. This is because the new design enables the haul truck/dump truck to increase their speed, which eventually reduces the dump truck population, thus suppressing the dump truck population. In addition, the safety of workers on the coal haul road is guaranteed because the new road can eliminate all unsafe conditions. Moreover, increasing production supply to the domestic market, and to the Electricity Power Company (PLN) can encourage sustainable development, especially development in South Borneo [1, 3].

The project's first phase covered the road length of 15 km and was divided into 2 segments. This project was unique compared to the public road project. Coal haul road is special road based on government regulation. The differences lie on the regulations applied, the units that pass, the safety rules, the materials used, and the production targets. This unique feature has a significant risk at the implementation stage of the construction process. If this risk is not appropriately anticipated by the project implementing party, it may hamper the project implementation or even result in losses [4, 5].

Therefore, it is necessary to conduct a study on risk management analysis at the construction stage of the coal haul road project. This research can minimize the arising risks and ensure that later development aligns with the concept of sustainable development. For that purpose, identifying risks and the mitigation strategy, especially for risks that fall into the dominant category, can be used as a basis for decision making by the parties concerned to overcome the negative consequences that may occur in the project implementation. The final objective of the project is to achieve the goal in terms of cost and time. It can also ensure that the project development is carried out based on a sustainable concept [6, 7, 15].

In this study, a risk analysis was carried out at the implementation stage of the coal haul road improvement work with a descriptive qualitative method using the Godfrey method. The most dominant risk was obtained from the multiplication between likelihood and consequence. Then the most dominant risks were mitigated based on the recommended mitigation strategies [8, 13, 14].

2. Research method

This study analyzes the implementation risk of PT. Borneo Indobara coal haul road project. The approach used was a descriptive qualitative risk management method. Qualitative analysis was carried out by looking at the level of risk acceptance which depends on the risk value, namely the result of multiplying the trend with the risk consequences [8]. Data collection was carried out through an interview mechanism and questionnaires distributed to respondents from contractors and project owners who carried out projects and also to similar contractors who carried out maintenance of coal haul roads at PT. Borneo Indobara.

2.1. Research Location

The research was conduted at the coal haul road improvement project carried out in the concession area of PT. Sebamban Indobara Borneo Village, Loban River, Tanah Bumbu District. Haul Road Project Location PT. BIB can be seen in figure 1.

2.2. Research Flowchart

This study was started with a preliminary study and literature review by reviewing sources from journals, books, and articles, as well as regulations, and taking into account all related aspects to achieve the goals and objectives. The next stage was to collect both primary and secondary data. The primary data was collected by questionnaires [9]. The data obtained from the questionnaire were then processed using descriptive analysis to obtain the most dominant risk results. Based on these risks, a mitigation strategy was developed. In addition, validation was carried out based on strategies that have been determined. The flowchart of the research is as shown in Figure 2.

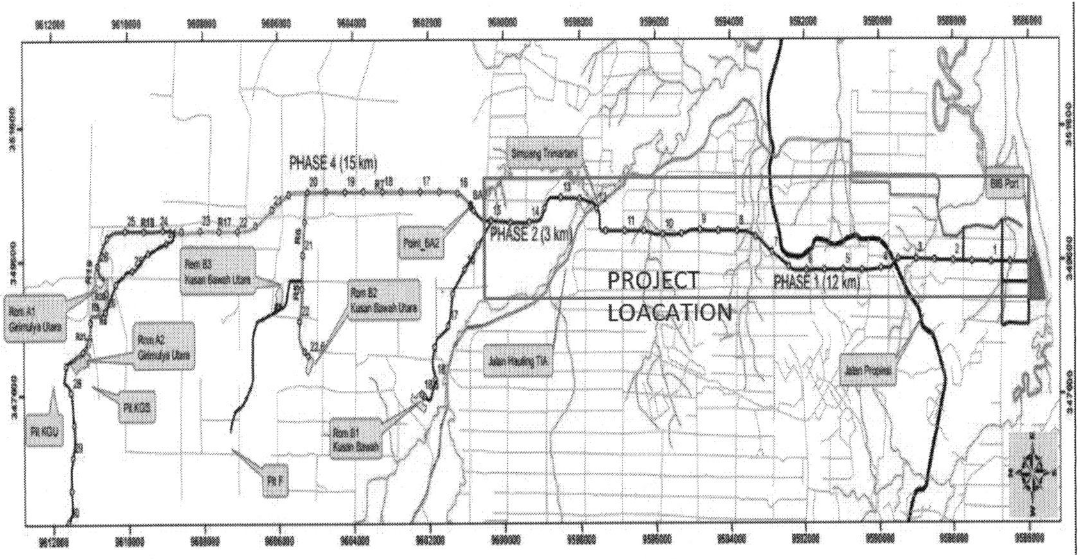

Figure 1. Site Project of Coal Haul Road PT. Borneo Indobara

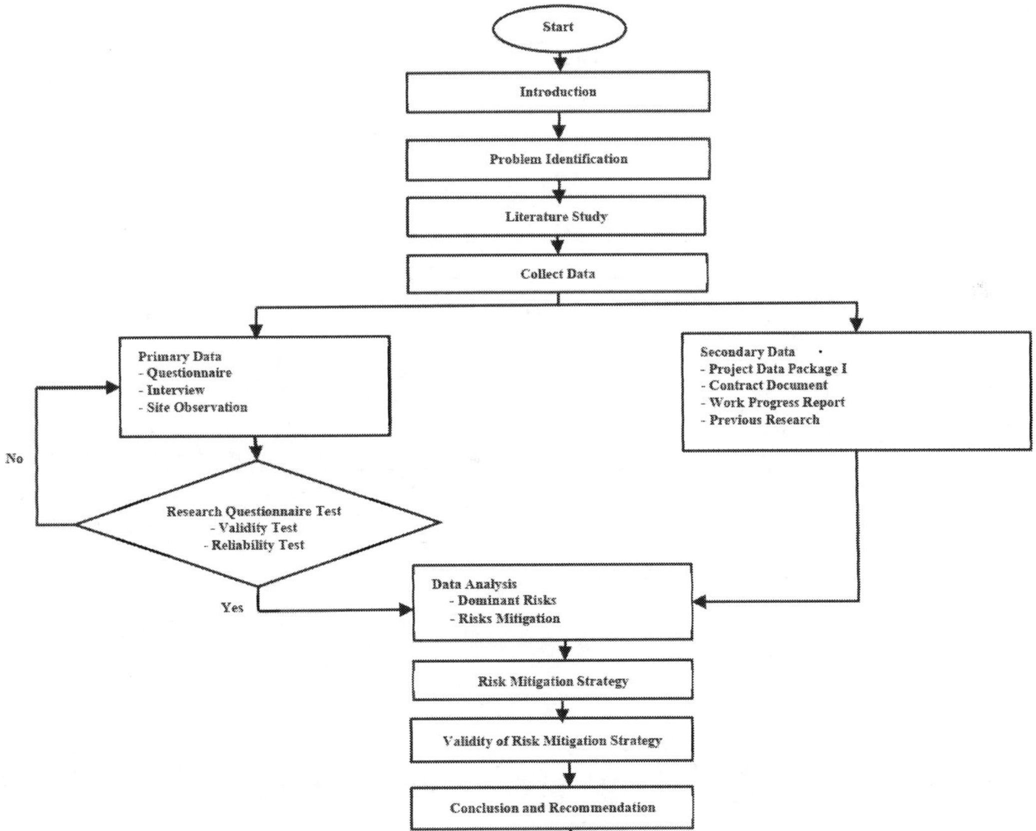

Figure 2. Research Flowchart

3. Results and discussion

3.1. Data collection technique

The data collected for this research were primary and secondary data. The primary data were collected by questionnaires and observation at the project site. The secondary data derived from literature studies, previous similar studies, and historical data in the form of risk data from similar projects [11, 12].
In this research, the first stage was a preliminary survey to determine the risk variables that occured in the project. These were then added to the appropriate previous research risk variables. The second stage was the distribution of questionnaires to obtain the frequency and impact.
The sources of risk obtained in this study were identified from several risk sources, namely: Environmental, Economic, Financial, Natural, Project, Technical, Human, Criminal, and Safety. The initially identified risk sources found 42 risk variables as shown in Table 1.

Table 1. Risk Variables

Risk Sources	**Risk Variables**		**Risk Sources**	**Risk Variables**	
	Code	**Variables**		**Code**	**Variables**
Environment	X1	Potential for asphalt spills	Project	X22	Different volume calculation
	X2	Noise pollution		X23	Violation of prime coat area
	X3	Coal spillage	Technical	X24	Different design
Economy	X4	The increase of fuel price		X25	Lack of traffic management implementation
	X5	Material price escalation		X26	Drawing vs technical spec.
Finance	X6	Delay in payments from contractor to vendors		X27	Quantity work vs actual condition
	X7	Delay in payments from owner to contractor		X28	Availability of heavy equipment
Natural	X8	Weather factor		X29	Repair speed of heavy equipment
	X9	Force major	Human	X30	Delay in contract sign
Project	X10	Contract document misinterpretation between owner and contractor		X31	Delay in BAPP
	X11	Mismatch of Bill of Quantity vs actual		X32	Low productivity
	X12	Delay in material supply		X33	Ied Mubarak effect
	X13	Design changes		X34	Fatigue of the workers
	X14	Unload material via sea/river		X35	Covid effect
	X15	Work not up to specification	Criminal	X36	Material and equipment losses
	X16	Lack of safety equipment		X37	Destruction of project facility
	X17	Insufficient manpower		X38	Incident occur
	X18	Incompetent manpower		X39	Lack of safety system
	X19	Poor coordination between owner and contractor		X40	Healthy worker
	X20	Delay in work activity		X41	Safety tools unapplied
	X21	Allocation of funds outside the project		X42	Lack of sanitary facility

3.2. Validity and Reliability Test

The respondents in this study were 33 people. The validity test results showed that 1 variable out of 42 risk variables was not valid. Thus only 41 risk variables were analyzed. The invalid risk was X2 (noise pollution) due to the operation of heavy equipment (r count= 0.183 < r table=0.334). The reliability test was carried out for the Frequency and Impact data. The results of each Cronbach's Alpha value were frequency = 0.959, the impact on time delay = 0.972, and the impact on cost loss = 0.976. Cronbach's alpha coefficient values from all data were greater than 0.6. Thus, the instrument used in the data collection can be categorized as reliable [9, 10, 16].

3.3. Descriptive risk analysis

The questionnaire data in this study were tabulated using the mode value, then analyzed by referring to Godfrey [8]. The dominant risk analysis results can be seen in Table 2 and Table 3.

Table 2. The results of the dominant risk analysis of the impact of time delay

Var.	Freq.	Time Impact	Risk Index	Category
X1	1	2	2	Negligible
X3	5	2	10	Undesirable
X4	4	2	8	Undesirable
X5	3	2	6	Undesirable
X6	1	1	1	Negligible
X7	2	1	2	Negligible
X8	4	4	16	Unacceptable
X9	2	1	2	Negligible
X10	2	1	2	Negligible
X11	2	2	4	Acceptable
X12	2	3	6	Undesirable
X13	2	3	6	Undesirable
X14	2	2	4	Acceptable
X15	1	2	2	Negligible
X16	3	2	6	Undesirable
X17	3	2	6	Undesirable
X18	1	2	2	Negligible
X19	1	1	1	Negligible
X20	2	2	4	Acceptable
X21	1	3	3	Acceptable
X22	1	3	3	Acceptable
X23	3	2	6	Undesirable
X24	2	3	6	Undesirable
X25	2	3	6	Undesirable
X26	1	2	2	Negligible
X27	2	2	4	Acceptable
X28	2	2	4	Acceptable
X29	2	3	6	Undesirable
X30	2	2	4	Acceptable
X31	2	2	4	Acceptable
X32	2	3	6	Undesirable
X33	2	3	6	Undesirable
X34	1	3	3	Acceptable
X35	2	2	4	Acceptable
X36	1	1	2	Negligible
X37	2	2	4	Acceptable
X38	1	1	1	Negligible
X39	1	1	1	Negligible
X40	1	2	2	Negligible
X41	1	1	1	Negligible
X42	1	1	1	Negligible

Table 3. The results of the dominant risk analysis of the impact of cost losses

Var.	Freq.	Cost Impact	Risk Index	Category
X1	1	2	2	Negligible
X3	5	2	10	Undesirable
X4	4	4	16	Unacceptable
X5	3	5	15	Unacceptable
X6	1	1	1	Negligible
X7	2	1	2	Negligible
X8	4	4	16	Unacceptable
X9	2	1	2	Negligible
X10	2	1	2	Negligible
X11	2	2	4	Acceptable
X12	2	3	6	Undesirable
X13	2	3	6	Undesirable
X14	2	2	4	Acceptable
X15	1	2	2	Negligible
X16	3	2	6	Undesirable
X17	3	1	3	Acceptable
X18	1	2	2	Negligible
X19	1	2	2	Negligible
X20	2	2	4	Acceptable
X21	1	2	2	Negligible
X22	1	2	2	Negligible
X23	3	2	6	Undesirable
X24	2	2	4	Acceptable
X25	2	2	4	Acceptable
X26	1	2	2	Negligible
X27	2	2	4	Acceptable
X28	2	2	4	Acceptable
X29	2	2	4	Acceptable
X30	2	2	4	Acceptable
X31	2	2	4	Acceptable
X32	2	2	4	Acceptable
X33	2	2	4	Acceptable
X34	1	3	3	Acceptable
X35	2	2	4	Acceptable
X36	1	1	1	Negligible
X37	2	1	2	Negligible
X38	1	1	1	Negligible
X39	1	1	1	Negligible
X40	1	1	1	Negligible
X41	1	1	1	Negligible
X42	1	1	1	Negligible

The level of risk acceptance was obtained based on the risk index calculation, the multiplication between frequency and impact in Table 2 and Table 3. The dominant risk is the risk that falls into the unacceptable and undesirable categories, while the acceptable and negligible categories are not the dominant risk. The dominant risks in each impact of the risk source were:

1. Dominant risks to the impact of time delay:
   a. Environment: X3
   b. Economy: X4, X5
   c. Natural: X8
   d. Project: X12, X13, X16, X17, X23
   e. Technical: X24, X25, X29
   f. Human: X32, X33

2. Dominant risks to the impact of cost losses
   a. Environment: X3
   b. Economy: X4, X5
   c. Natural: X8
   d. Project: X12, X13, X16, X23

3.4. Risk mitigation discussion

As we could see from the results, the dominant risk of the impact of time delay and cost losses were generally the same. Only for the dominant risk of time delay, there were other dominant risks, namely: X24, X25, X29 (Technical), and X32, X33 (Human). Thus the discussions for risk mitigation were carried out for both dominant risks.

The risk mitigation was obtained from the results of field observations and interviews with personnel who were experts in their fields. The results of observations and interviews with experts regarding mitigation actions that can be taken for each of the most dominant risk impacts could be described as follows:

1. Environment
   - Coal spillage (X3)
     Making a crossbar system as a levelling tool in the ROM (Run of Mine) area before the Dump Truck unit goes out to the haul road. Implementing the operator's ID holes sanction even up to the revocation of upstream loading equipment. Carrying out inspections once a week on the road and implementing sanctions with all stakeholders. This mitigation is carried out so that the load does not exceed the planned load and does not pollute the environment.
2. Economy
   - The increase of fuel price (X4)
     Determining options for implementation work methods for all resources to reduce operational costs. Material supply through sea transportation.
   - Material price escalation (X5)
     Using a contract system (base) with material vendors per 6 months (contract binding). Bring in materials early. Preparation of additional material stockpile area to accelerate volume fulfilment. Use of material supply through sea or river routes. Direct supply of materials on site or work segment to reduce double handling cost.
3. Natural
   - Weather factor (X8)
     Determining the method of carrying out the work following field conditions, especially those that can synergize with nature, do not damage the environment, and utilize the natural potential. The starting and finishing time for the working hours need to be effective. A stabilization system needs to be used to accelerate the stability of the pavement layer. Weather modification needs to be applied in the nearest area. Conducting work addendum (rescheduling).

4. Project
    - Delay in material supply (X12)
      Bringing materials early to the site project. Tighten monitoring of the entry and departure of the material used for daily project activities. Combination of material supply through land and sea transportation. Conducting incidental meetings or show cause meetings if there are project delays (referring to the contract).
    - Design changes (X13)
      Coordination between project owners and contractors for design adjustments and work planning.
    - Lack of safety equipment (X16)
      Management of Change (MOC) is determined and approved by the Head of Mining Engineering (HME) before the project starts. Fulfillment of safety equipment before the project starts. Checking the beginning of work shifts must comply with MOC.
    - Insufficient manpower (X17)
      Creating work schedules and methods to determine manpower needed from the resource planning stage. Implementing a long shift or overtime work system to cover the delayed progress
    - Violation of prime coat area (X23)
      Traffic management settings are set by a dedicated traffic man. Using the warning signs approaching the work segment. Using the portable traffic lights *(stop and go lights)*.
5. Technical
    - Different design (X24)
      Early identification in the field before the start of work in the road segment. Coordination between project owners and contractors for design adjustments and implications of works. The environmental impact must be kept to a minimum.
    - Lack of traffic management implementation (X25)
      Traffic management preparation work, especially regarding resource adequacy. Socialization and blast in groups of coal haul road users for the work area segment. Adequate barricades in the work area.
    - Repair speed of heavy equipment (X29)
      Availability of proper workshops with adequate facilities. Determining equipment maintenance schedule periodically and monitor its implementation. Maintaining the availability of spare parts at the job site, particularly for special spare parts (imports). Accelerate repairs and overtime if needed. Finding or providing a replacement unit if the unit or equipment has been damaged for a long time.
6. Human
    - Low productivity (X32)
      Setting performance targets for each work item/activity. Strengthening and tightening supervision in the field so that work is carried out according to the schedule. Implement an overtime work system.
    - Effect of Eid celebrations (X33)
      Arrangement of special work schedule for Eid holidays. Special incentives when working on holidays and on-time arrival after the specified holiday. Pick-up system with pick-up point assignment or travel expense cover.

3.5. Risk mitigation strategy
The formulation of a risk mitigation strategy was based on the availability of change management at the beginning, work methods and options available, supply chain management, technology application, facilities and infrastructure maintenance management, supervision and coordination, as well as human resources. The purpose of establishing the risk mitigation strategy is to focus more on mitigating and minimizing or even avoiding these risks during project implementation.

The discussions on the previous section show the correlation link among several risk mitigations. Thus, mitigation actions can be carried out simultaneously. The proposed risk mitigation strategy is outlined as follows:

1. Implementation of Management of Change (MoC)
   The strategy mitigates for the risk variables X3, X16, X23, and X25
2. Determination of the implementation method that best suits the needs and conditions in the field, both technical and environmental.
   The strategy mitigates for the risk variables X4, X8
3. Strengthened supply chain management
   The strategy mitigates for the risk variables X4, X5, X12, and X29
4. Construction technology application
   The strategy mitigates for the risk variable X8
5. Strengthened material and equipment maintenance management
   The strategy mitigates for the risk variable X29
6. Tightened supervision and monitoring of implementation activities
   The strategy mitigates for the risk variables X3, X12, X16, and X23
7. Improved coordination between the owner, contractors including interested stakeholders
   The strategy mitigates for the risk variables X13, X24
8. Strengthened manpower resource management and implementation
   The strategy mitigates for the risk variables X17, X32, X33

4. Conclusions and recommendations

4.1. Conclusions

Based on the results of the analysis, it can be concluded that:

1. There are 9 risk sources that could occur on the project of improving the coal haul road at PT. Borneo Indobara, namely: environmental, economic, financial, natural, project, technical, human, criminal, and safety. 41 risk factors were identified in these risk sources.
2. The analysis of the level of risk acceptance shows the dominant risks that may occur at the project implementation stage. Dominant risks are divided into 2 (two) impacts, namely impacts on project delays and impacts on loss of project costs. Overall, there are 22 (twenty-two) dominant risks with the following details: 14 (fourteen) or 34.15% dominant risks to the impact of project delays and 8 (eight) or 19.51% dominant risks to the impact of project cost losses.
3. Mitigation strategies are carried out on the dominant risks that arise so that they are more focused on their implementation. A total of 8 (eight) risk mitigation strategies are carried out as follows.
   a. Implementation of Management of Change *(MoC)*
   b. Determination of the implementation method that best suits the needs and conditions in the field, both technical and environmental.
   c. Strengthened supply chain management
   d. Construction technology application
   e. Strengthened material and equipment maintenance management
   f. Tightened supervision and monitoring of implementation activities
   g. Improved coordination between the owner, contractors including interested stakeholders
   h. Strengthened manpower resource management and implementation

4.2. Recommendations

Based on the analysis and discussion, several suggestions must be considered for similar research. The suggestions are as follows.

1. The risks that fall into the unacceptable category must get more attention to reduce the negative impacts they may cause. Therefore, time delay or loss of project costs can be eliminated, especially for the contractor.
2. Study the impacts of risk on other project stages, especially those affecting the implementation stage.

3. Continuous improvement of the mitigation strategies proposed for similar research, particularly in the mining industry.
4. The results of this study are expected to be a guideline or baseline for further research and can be valuable information for companies and related parties to improve coal-hauling roads in the mining industry.

5. References

[1] PT. Borneo Indobara 2022 *Dokumen Rencana Kerja dan Anggaran Biaya Tahun 2022* (Jakarta: PT. BIB) chapter II p 21-22

[2] Simanjuntak, Manlin Ronald A 2009 *Analisis Risiko Pelaksanaan Konstruksi untuk Meningkatkan Kinerja Biaya pada Proyek Jalan Tol* (Jakarta: Jurnal Universitas Pelita Harapan-KoNTekS 3 Jakarta) p M-171

[3] Haris J 2016 *Konsepsi Suistainable Development Terhadap Tata Kelola Pertambangan* vol 2 (Jakarta: Selisik 2-Univ. Pancasila Jakarta) p 43

[4] Menteri Pekerjaan Umum Republik Indonesia 2011. *Peraturan Menteri Pekerjaan Umum Nomor 11/PMT/M/2011, Tentang Pedoman Penyelenggaraan Jalan* (Jakarta: Kementrian PUPR) p 2

[5] Direktorat ESDM 2018 *Keputusan Menteri ESDM, Nomor 1827K/30/MEM/2018, Tentang Pedoman Pelaksanaan Kaidah Teknik Pertambangan yang Baik* (Jakarta: Kementrian ESDM) p 98

[6] Australian/New Zealand Standards 1999 *Risk Management AS/NZS 4360: 1999* (Australia/New Zealand: the council of Standards Australia) p 16

[7] Thompson, P.A. Perry, J.G. 1991 *Engineering Construction Risk* (London: Thomas Telford Ltd) p 30

[8] Godfrey, P.S., Sir William Halcrow and Partners Ltd. 1996 *Control of Risk A Guide to Systematic Management Of Risk from Construction* (Westminster London: Construction Industry Research and Information Association-CIRIA) p 40

[9] Wirawan, Nata, 2001 *Cara Mudah Memahami Statistik Deskriptif & Inferensia* (Denpasar: Keraras Emas) p 15

[10] Priyatno 2010 *Paham Statistik Data dengan SPSS* (Yogyakarta: Media Komp) p 25

[11] Flanagan, R. and Norman, G. 1993 *Risk Management and Constructions* (British England: Blackwell Science Ltd. Oxford) p 40

[12] Project Management Institute 2013 *A Guide to the Project Management Body of Knowledge – PMBOK Guide – Fifth Edition* (Pennsylvania: Project Management Institute) p 18

[13] Akintoye, A.S, MacLeod, M.J. 1997 *Risk Analysis and Management in construction* vol 15 (Great Britain: Elsevier) pp 31-38

[14] Al-Bahar, J.F. and Crandall, K.C. 1990 *Systematic Risk Management Approach for Construction Projects* vol 116 (Calilfornia-Barkeley: Journal of Construction Engineering and Management) pp 533-546

[15] Fisk, E. R. 1997 *Construction Project Administration 5th edition* (New Jersey: Prentice Hall, Upper Saddle River, N.J) p 225

[16] Azwar, S. 2001 *Reliabilitas dan Validitas* (Yogyakarta: Pustaka Pelajar) p 86

IOP Conf. Series: Earth and Environmental Science 1184 (2023) 012021 doi:10.1088/1755-1315/1184/1/012021

Important Factors for Inclusive Pedestrian Ways Development in Pecinan Semarang City

G L Wungo[1*], Mussadun Mussadun[1] and A Novianto[1.]

[1] Department of Urban and Regional Planning, Universitas Diponegoro, Semarang, Indonesia

Email: grandywungo@live.undip.ac.id

Abstract. Urban Planning is strongly related to public spaces. Open public spaces are an essential element in sustaining community activity spaces. Social, economic, and other activities regularly happen in this area. Pecinan (Chinatown) is one of the heritage areas in Semarang that still has high community activity. Narrow and crowded street in this area becomes one the issues for pedestrian. The unsafe feeling occurs when people cross the street. The store's activity on the side of the street also creates a fuzzy condition for pedestrians. The study aims to the sustainability and inclusivity of pedestrian ways in Pecinan. Data Collection is done by observing the actual condition of 1380 samples in the field through interviews and questionnaires for the pedestrian. The method of this study uses quantitative methods and use behavioral mapping for the analysis methods. The result of the study shows that pedestrians tend to rank accessibility as the most important factor, followed by transit points, infrastructure, and parking areas. The result also shows that continuous arcade shade and vegetation coverage as other factors that affect pedestrian comfort in Southeast Asian Country climates condition. The result also reveals that an active frontage of a store creates a more lively atmosphere. The recommendation based on the study is to relocate the street vendors' activity alongside the narrow road. The relocation also should still be near the area of the Pecinan.

Keywords: *Important Factor, Pedestrian Ways, Pedestrian Preferences*

1. Introduction

The provision of urban open spaces in recent years has become one of the main issues of urban or city life. Many cities in the world are shifting their focus not only on the people's welfare and improvement in economic and infrastructure sectors, now many cities tend to focus on the community's everyday life needs such as recreation, social activity, etc. One of the ways that a city could provide that needs can be done by designing quality urban public spaces. According to The Law of Republic Indonesia No. 26 of 2007 on spatial planning, a city consists of two main aspects: built-up space and open space [1]. This means that open space in a city, especially in Indonesia, has to provide open space for the community. It is mandatory for any city in Indonesia to achieve open space needs.

In recent studies, the fulfillment of urban public spaces that accommodate the social and economics of the community can affect people's happiness and well-being. Furthermore, when the community feels happy, they tend to be less sick and have lower crime rates which also can lead to achieving a welfare life [2]. Public open spaces in a city can be formed in various ways, such as parks, plazas, sports fields, etc. One of the forms of urban public spaces is the sidewalk or pedestrian ways. Sidewalks can be categorized as urban public spaces as they accommodate the community social activity where they can socialize with other individuals or wander around the areas [3]. Open public spaces play a vital role to creates a more habitable environment [4].

Semarang city is one of many cities that are dealing with this problem. As the city's population grows 6,64% per year [5], the complex and various needs of each individual, including the need for public spaces, create a fuzzy situation. Pecinan district (Chinatown) is one of the most bustling areas in Semarang City. This district is a mix of residential, heritage, tourism, and economic areas in Semarang with a strong Chinese ethnical background strongly related to trade and service activity [6]. Based on The Semarang City Spatial Plan 2011-2031 [7], the Pecinan district (Chinatown) is included as one of the strategic locations of Semarang for tourism activity and cultural preservation. It is essential to manage the urban public spaces in this area.

One of the main issues of the Pecinan District (Chinatown) related to the urban public spaces is the poor quality of pedestrian ways. The high traffic flow with the mobility of parking vehicles in the busy trade and service area added with other activities such as street vendor activity that occupies the sidewalk and roadside service generate an unsatisfactory condition for pedestrian ways users. The impact of this phenomenon affects pedestrians in many ways. Pedestrians feel unsafe when they cross the streets and walk around at night. This condition happens as there is no sign for the crossing path and the bad quality of pedestrian utility, such as the broken streetlight and broken pavement. The other major problem in the Pecinan district sidewalk is the absence of utility for disabled people. There are no guidance paths, wheelchair paths, etc. This condition makes the sidewalks as urban space inaccessible and does not utilize the community needs as it should.

Certain problems stated before shows that there is no inclusivity within the sidewalks. The inclusive value in an urban space is essential. Inclusive urban spaces refer to a condition where an urban space can accommodate all communities with no exception or difference whether physically or socially [8]. The inclusive value of an urban spaces should bring safety, flexibility, and quality so that it can benefits the community at best [9]. The broken pavement, less utility and the lack of support for disabled people to use the sidewalks become the evidence of uninclusive urban spaces, because this condition will complicate and harm the elderly people, children and the disabled people. Accordingly, this study aims to reveal the important factors of pedestrian ways that will be used to determine its development and bring inclusiveness. This study also aims to show the pedestrian users' preferences for ideal sidewalks condition.

2. Literature Review

2.1. *Inclusive Design*

An inclusive design can be explained as the creation of a place that all people can use without any barriers related to each individual ability and condition so that every people can participate in an activity equally [10]. This inclusiveness has to certify that any barriers that interrupting some community group with buildings or space with its environment are cleared. The focus on the community also included women, child and elder group that is vulnerable to some harm such as health condition, harassment or any criminal activity [11]. Commission for Architecture and the Built Environment (CABE) stated there are five principles of inclusive urban spaces and that is (1) Inclusive design put community as the core of the design process (2) Inclusive design acknowledge diversity (3) Inclusive design offers various accommodation regarding the people need (4) Inclusive design provide flexibility (5) Include design has to creates comfort and joyful ambiance for any people.

2.2. *Inclusive Sidewalks/Pedestrian Ways*

In the sense of pedestrian ways, an inclusive pedestrian way can be described as a condition where a sidewalk or pedestrian ways can accommodate the mobility of the community as its main purpose and provides the value of equality, accessibility, safety, and comfort. An appropriate or good sidewalks should have the attributes of (1) accessible for all users (2) Has proportional width (3) Safe to use (4) Continuity and Connectivity (5) Landscaping with vegetation to provide shading and create a border for pedestrian and traffic (6) Social space for social activity [12]. In more detailed manners, the ideal condition for inclusive sidewalk/ pedestrian has certain features [Table 1[

Table 1. Ideal pedestrian ways condition

Aspect	Design	Description
Clear Zoning	Sidewalks divided into three clear Zone 1. Frontage Zone 2. Pedestrian Clear Zone 3. Furnishing zone	1. Frontage Zone used as the zone for building activity for doors, railings and bench 2. Pedestrian clear zone used as the main path for people to walk 3. Furnishing used as the barrier for pedestrians with road traffic
Accessibility	Sidewalk Design: Curb, railings, safe surface (unslippery), bollards, Disabled people feature	The sidewalk design has to ensure that all pedestrian can access it easily. such. The ground has to be unslippery for pedestrian safety. Bollards, curbs and railings can be used to protect pedestrians from vehicles harming the pedestrian. The disabled path becomes one of the essentials aspects, they can be formed as guidance path, and wheelchair entrances.
Safety	Signage: Crossing path, Road sign, street marks, Street names, directions, assembly point	Signage like crossing paths, road signs, and street marks will direct the flow of pedestrian and control the vehicle mobility. Street names and direction will help people to understand the location in the area.
	Utility and Furniture: Bench, street lamp, trash bin	Utility functioned as to support the pedestrian and can be used as transit point to rest or to gather. This also help the vulnerable group such as elderly, disabled people, pregnant women and child for their limitation.
Comfort and Beauty	Vegetation and Landscaping	Vegetation and landscaping create shading and building façade to improve the attractive ambience and comfort for pedestrian users.

Source: [10] [13] [14] [15]

3. Methods

3.1 Behavioral Mapping

Behavioural mapping is a method to tell the behavior tendencies of pedestrian users in the Pecinan District in the sense of spatial analysis. This method will capture whether the pedestrian or other physical environment supports or disturb their activity in a specific location [16]. This will help the researcher compose the prioritized location for sidewalks/pedestrian development in the study area. This method reveals the location of where the community within each category such as gender, age group, destination activity, and posture, mostly happened. The data used for this analysis is collected by filling out a questionnaire in some locations in the study area.

3. 2 Simpson Diversity Index

Being able to be used by all of the community without any discrimination becomes the purpose of inclusivity, this aspect can be measured using Simpson's Diversity Index. Simpson's Diversity Index was actually an index that used in environmental or ecological studies to measure the degree of biodiversity of a population of animals or plantation in an ecosystem [17]. This method also has been used in other studies related to the community to reveal the richness of some aspects related to the community [18]. In the sense of public spaces, the diversity in a sidewalk, this study uses gender, age group, people's activity, and social classes based on people's income to rate the diversity in the study area. This Simpson's Diversity is also an input for the Public Space Index analysis. The equation of Simpson's Diversity Index can be explained that n refers to the amount of sample measured from each variable and N refers to the total population. The closer D value toward 1 means more diversity.

$$D = 1 - (\frac{(\Sigma n(n-1))}{N(N-1)})$$

3.3 Public Space Index

To measure the quality of the pedestrian ways/sidewalk, this study uses a quantitative method through Public Space Index (PSI) analysis. Public Space Index is a method to evaluate the public space related to the aspect of the public space like : (1) purposeful in its design and can accommodate activities (2) create a sense of safety (3) Convenience and comfortable either physically or environmentally (4) has a sense of control and (5) giving a sensory pleasure for the community [19]. These 5 aspects of the Public Space Index will be analyzed based on field observation. Field observation aims to obtained the data on pedestrian users' perception on Pecinan District's sidewalk. The field observation is done by giving the pedestrian users in the Pecinan area a questionnaire. The questionnaire asked the pedestrian users' perception of the five dimensions of the Public Space Index (PSI) which are (1) Inclusiveness (2) Meaningful Activity (3) Comfort (4) Safety and (5) Pleasurability.

The criteria used for the Public Space Index (PSI) heavily linked to the usage to analyze public parks. Therefore, for this study, this method has been modified and simplified based on the sense of pedestrian users.

Table 2. Public Space Index (PSI) assessment

Dimension	Variable	Weight	Source of Indicator
Inclusivity	Gender	0.8	Simpson's Diversity Index Analysis
	Age-Group	0.6	Simpson's Diversity Index Analysis
	Economic Classes (Income)	0.6	Simpson's Diversity Index Analysis
	Accessibility	2	Questionnaire
	Presence of Surveillance cameras, security and etc. disturb pedestrian privacy	1	Questionnaire
	Presence of disabled people and disabled people's utility	2	Field observation
	Transit Point for elder and disabled community	1	Field observation
	Control of Entrance point	2	Field observation
Meaningful Activities	Perceived benefit for pedestrian	3	Questionnaire
	Perceived usefulness for pedestrian activity	3	Questionnaire
	Reachable food stall	2	Field observation

Dimension	Variable	Weight	Source of Indicator
	Flexibility for various activities	2	Field observation
Safety	Perceived security from cameras, guards etc.	2	Questionnaire
	Signage for vehicle and pedestrian	2	Field observation
	High people accumulation disturbs safety	2	Questionnaire
	Sidewalk safeness from vehicle mobility	3	Questionnaire
	Streetlights at night (proper brightness at night)	1	Field observation
Comfort	Street attribute (Shading and Furniture) quality	3	Field observation
	Street vendors disturbance	3	Questionnaire
	Homeless people disturbance	2	Questionnaire
	Parking activity disturbance	2	Questionnaire
Pleasurability	Presence of design integration with local context	5	Questionnaire
	User's experience	5	Questionnaire

3.4 Study Area

The study area in this study is the Pecinan area, divided into 6 corridors. The areas chosen are based on the number of pedestrian users in the corridors and the various activity that occurs in the area. The 6 corridors are as follows:

1. Pasar Baru Corridor
2. Semawis Corridor
3. Gang Pinggir Corridor
4. Gang 5 Pasar Corridor
5. Plampitan Corridor
6. KH. Wahid Hasyim Corridor

The 6 study areas also have their own characteristics defined by their main activities. The Gang 5 Pasar main activity is a traditional market that makes a high accumulation of people's flow. The mobility of the conventional market traders that also move their goods and mixed up with a lot of street vendors' activity makes this corridor very busy.

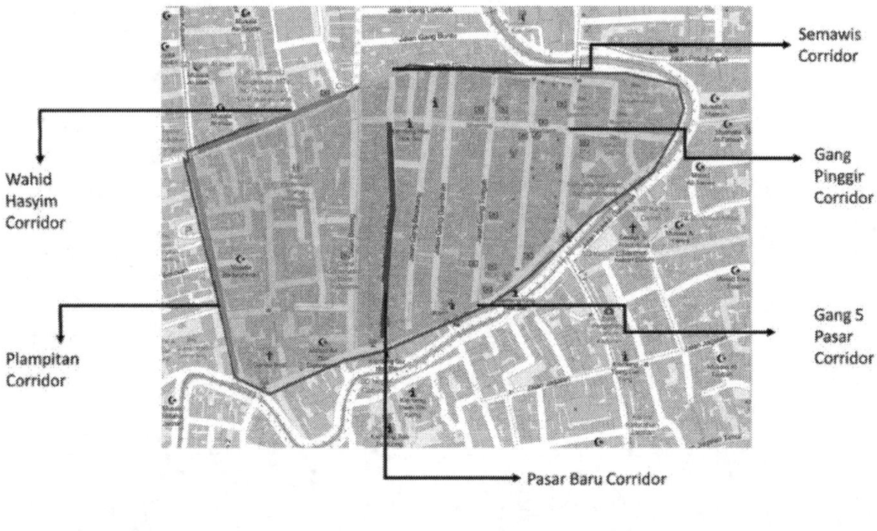

Figure 1. Map of Study Area.

IOP Conf. Series: Earth and Environmental Science 1184 (2023) 012021 doi:10.1088/1755-1315/1184/1/012021

4. Result and Discussion

4.1 Behavioural Mapping Result

The analysis of the behavioral mapping of the pedestrian users on each of the 6 corridors was divided into 4 categories such as gender, age group, destination activity, and posture.

4.1.1 Behaviour mapping of gender

The male gender is ranked as the most pedestrian ways user in 5 Corridor; only Gang 5 Pasar corridor shows women as the highest pedestrian ways user. This correlated to the main activity in each corridor. The corridor that shows men as the most user has a trade area (shops and warehouse), accommodation (Hotel), and street vendors as primary activities. This means that men do the activity of the trade and service area, trade and service area activity usually includes the load and unloading of stuffs, assembly, and warehouse activity. The corridor mainly used for women has the main activity of a traditional market that sells vegetables and common groceries [see Figures 2]

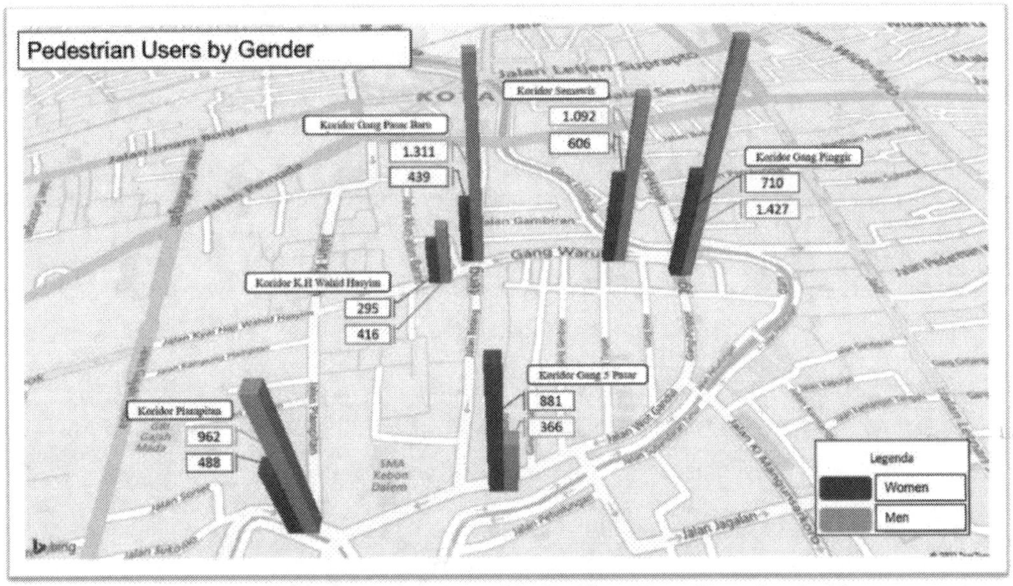

Figure 2. Behavioral Mapping of Pedestrian Ways Users by Gender

4.1.2 Behaviour mapping of age group

All corridor shows "adults" as the primary pedestrian way users. Related to the main activity of the Pecinan Area, economic activity such as trade area, market, and street vendors can be considered as the local community occupation. Therefore, "adult" is the most dominant pedestrian ways user as they are categorized as the working-class age group of the community.

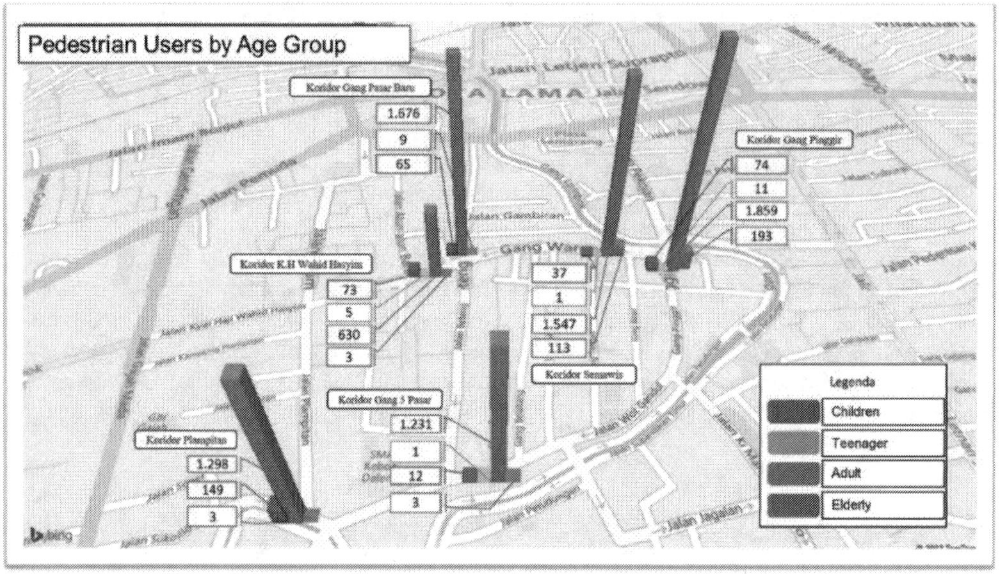

Figure 3. Behavioral Mapping of Pedestrian Ways Users by Age Group

4.1.3 Behavioral mapping of destination activity

Corridor gang 5 pasar that shows common activity can be explained as the activity for daily living needs such as traditional market. The specific corridor activity can be described as a more purposeful economic activity, such as (Electronic shops, warehouses, grocery shops, household furniture shops, etc. The social activity corridor referred to the everyday life activity of the community, such as religious activity, school and etc.

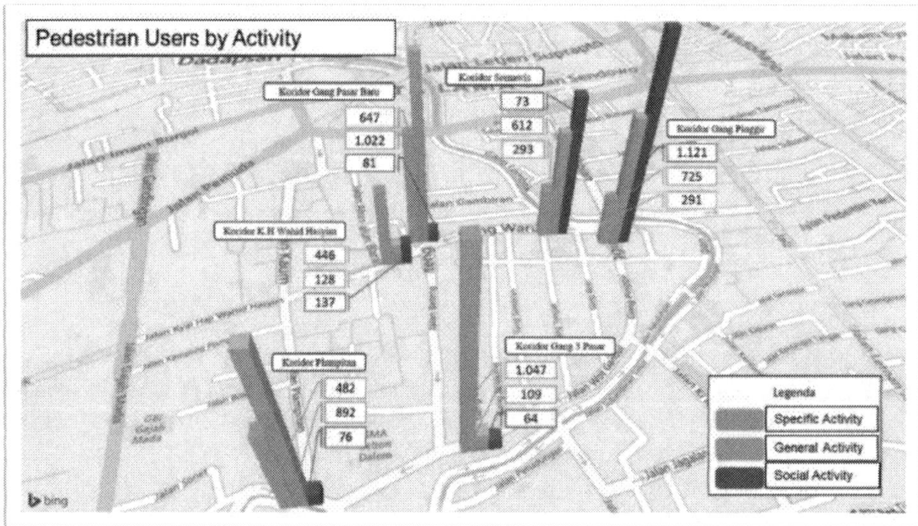

Figure 4. Behavioral Mapping of Pedestrian Ways Users by Activity

4.1.4 Behavioral mapping of posture

The corridor with specific activities such as grocery shops and warehouse tend to have more posture stationary. The corridor that has standard and social activities tend to have more posture moving activity.

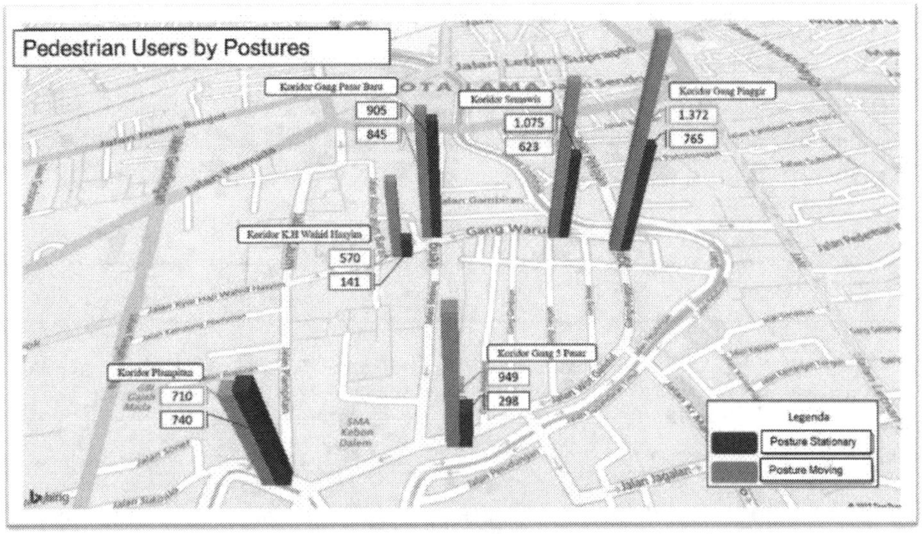

Figure 5. Behavioral Mapping of Pedestrian Ways Users by Postures

4.2 Simpson's Diversity Index

Simpson's Diversity Index analysis that measured the degree of diversity based on the pedestrian user's gender, age group, economic classes and activity shows good diversity in the study area. Simpson's diversity index analysis explained that the closer the value to the value of 1 means that there are more diverse. The average of all Pecinan Districts shows a value of 0.73. This value indicates richness in Pecinan's sidewalk regarding gender, age group, economic classes, and activity. The mixed-use of residential, trade and service, traditional market, and tourism is the main reason that many people use the sidewalk. Even though, every 6 corridors also have their diverse characteristic.

Table 3. Simpson's Diversity Index of 6 Corridors

No	Corridors of Study Area	User Diversity (Gender & Age Group)	Economy Classes (Income)	Activity Diversity	Average
1	Semawis	0.65	0.74	0.64	**0.68**
2	Gg Pinggir	0.70	0.74	0.64	**0.69**
3	Gang 5 Pasar	0.63	0.79	0.70	**0.71**
4	KH. Wahid Hasyim	0.70	0.77	0.68	**0.71**
5	Pasar Baru	0.68	0.81	0.68	**0.72**
6	Plampitan	0.83	0.85	0.83	**0.84**
	Average	**0.70**	**0.78**	**0.69**	**0.73**

The Plampitan corridor shows the highest diversity out of other corridors, which is related to the various actors in the Plampitan corridors. Plampitan corridors have a mix of residential activity, school activity, accommodation (Hotel) activity, and trade and service areas. The combination of pedestrian users that consist of students that go to the school and the trade and service area worker makes this corridor diverse. Semawis corridor has the lowest rate of diversity, the low rank of this corridor is heavily linked with the establishment of Semawis corridors as the center of culinary tourism in Semarang City. The activity food stall that sells unique food and beverages attracts many young adults and teenagers to

IOP Conf. Series: Earth and Environmental Science 1184 (2023) 012021 doi:10.1088/1755-1315/1184/1/012021

the area. This also means that the Semawis corridor has not had many activities since its establishment, with people in the middle income category dominating the economy classes in this corridor.

4.3 Public Space Index

Public Space Index as mentioned before that used to measure how public space has a good quality in 5 dimensions. Public Space Index analysis is done in every 6 corridors and then accumulated to show the entire Pecinan Area index.

Table 4. Public Space Index of Each 6 Corridors.

No	Corridors of Study Area	Inclusivity	Meaningful Activity	Safety	Comfort	Pleasurability	Average
1	KH. Wahid Hasyim	41.56	74.00	48.00	40.67	57.78	**52.40**
2	Gang Pinggir	49.33	78.00	49.62	46.79	50.32	**54.81**
3	Plampitan	47.38	87.69	51.03	44.62	43.59	**54.86**
4	Semawis	47.38	81.67	51.89	56.44	43.89	**56.25**
5	Gang 5 Pasar	46.29	86.43	51.17	50.71	54.76	**57.87**
6	Pasar Baru	49.33	89.23	55.64	54.62	48.72	**59.51**
	Average	**46.88**	**82.84**	**51.22**	**48.97**	**49.84**	**55.95**

The Public Space Index analysis shows that the average of the study area can be considered as a moderate public space index with a value of 55.95. The dimension of inclusivity becomes the lowest dimension on the public space index analysis. This finding is heavily linked with the absence of utility for disabled people and elderly people variable [see Table 1]. The sidewalk on the whole Pecinan Area does not have any disabled people utilities such as a guidance path, wheelchair entrance, etc. The sidewalk of the Pecinan area also does not have any transit point for elderly people or disabled people to take a rest because of their limitations. This situation makes it clear that the sidewalk in the study area does not accommodate the activity of the community.

Meaningful Activity becomes the highest dimension compared to others. This finding is related to the various activity in the Pecinan Area. The mixed-use of residential, trade and service activity, tourism, traditional market and other activities actually helped by the sidewalk. People that are going to work, shop, or do other activities are still utilizing the sidewalk for their mobility. The local people also use the sidewalk to make a short trip destination. Therefore, the sidewalk still benefits the community as their space to walk to their destination and accommodate their needs.

KH. Wahid Hasyim corridor has the lowest rate of Public Space index of all 5 dimensions. This corridor has less inclusivity and safety compared to other corridors. KH. Wahid Hasyim corridor is dominated by trade and service activity, so this corridor has high traffic mobility added with the parking activity to unload and load groceries for shops and warehouses. The high traffic mobility with parking activity makes the pedestrian feels unsafe. The sidewalk in this area is also occupied by the illegal parking of motorcycle and filled with street vendors' activity. Therefore, this corridor makes the pedestrian feel unsafe, and not all communities can easily access the sidewalk.

Pasar Baru corridor has the highest rate of Public Space Index. This result is linked with the main activity that happens in this corridor. The Pasar Baru main activity is traditional market and street vendor activity that is also connected with the Gang 5 Pasar corridor. This corridor has low mobility of traffic since this area is filled with market activity that occupies the road spaces. This area tends to have high mobility of pedestrian users from market sellers and consumers. However, the whole study area still needs development for the sidewalk. The lack of inclusivity in the pedestrian ways as a public space for the community has to be solved in the future. People with certain limitations are still facing issues in utilizing the sidewalks.

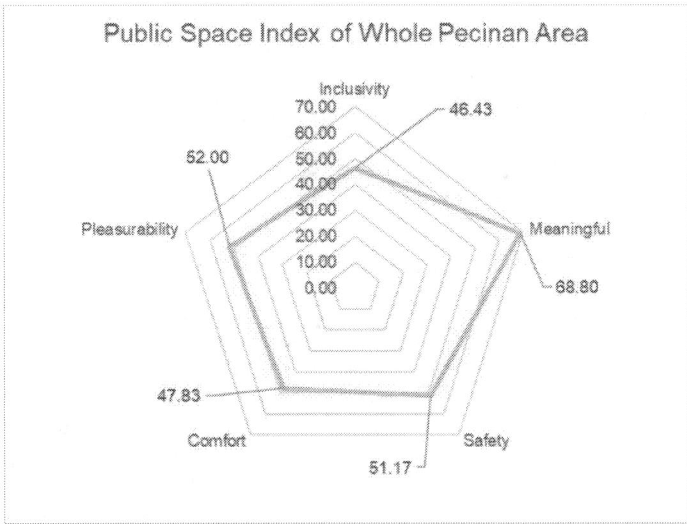

Figure 6. Public Space Index of Whole Pecinan Area

4.4 Existing Condition of Pecinan District Pedestrian Ways

The sidewalk condition in Pecinan District is actually still far from the ideal condition. Broken pavement, no signage for pedestrian crossing, no entrance point, the absence of street furniture for transit points and the absence of disabled people's utility become the issues that need to be solved. Based on the field observation, from 6 corridors that have been observed, there are some additional problems that occur in each corridor. Parking violations from vehicles and street vendor activity that placed their barrow or wheelbarrow on the sidewalks disturb the pedestrian. Parking violation becomes the primary issue that occurs in the study area, this issue is found in 4 out of 6 corridors that are Pasar Baru Corridor, Semawis Corridor, Gang 5 Pasar Corridor and Plampitan Corridor. The street vendors' activity that occupies the sidewalk found in 2 out of 6 corridors which are Gang Pinggir Corridor and KH. Wahid Hasyim Corridor. These problems make the pedestrian spaces become lesser, which means the community's need for public space is yet to be fulfilled.

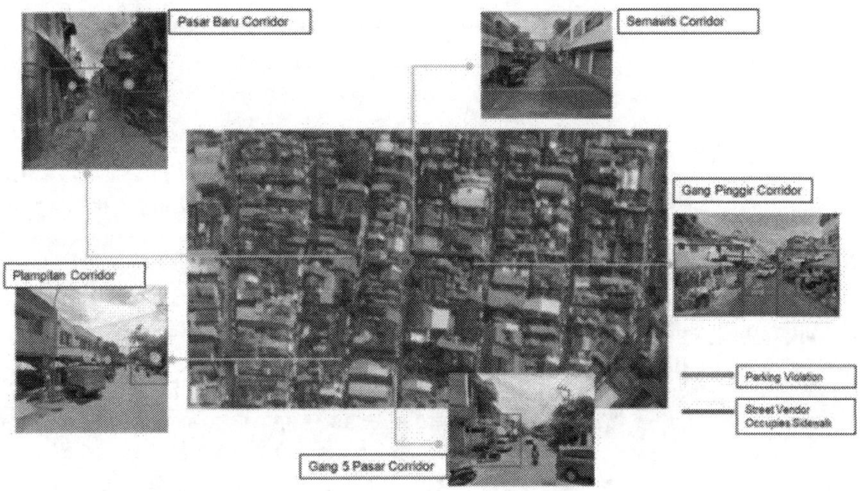

Figure 7. Existing Problems of each 6 Corridor

4.5 Pedestrian Preferences for Ideal Sidewalk

The field observation process that has been done through giving questionnaires to pedestrian users in the study area also reveals some community perceptions, the perception asked of the respondent is about the ideal sidewalk condition. The community expresses that the sidewalk in the Pecinan Area still needs to be developed in many aspects. The community's preference for the ideal sidewalk can be seen in Table 4

Table 5. Pedestrian Preferences for Ideal Sidewalk

Important Factor	Variable	Pedestrian Users Preference
Inclusivity	Accessibility	Pedestrian users in the study area put the "easy to access" aspect as the most important factor for pedestrian ways. This includes the need for distinct quality Pavement and sidewalk material
	Transit Point	Pedestrian users suggest that there is a spot for them to stop and rest, this spot needs to be comfortable and easy to access. This thing correlated with the limitation of disabled people and the elderly age group.
	Entrance	Pedestrian users suggest a better sidewalk entrance that distinct and safe.
	Disabled people utility	Pedestrian users feel that the utility for disabled people that is still non-existent in the study area needs to be developed.
Safety	Signage	Pedestrian users in the area need a better signage such as pedestrian crossing sign, and assembly points in case of an emergency that may occur in the market or shops. Pedestrians also advise that parking space for shops has a visible sign
	Street furniture	Related to the needs for transit point. Pedestrian users suggest that there is a bench in some spots and recommend better lighting at night. Pedestrians also suggest there are some trash bins at some distance
	Parking Area	Parking activity in trade and service areas is one of the unpleasant things for pedestrians. The pedestrian urge that parking activity needs to have a clear border and better settings.

5. Conclusion and Recommendation

Pedestrian ways in the Pecinan area are still far from inclusive pedestrian ways. Located in a busy economic area, including a market, grocery shops, and warehouse area, problems like parking violations and street vendors activity that occupied the pedestrian ways become the main reason for the bad pedestrian ways quality. An inclusive pedestrian ways design requires much effort to satisfy the community. Pedestrian ways must be easy to access for all communities, and this also has to be supported by upgrading better street elements such as signage, street furniture, etc. The main goal of the proper pedestrian ways is to ensure the pedestrian's safety, comfort and lively atmosphere.

Based on the field observation, Community prefers a better building shading by giving continuous arcade shade of the building from the 2nd story of building and vegetation covers. This is correlated with the temperature comfortness for pedestrian in the southeast Asian climate. Pedestrian users will also feel safer by encouraging shops to be more lively by having an active building frontage. The recommendation for this study is to relocate the street vendors activity to a more compact area near the

Pecinan District. This will create a more ordered and centered activity for the street vendors, so they do not occupy the sidewalk as their activity space.

References

[1] Law of the Republic Indonesia about Spatial Plan No 26/ 2007

[2] Montgomery C 2013 *Happy City : Transforming Our Lives Through Urban Design*

[3] Mauliani L, Purwantiasning A W and Aqli W 2013 Kajian Jalur Pedestrian sebagai Ruang Terbuka Pada Area Kampus *NALARs J. Arsit.* **12** 1–9

[4] Faragallah R N 2018 The impact of productive open spaces on urban sustainability: The case of El Mansheya Square – Alexandria *Alexandria Eng. J.* **57** 3969–76

[5] Regional Statistic Agency of Semarang City 2020 *Kota Semarang Dalam Angka 2020*

[6] Dewi S P and Debby T R 2019 Revealing Community Awareness in Semarang China Town Area **21** 90–9

[7] Semarang City Spatial Plan 2011-2031

[8] Landman K 2020 Inclusive public space: rethinking practices of mitigation, adaptation and transformation *Urban Des. Int.* **25** 211–4

[9] Zhou S L 2019 Understanding 'Inclusiveness' in Public Space: Learning from Existing Approaches

[10] CABE 2006 The principles of inclusive design (They include you.)

[11] Ouf A and El-Zafarany N 2018 Diversity and Inclusion in the Public Space As Aspects of Happiness and Wellbeing *J. Urban Res.* **28** 109–29

[12] U.S. Department of Transportation 2004 Accessible Sidewalks and Street Crossings an informational guide 28

[13] Minneapolis City 2021 City of Minneapolis Street Design Guide *Isl. Press* 426

[14] Department for Transport 2021 *Inclusive Mobility A Guide to Best Practice on Access to Pedestrian and Transport Infrastructure* (London)

[15] Tanuwidjaja G, Levina C, Tandiono C and Tandiono C 2018 Service Learning on Inclusive Design: Sidewalk Redesign for Siwalankerto, Surabaya, Indonesia *SHS Web Conf.* **59** 01013

[16] Loebach J, Cox A and Little S 2020 Behavior mapping to support the development of youth-friendly public outdoor spaces *Routledge Handb. Des. Public Spaces Young People Process. Pract. Policies Youth Incl.* 308–26

[17] Kumar P, Dobriyal M, Kale A, Pandey A K, Tomar R S and Thounaojam E 2022 Calculating forest species diversity with information-theory based indices using sentinel-2A sensor's of Mahavir Swami Wildlife Sanctuary *PLoS One* **17**

[18] McLaughlin J E, McLaughlin G W, McLaughlin J S and White C Y 2016 Using Simpson's diversity index to examine multidimensional models of diversity in health professions education *Int. J. Med. Educ.* **7** 1–5

[19] Mehta V 2014 Evaluating Public Space *J. Urban Des.* **19** 53–88

IOP Conf. Series: Earth and Environmental Science 1184 (2023) 012022 doi:10.1088/1755-1315/1184/1/012022

Black Spot Identification and Road Accident Prediction Model on Cebu South Coastal Road (CSCR)

A P Sisbreño[1], J A Sumilhig[1], T J C Venci[1,2], H A R Canseco-Tuñacao[1,2], D B M Cavero[1]

[1] Civil Engineering Department, Cebu Technological University-Main Campus, Cebu 6015, Philippines
[2] Corresponding author: dyannebrendalyn.cavero@ctu.edu.ph

Abstract. Millions of people die in road accidents yearly, causing significant damage to the economy and humanity. The main focus of the study is to analyze road accidents through Black Spot Identification and to establish a Prediction Model on the Cebu South Coastal Road (CSCR)-N840 Highway. The data gathered from government agencies were digitized using SPSS and MS Excel. The study revealed that the year 2018 has the highest number of accident cases with 555 (46.52%) and the lowest severity ratio of 0.41. In contrast, 2019 has the lowest frequency of accident cases with 159 (13.33%) and a severity ratio of 0.72. Through the APW system, the CSCR-N840 Highway's top black spot locations were at 2.6 km, 1.2 km, 2.3 km, 0.9 km,2.2 km, 5.3 km, 4.6 km, 4.7 km, 3.7 km, 5.9 km, 7.6 km, 7.7 km, 8.2 km, 8.5 km, 6.5 km, 10.2km, 11.4 km, 9.4 km, 12.1 km, and 12.0 km. Also, the researchers presented a heat mapping through QGIS for visual analysis, where severe road traffic accidents usually occur on the Talisay side of CSCR N840 Highway due to 13 junctions, 7 intersections, and 6 curvatures. Lastly, in the developed regression model the accident rate decreases gradually with an increase in the traffic volume. Also, Poisson prediction model, as the accident probability increases, the accident rate may either increase or slowly decrease.

1. Introduction

The global population has balanced the number of fatalities in the past few years, and one of the major global causes of these fatalities is road traffic accidents. According to the Global Status Report on Road Safety of the World Health Organization (2018), over 1.35 million people die on the world's roadways yearly. It became a severe economic concern, particularly in low- and middle-income countries, where road fatality rates are three times greater than in high-income countries (WHO, 2018). The Philippines reached 13, 017 road traffic accidents or 1.93% of total fatalities, ranking at 109th globally (Road Safety in the Philippines, 2016). It is one of the South-East Asian countries with the highest road traffic fatality rate, reaching 1.87 percent of all fatalities (WHO, 2014). Different factors can contribute to a road traffic accident, such variables on accident frequency varies significantly on a particular instance basis, both behavioural and non-behavioural factors such as road geometry, traffic flow conditions, vehicle type, and surroundings (Caliendo et al., 2007). They account for 54% of all road traffic fatalities.

Cebu City is the second-largest development center in the Philippines due to its strategic location in Visayas and its developed ports and road networks (Metro Cebu: Integrating Land Use and Transport Development, 2007). Cebu South Coastal Road (CSCR)- N840 Highway, one of Cebu City's road networks has a 300-hectare island-type reclamation area with a six-lane two-way national highway connecting Talisay City to Cebu City (Filipinas Dravo | Projects: Roads & Bridges, 2018). During the last decade, significant statistics on road traffic accidents have been compiled. According to Sun Star and The Freeman, most cases were car accidents that caused fatalities and severe and minor injuries in the ensuing years.

Researchers employed numerous techniques and statistical models to address the alarming number of global road traffic accidents in the past years. Geurts et al. (2004) conducted a sensitivity analysis to determine whether a change in traffic safety policy would have a significant impact not only on the number of accident locations affected by the change in selection and ranking of black spots but also on the types of accident locations. In the same way, R. K. and Suman (2012) developed an accident prediction model based on Average Annual Daily Traffic (AADT) and road conditions to predict road traffic accidents in the future years. The accident prediction model indicates that the number of accidents per km-year increases with AADT and reduces with road/shoulder condition improvement.

The main focus of the study is to analyze road traffic accidents through Black Spot Identification and to establish a Prediction Model on the Cebu South Coastal Road (CSCR)-N840 Highway. The CSCR N840 Highway was recorded with an average of 10 major accidents in the span of 3 years, 2017-2019. Based on the DPWH Accident Black Spot Investigation Handbook (2004), CSCR N840 Highway passed the criteria for black spot identification, stating that a road section must contain an average of 3 major accidents per year per km of road length over the past 2 or 3 years. These recorded statistics made CSCR N840 Highway qualified for black spot identification analysis. The first phase of this study will analyze the Black spots using the Accident Point Weightage to quantify the effect by incorporating the number of road traffic accident cases and assigning weight to its severity. After ranking the black spots, the data will be analyzed using heat mapping analysis. In the second phase of this research, a road accident prediction model will be developed using Regression and the Poisson distribution method to quantify the predicted number of accidents given by the historical count data.

2. Materials and Methods

2.1. Site Selection

The researcher decided to focus on Cebu South Coastal Road (CSCR) specifically the N840 Highway as per the researcher's interest and prominent accident related cases on the road network. As shown in CSCR Road Map in *Figure 1,* the researcher established a four divided stretch from a 12.1 km span of the road network, with a 3.025 km per stretch categorized as Cebu Stretch A, Cebu Stretch B, Talisay Stretch A, Talisay Stretch B.

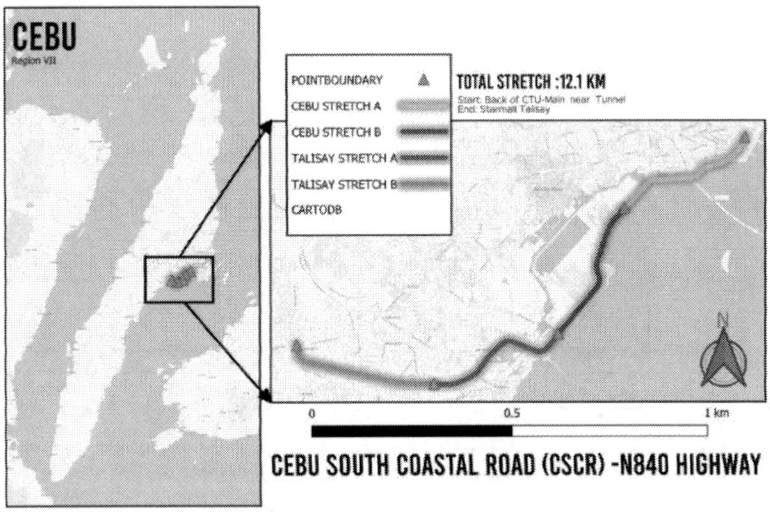

Figure 1 – Stretch of CSCR- N-840 Highway.

2.2. Data Collection

The three years of CSCR road traffic accidents data from 2017-2019 were from the collaboration of all the agencies, including the Department of Public Works and Highway RO7, assessed by the Planning and Design Section, South Road Properties-Management Office, Philippine National Police, and Disaster and Risk Reduction Management Office of Talisay City. These data sets were gathered before the pandemic. Thus, a factor like the pandemic will not affect the data's behaviour. Data collection was justified with a formal letter and supported by ethical permission (*see Appendix Figure A1*). Furthermore, the Department of Public Works and Highway RO7, Planning and Design Section

IOP Conf. Series: Earth and Environmental Science 1184 (2023) 012022 doi:10.1088/1755-1315/1184/1/012022

provided the Annual Average Data (AADT) in 3 years using an Automated Traffic Classifier monitoring system device (*see Figure 2*). The ATC is located in front of Plaza Talisay. It is an embedded sensor device counter that accurately counts the number of passing vehicles and categorizes them according to type along CSCR N840 Highway.

Figure 2 – ATCC (Automatic Traffic Counter/Classifier)

2.3. Methodology Flow

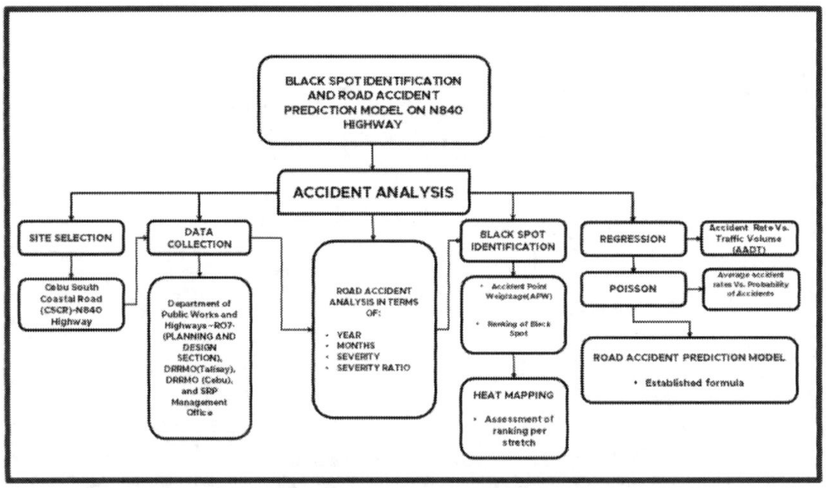

Figure 3 –Methodology flow of CSCR N-840 Highway.

As shown in Figure 3, Cebu South Coastal Road (CSCR) was used as the study stretch for the Black Spot Identification and Road Accident Prediction Model on N840 Highway. The Department of Public Works and Highways- RO7 (Planning and Design Section), the Disaster and Risk Reduction Management Office of Talisay City, Philippine National Police, and the South Road Properties-Management Office were addressed to acquire the data. The data variations for accident analysis included the road traffic accidents sorted yearly and monthly, their respective severity, and severity ratio. The data was digitized and transcribed using MS Excel and SPSS.

The values gathered were employed in two statistical approaches; the Black Spot Identification and Accident Prediction Model. The Black spot locations were analyzed using the Accident Point

Weightage (APW) method to quantify the effect by incorporating the number of road traffic accident cases and assigning weight to its severity. After arriving at the APW values, the spans were ranked per stretch based on the set criteria, and then a heat map analysis was used through Quantum Geographic Information System (QGIS). Consequently, the road accident prediction model was established utilizing the Regression and Poisson distribution methods by incorporating Accident Rate to Traffic Volume and Average Accident Rate to Probability of Accidents, respectively.

Data Analysis

2.4. Black Spot Identification (BSI)

Using the Accident Point Weightage, the researchers used a classification of black spots in road traffic accidents. According to Rahim et al. (2013) this weightage system is adopted from 'A Guide for Accident Black Spot Identification and Road Safety Countermeasure' (Road Engineering Association of Malaysia 1998). Moreover, the same weightage system employed in the study of Omar, N. et al (2017) and Ab. Ghani, A.R. et. al (2019). The accident point weightage formula was used to determine the ranking of accidents for every kilometer. Each accident severity level has its own weightage. For accidents involving fatalities, the numbers need to be multiplied by 6.0. For major injury, minor injury and property damaged, the numbers need to be multiplied by 3.0, 0.8 and 0.2, respectively. Following the calculation of APW, the top ranks were chosen to identify black spots. The equation used:

$$\mathbf{APW = 6(A) + 3(B) + 0.8(C) + 0.2(D)} \qquad \textit{(Equation 1)}$$

2.5. Road Accident Prediction Model (RAPM)

2.5.1. Regression (Accident Rate Vs. Traffic Volume in terms of AADT).

Linear regression models were utilized in Zhang et. al (2019) study, where the dependent variable (either the number of accidents or the rate of accidents) was a function of a series of independent variables, such as speed or traffic volume, in these models. These models made the assumption that accident frequency is regularly distributed. The researcher focuses on the rate of accidents, and for the dependent variable was the number of vehicles in terms of Annual Average Daily Traffic (AADT). The process may be stated as follows:

$$z = \propto + \propto x_1 + \propto x_2 \ldots + \propto x_n \qquad \textit{(Equation 2)}$$

2.5.2. Poisson Distribution Method (Average Accident Rate Vs Accident Probability)

Several researchers, like Ünlü et. al (2020) , Sapuan et. al (2016), Zha et al (2014) and Ma et. al (2014), have employed the Poisson log linear model to explain fluctuations in accident rates in order to account for the probabilistic character of accident occurrence. This Poisson regression model was particularly well suited for dealing with data that has a substantial number of zero counts. The researcher considered this method by observing the average accident rate as the dependent variable for/to the behavior of the data. The method can be expressed as:

$$P(r) = \frac{e^{-m}(m^r)}{r!} \qquad \textit{(Equation 3)}$$

Data Processing and Visual Analysis

2.6. Road Accident Analysis

The road traffic accident data were entered and digitized in MS Excel.

2.6.1. Frequency count. It will be used to determine the number of accident occurrences belonging to a category such as for Years, Months, and Severity.

2.6.2. Severity Ratio. It will be utilized to be the basis for the annual extent of road safety. It can be expressed as:

IOP Conf. Series: Earth and Environmental Science 1184 (2023) 012022 doi:10.1088/1755-1315/1184/1/012022

$$\mathbf{SV} = \frac{\textbf{Fatal+Major+Minor}}{\Sigma(\textbf{Total number of all Severity Classification})} \qquad \textit{(Equation 4)}$$

2.7. Black Spot Identification

The Accident Point Weightage (APW) was employed to identify the black spots and rank the top five black spots per defined stretch. The heat mapping was presented through QGIS analysis and incorporated with the black spot identification to visualize the severity and concentration of the accident-prone areas. The area with the highest concentration in a specified stretch represents more fatal injuries involved. In contrast, the area with the lowest concentration represents property damage. Notably, as shown in *Table 1,* a scale was established to visualize the severity concentration of the road traffic accident data, respectively.

Table 1-Concentration Scale of Accident Severity

CONCENTRATION VALUE	COLOR	SEVERITY
4	Red	Fatal
3	Orange	Major Injury
2	Yellow	Minor Injury
1	Blue	Property Damage

2.8. Prediction Model

The MS Excel and Statistical Package for Social Science (SPSS) software will be utilized in processing the gathered data. The study will be subjected to computer-processed statistics, which included the following statistical tools:

2.8.1. Regression. It is used to determine the relationship between:

- AADT and Accident Rate
- Average Accident Rate and Accident Probability

2.8.2. Poisson Distribution. It is used to formulate the values of Accident Probability.
In computing the accident rate and its average, it can be expressed as:

- Accident Rate (z) is expressed as:

$$\mathbf{z} = \frac{\textbf{number of accidents}}{\textbf{length of the stretch}} \qquad \textit{(Equation 5)}$$

- Average Accident Rate (Z') is expressed as:

$$\mathbf{z'} = \frac{\mathbf{z}}{\textbf{\# of observed year}} \qquad \textit{(Equation 6)}$$

IOP Conf. Series: Earth and Environmental Science 1184 (2023) 012022 doi:10.1088/1755-1315/1184/1/012022

3. Results and Discussion

Table 2 –Road Accident Analysis of N840 Highway.

VARIABLE	CATEGORIES	ACCIDENT FREQUENCY IN TERMS OF	
		NUMBER OF CASES	PERCENTAGE VALUE
Years	2017	479	*40.15*
	2018	555	*46.52*
	2019	159	*13.33*
Months	Jan	130	*20.89*
	Feb	63	*5.28*
	March	106	*8.89*
	April	108	*9.05*
	May	111	*9.30*
	June	114	*9.56*
	July	123	*10.31*
	Aug	101	*8.47*
	Sept	53	*4.44*
	Oct	59	*4.95*
	Nov	105	*8.80*
	Dec	120	*10.06*
Severity	Fatal	30	*3.00*
	Major Injury	167	*13.93*
	Minor Injury	321	*26.77*
	Property Damage	675	*56.30*
Severity Ratio	2017	0.41	*27.33*
	2018	0.37	*24.67*
	2019	0.72	*48.00*

Table 2 shows the N840 Highway Road Accident Analysis by Year, Month, Severity, and Severity Ratio from 2017 to 2019. The highest frequency value by year is in 2018, with 555 (46.52%) road traffic accident cases, while the lowest is in 2019, with 159 (13.33%) road traffic accident cases. The highest number of road traffic accidents with respect to months occurred in January, with 130 (10.89%) cases, and the lowest number was in September, with 53 (4.44%) cases. There were 30(3%) cases of fatal road traffic accidents, 167 (13.93%) cases of major accidents, 321 (26.77%) cases of minor accidents, and 675 (56.30%) cases of property damage. In 2019, the severity ratio reached an equivalent of 0.72 (48 %), and 0.37 (24.67%) in 2018. This reveals that 2018 has the highest number of accident cases with 555 (46.52%) and the lowest severity ratio of 0.41. This is because most accidents are less severe, signifying a higher frequency of minor and property-damage cases. In contrast, 2019 has the lowest frequency of accident cases with 159 (13.33%). However, it shows the highest severity ratio of 0.72 because accident cases in 2019 have a lower frequency of minor and property-damage cases.

3.1.1. Graphical Illustration of the Statistical Data

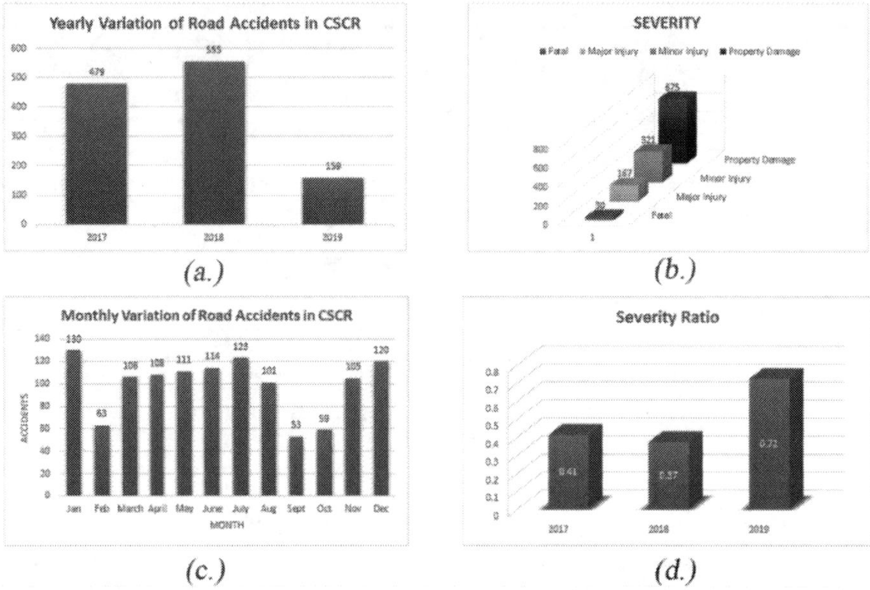

Figure 4 – Road Accident Analysis of CSCR N840 Highway

Figure 4 reveals the graphical representation of the numerical data from Table 2 for 2017-2019. In 2019, a low number of road traffic accident cases was depicted among all the years (see Figure 4-a). Still, its severity ratio reached the highest equivalent value (*see Figure 4-d)* due to the fatal occurrences in that specific year. Concerning the monthly variations of road traffic accidents (see Figure 4-b), a high frequency occurs in the year's first, middle, and last two months. Also, it justifies the low severity ratio for the years 2017, 2018, and 2019 (*see Figure 4-c)*, where the accident cases frequency is high. However, the severity is property damaged, thus resulting in a low severity ratio.

Table 3 - Ranking of Black spots

Stretch	Distance (km)	Accident Severity Fatal	Major	Minor	Damage to Property	APW	RANK
Cebu-A (0km-3.025km)	2.6	1	4	2	8	**18.4**	*1*
	1.2	1	1	4	3	**15.0**	*2*
	2.3	1	2	0	2	**9.4**	*3*
	0.9	1	0	3	8	**8.6**	*4*
	2.2	1	1	0	1	**8.4**	*5*
Cebu-B (3.025km-6.05km)	5.3	1	4	9	4	**18.6.**	*1*
	4.6	1	4	6	5	**17.8**	*2*
	4.7	0	4	3	13	**17.2**	*3*
	3.7	1	2	3	4	**15.4**	*4*
	5.9	0	3	0	5	**14.2**	*5*
Talisay-A (6.05km-9.075km)	7.6	5	23	35	61	**103.8**	*1*
	7.7	2	7	21	24	**43.2**	*2*
	8.2	0	14	12	15	**40.8**	*3*
	8.5	0	11	11	46	**40.0**	*4*
	6.5	0	9	4	7	**18.4**	*5*
Talisay-B (9.075km-12.01km)	10.2	3	23	42	57	**106.4**	*1*
	11.4	1	29	27	44	**101.4**	*2*
	9.4	3	14	25	21	**59.4**	*3*
	12.1	2	2	10	14	**23.4**	*4*
	12.0	0	5	3	4	**12.8**	*5*

Table 3 shows the N840 highway's top five black spot locations of the four established stretches: CEBU Stretch A, CEBU Stretch B, TALISAY Stretch A, and TALISAY Stretch B. An Accident Point Weightage (APW) System was used to identify the black spots. Black spots were identified quantitatively by counting the frequency of accidents based on severity for each 100-meter segment along each stretch. From 0 km to 3.025 km along the CEBU Stretch A section, the top five black spot locations were at 2.6 km, 1.2 km, 2.3 km, 0.9 km, and 2.2 km, in rank order. These findings comprised 2.6 km with an APW of 18.4, 1.2 km with an APW of 15.0, 2.3 km with an APW of 9.4, 0.9 km with an APW of 8.6, and 2.2 km with an APW of 8.4. Then, for the CEBU Stretch B from 3.025 km to 6.05 km distance of the highway, the top five black spots recorded were 5.3 km, 4.6 km, 4.7 km, 3.7 km, and 5.9 km, in sequential order. Five black spot locations were at 5.3 km with an APW value of 18.6, at 4.6 km having an APW value of 17.8, at 4.7 km with an APW value of 17.2, at 3.7 km with an APW value of 15.4, and at 5.9 km with an APW value of 14.2. Also, for the TALISAY Stretch A form 6.05 km to 9.075 km distance of the highway, the top five black spots recorded were 7.6 km, 7.7 km and 8.2 km, 8.5 km, and 6.5 km, in sequential order. Five black spot locations were at 7.6 km with an APW value of 103.8, at 7.7 km with an APW value of 43.2, 8.2 km both having an APW value of 40.8, at 8.5 km with an APW value of 40.0, and at 6.5 km with an APW value of 18.4. Lastly, for the TALISAY Stretch B from 9.075 km to 12.01 km distance of the highway, the top five black spots recorded were 10.2 km, 11.4 km, 9.4 km, 12.1 km, and 12.0 km, in sequential order. Five black spot locations were at 10.2 km with an APW value of 106.4, at 11.4 km having an APW value of 101.4, at 9.4 km with an APW value of 59.4, at 12.1 km with an APW value of 23.4, and at 12.0 km with an APW value of 12.8.

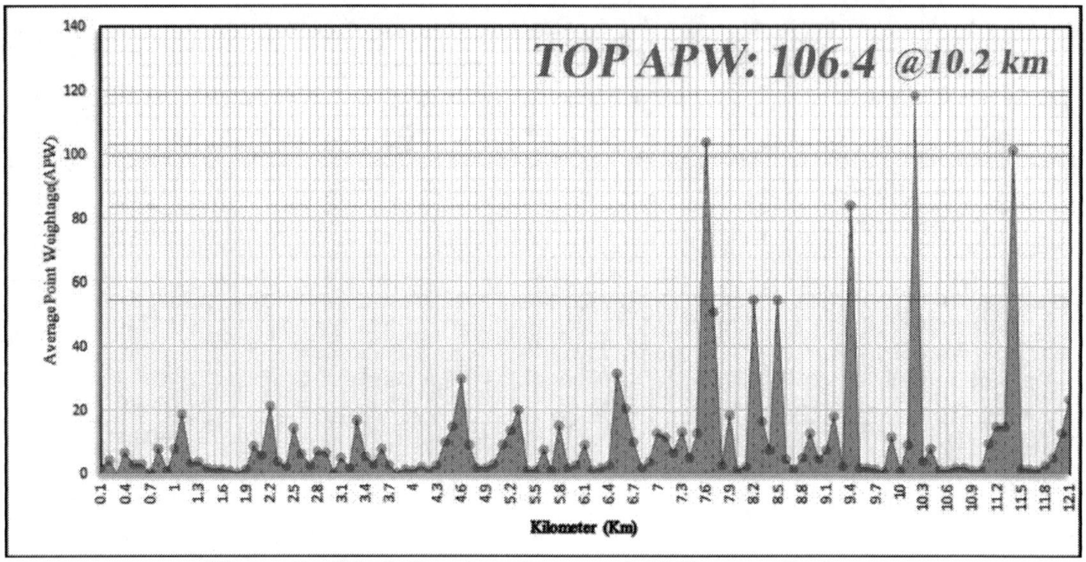

Figure 5 – Black Spot locations per 0.1 kilometer (km)

Figure 5 shows the computed Accident Point Weightage (APW) values in every established segment. The figure reveals three segments with higher APW values. The highest Accident Point Weightage (APW) value was observed at a 10.2 km distance, with an APW value of 106.4, implying that major accidents usually occur in that location. The recorded severity values significantly impact the APW values used to identify black spot locations, making 10.2 km the top black spot location along CSCR N840 Highway.

Black spot by Severity through Heat Mapping per Stretch

As shown in *Figure 6,* Cebu Stretch A was presented in zoomed-in spans (in horizontal point of view) with specified lengths of 1.008km each, where the span of 2.016 to 3.025 km reveals the highest concentration (red) of road traffic accident severity. The figure shows that the most concentrated area is across Ludo Corporation, indicating that multiple fatal accidents frequently occur in that portion. Also, the figure shows a moderate concentration (orange & yellow) of road traffic severity at the SRP Tunnel from 0 to 1.008km, indicating that major and minor accidents occur in that portion due to the long length of the tunnel. Lastly, the less concentrated (blue & yellow) portion of the stretch is from 2.016-3.025km, Pasil Viaduct. The low concentration of road traffic severity indicates that fewer minor accidents usually occur in that portion. Based on the figure above, the road's curvature was a factor in the occurrence of road traffic accidents in this stretch. Utilizing QGIS, the figure illustrates that in Cebu Stretch A, 1 junction, 2 curvatures, and a U-turn slot near the CSCR viaduct were identified as potential areas of road traffic accidents.

IOP Conf. Series: Earth and Environmental Science 1184 (2023) 012022 doi:10.1088/1755-1315/1184/1/012022

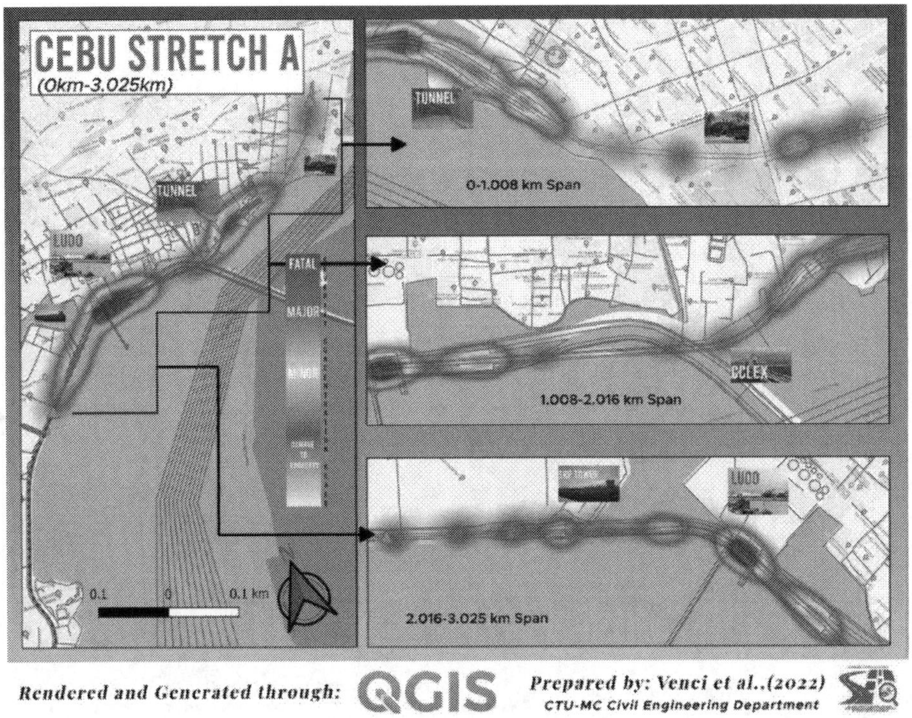

Figure 6 – Black spot by Severity through Heat mapping of Cebu Stretch

Furthermore, as shown in *Figure 7*, Cebu Stretch B was also presented into zoomed-in spans with specified lengths of 1.008km each, where the span of 4.033 to 5.042km reveals the highest concentration (red) of road traffic accident severity. The figure shows that the most concentrated area is near Nustar and Kawit Island, signifying that multiple fatal accidents frequently occur in that portion due to the road's curvature. Also, the figure shows a very high concentration (red & orange) of road traffic severity in front of Il Corso Malls, indicating that a few fatal accidents occur around the area. Lastly, this stretch's moderate concentrated (blue) portion is from 3.025 to 4.033km at the junction of SM Seaside City Cebu. It indicates that few major accidents also occur due to the vehicles crossing paths at that junction. Utilizing QGIS, the figure illustrates that in Cebu Stretch B, 7 junctions, 2 curvatures, and a U-turn slot near DPWH were identified as potential areas of road traffic accidents.

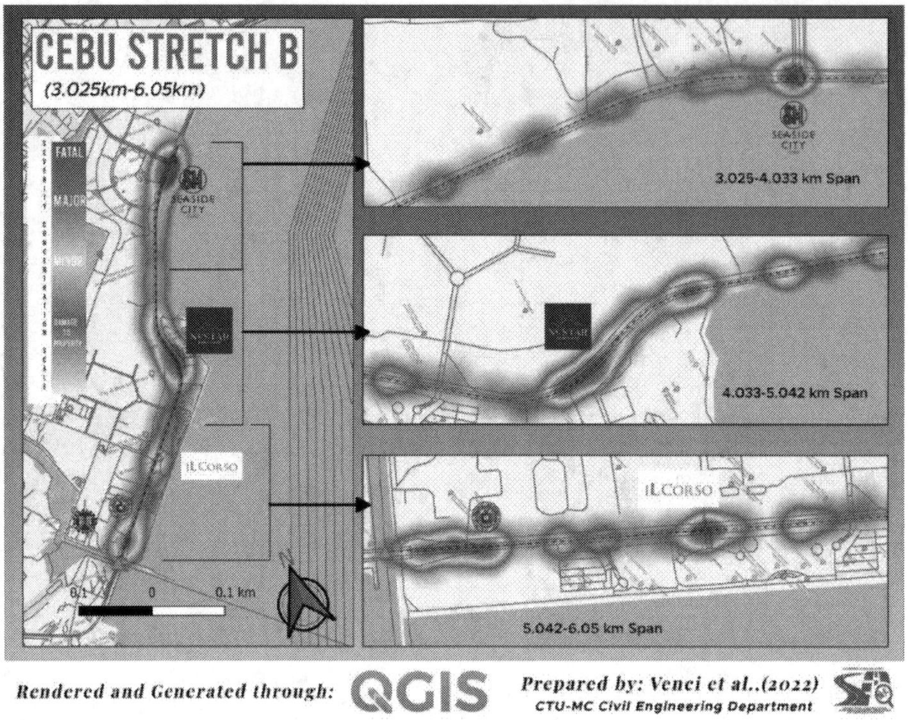

Figure 7– Black spot by Severity through Heat mapping of Cebu Stretch B

As shown in *Figure 8,* Talisay Stretch A presented in zoomed-in spans reveals that the stretch's most concentrated (red) portion is in front of Gaisano Capital SRP, 7.058 to 8.067 km. This indicates that several fatal accidents usually occur at its intersection due to the high number of vehicles passing and crossing paths at that portion. Also, the figure shows a moderate concentration (yellow) of road traffic severity beside PTT Gas station, signifying that few minor accidents occur in that portion. Lastly, this stretch's less concentrated (blue) portion is from 6.05 to 7.058km at the Talisay Fish Port, indicating fewer accidents occurred in that portion. Utilizing QGIS, the figure illustrates that in Talisay Stretch A, 5 junctions, 5 curvatures, and 4 intersections were identified as potential areas of road traffic accidents.

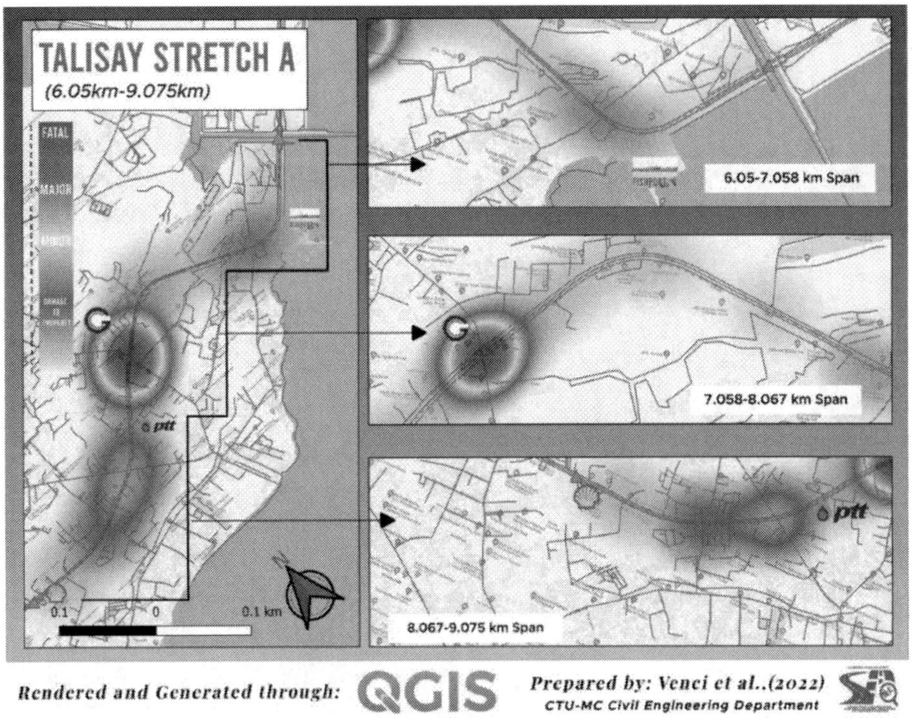

Figure 8 – Black spot by Severity through Heat mapping of Talisay Stretch A

Figure 9 shows the Talisay Stretch B presented in zoomed-in spans, revealing a very high concentration of road traffic accident severity. The most concentrated portion is at the intersection of San Isidro Road, a 10.083 to 11.092 km span, and the intersection near Talisay City Hall, an 11.075-12.1 km span, which indicates that multiple fatal accidents occur in those respective portions due to several vehicles passing and crossing paths at the intersections. Also, the figure shows a moderate concentration (yellow) of road traffic severity in front of the Flying V gas station, signifying that only a few minor accidents occur in that portion. Utilizing QGIS, the figure illustrates that in Talisay Stretch B, 8 junctions, 1 curvature, and 3 intersections were identified as potential areas of road traffic accidents.

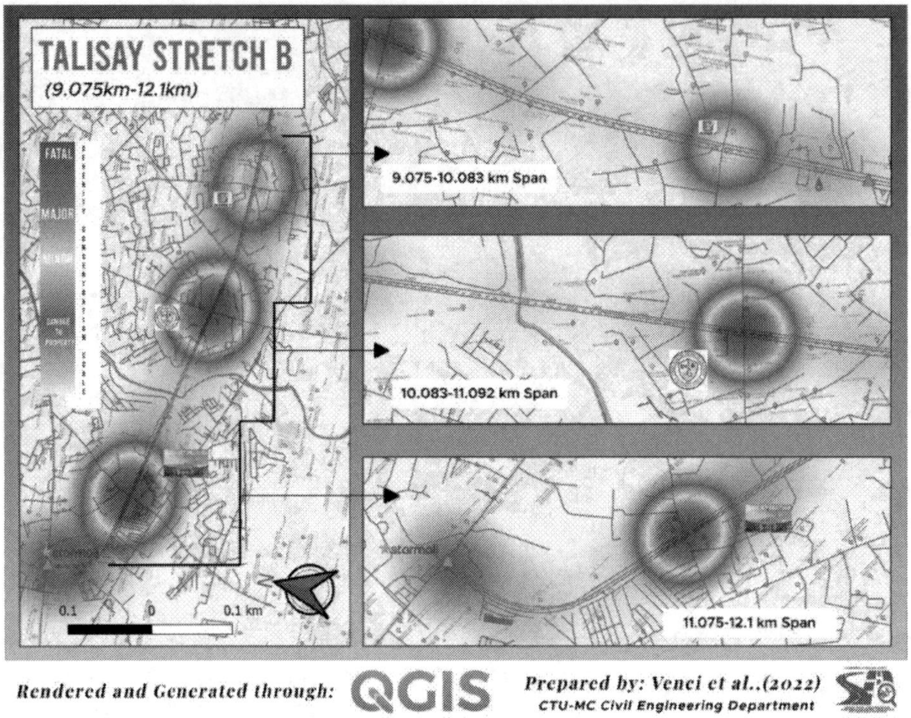

Figure 9 – Black spot through Heat mapping of Talisay Stretch B

3.2. Accident Rates Related to Traffic Volume (AADT)

As shown in *Figure 10,* the accident rate decreases gradually with an increase in the traffic volume (AADT). For 2017, the accident rate and AADT were as follows; 39.59 and 50957. Also, for 2018, the accident rate and AADT were 45.87 and 52889, respectively. Lastly, for 2019, the accident rate and AADT were 13.14 and 59035, accordingly. The factor contributing to it is the influence of traffic volume on the accident's severity. When traffic volume decreases, it results in major and fatal severity of road traffic accidents. Furthermore, the increase in traffic volume may result in minor and property-damaged road traffic accidents. However, this statement disaffirms the data's behavior in 2019, where the traffic volume increased yet the severity ratio was high, indicating fatal and major road traffic accident frequencies. This behavior may be affected by factors other than Traffic Volume (AADT) only, such as the design and operational features of the CSCR N840 Highway.

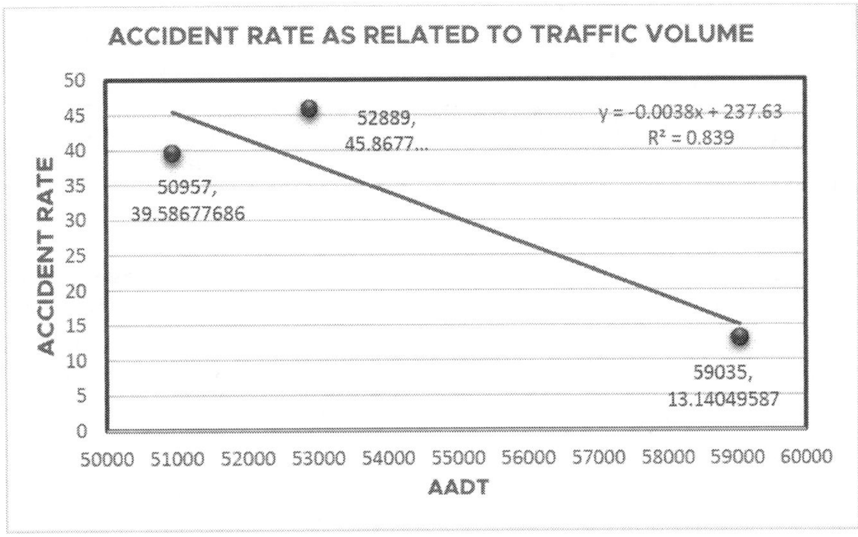

Figure 10 – Accident rate as per Traffic Volume

The researchers developed a regression model concerning the road traffic accidents, as indicated in *Equation 7*.

$$\text{Accidents Rate (Z)} = -0.0038(\text{AADT}) + 237.63 \qquad \textit{(Equation 7)}$$

The coefficient of determination, R^2, was derived from the regression statistic. The value R^2 was found to be 0.839, which indicates that the regression line explains 83.90% of the variance in the number of accidents. This also creates a significant and positive relationship between accident rate and AADT. Refer to the study's *Appendix Figure A8* for detailed information regarding the data.

3.3. Accidents Related to Accident Probability

Figure 11 demonstrates that as the accident probability increases, the accident rate may either increase or decrease due to the implemented countermeasures. For 2017, the accident probability and average accident rate were as follows; 4.60E-09 and 39.587. Also, for 2018 the accident probability and average accident rate were 5.13693E-08 and 22.934, respectively. Lastly, for 2019 the accident probability and accident rate were 3.8185E-07 and 4.380, accordingly. The data's behavior justified in 2019 reveals a high accident probability and a low accident rate. This indicates that the road interventions established by the local government may contribute to the low accident rate despite the high accident probability taken. Road interventions include road resurfacing by the DPWH for the roads along CSCR- N840 Highway.

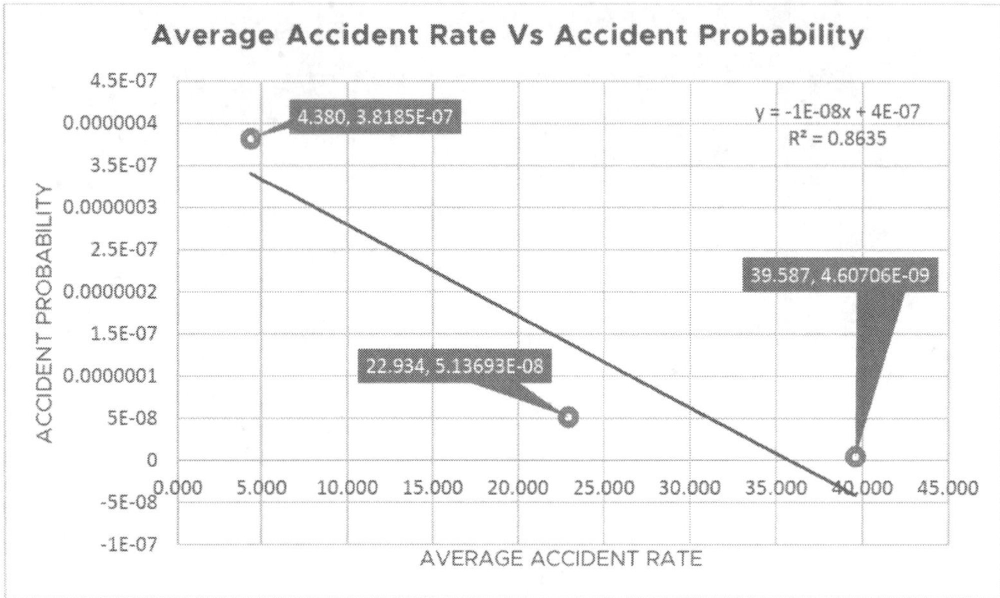

Figure 11 –Accident Rates as per Accident Probability

The coefficient of determination, R^2, was derived from the regression statistic. The value R^2 was found to be 0.8635, which indicates that the trend line explains 86.35% of the variance in the number of accidents. This also creates a significant positive relationship between accident rate and accident probability.

The researchers developed a Poisson distribution model, as indicated in Equation 8, influencing a behavior between the accident rate and accident probability.

$$P(r) = \frac{e^{-22.3}(22.3^{r})}{r!} \qquad \textit{(Equation 8)}$$

Figure 12 reveals the predictive output of the established Poisson distribution model. The plot shows a normally distributed trend in which the model can anticipate a maximum accident probability of 0.845 of 22nd year beyond the years established by the researchers. For example, the year 2017-2019, which follows 2019 by one year, serves as the starting - point for the model's predictive output. With the aid of this model, local government agencies can predict future countermeasures and road interventions for road traffic accident reduction. For instance, the Department of Public Works and Highways (DPWH) intervenes by resurfacing the road of the whole road section of CSCR N840 Highway.

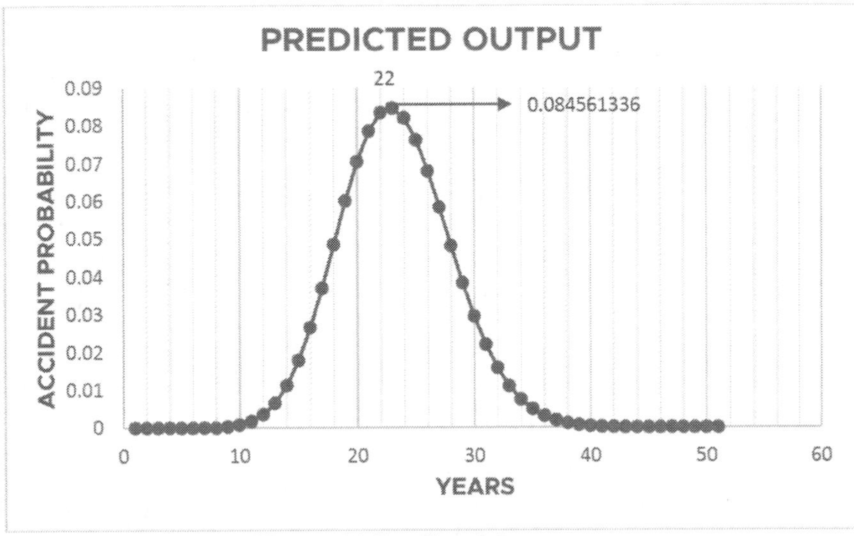

Figure 12-Predicted Output.

4. Conclusions

Road traffic accidents are a significant public safety concern, and a number of studies have invested into analyzing and forecasting these unpredicted occurrences. In this study, black spots were identified through an accident point weightage system and an accident prediction model by utilizing regression and Poisson distribution methods.

1. The study revealed that the year 2018 has the highest number of accident cases with 555 (46.52%) and the lowest severity ratio of 0.41. This is because most accidents are less severe, signifying a higher frequency of minor and property-damage cases. In contrast, 2019 has the lowest frequency of accident cases with 159 (13.33%). However, it shows the highest severity ratio of 0.72 because accident cases in 2019 have a lower frequency of minor and property-damage cases.

2. The study revealed the N840 highway's top five black spot locations of the four established stretches: CEBU Stretch A, CEBU Stretch B, TALISAY Stretch A, and TALISAY Stretch B.

From 0 km to 3.025 km along the CEBU Stretch A section, the top five black spot locations were at 2.6 km, 1.2 km, 2.3 km, 0.9 km, and 2.2 km, in rank order. These findings comprised 2.6 km with an APW of 18.4, 1.2 km with an APW of 15.0, 2.3 km with an APW of 9.4, 0.9 km with an APW of 8.6, and 2.2 km with an APW of 8.4.

Then, for the CEBU Stretch B from 3.025 km to 6.05 km distance of the highway, the top five black spots recorded were 5.3 km, 4.6 km, 4.7 km, 3.7 km, and 5.9 km, in sequential order. Five black spot locations were at 5.3 km with an APW value of 18.6, at 4.6 km having an APW value of 17.8, at 4.7 km with an APW value of 17.2, at 3.7 km with an APW value of 15.4, and at 5.9 km with an APW value of 14.2.

Also, for the TALISAY Stretch A from 6.05 km to 9.075 km distance of the highway, the top five black spots recorded were 7.6 km, 7.7 km and 8.2 km, 8.5 km, and 6.5 km, in sequential order. Five black spot locations were at 7.6 km with an APW value of 103.8, at 7.7 km with an APW value of 43.2, 8.2 km both having an APW value of 40.8, at 8.5 km with an APW value of 40.0, and at 6.5 km with an APW value of 18.4.

Lastly, for the TALISAY Stretch B from 9.075 km to 12.01 km distance of the highway, the top five black spots recorded were 10.2 km, 11.4 km, 9.4 km, 12.1 km, and 12.0 km, in sequential order. Five black spot locations were at 10.2 km with an APW value of 106.4, at 11.4 km having an APW value of 101.4, at 9.4 km with an APW value of 59.4, at 12.1 km with an APW value of 23.4, and at 12.0 km with an APW value of 12.8.

Most black spot locations from the Cebu and Talisay stretches prevailed at high concentrations along the area, revealing that curvatures, intersections, and junctions contribute significantly to road traffic accidents. Utilizing QGIS, on the Cebu side (Cebu Stretch A and Cebu Stretch B), a total of 4 curvatures, 8 junctions, and 2 U-turn slots were recognized; on the Talisay side (Talisay Stretch A and Talisay Stretch B), a total of 13 junctions, 7 intersections, and 6 curvatures were identified. (Refer to Figures 6, 7, 8, & 9 for detailed visual representation regarding the road features mentioned above). Severe road traffic accidents usually occur on the Talisay side of CSCR N840 Highway. Thus, the top black spot locations were at the Talisay side of CSCR N840 Highway.

3. In establishing the accident prediction model, based on the regression model results, the accident rate decreases gradually with an increase in the AADT, indicating that the factor contributing to it is the influence of traffic volume on the accident's severity. Furthermore, the data's behavior may be due to the design and operational features of the highway. On the other hand, based on the prediction model in Poisson distribution results, as the accident probability increases, the accident rate may either increase or slowly decrease due to the countermeasures implemented by the local government. Their coefficient of determination assessed the predictive capacity of the models. From the regression statistic, R^2 was calculated as the coefficient of determination. The value R^2 for the regression model was 0.839, indicating that the regression line explains 83.90% of the variance. The researchers developed regression model concerning the road traffic accidents is expressed as:

$$Accidents\ Rate\ (Z) = -0.004(AADT) + 237.631$$

For the Poisson distribution model, the value R^2 was 0.8635, indicating that the trend line explains 86.35% of the variance in the number of accidents. The researchers developed Poisson distribution model concerning to the road traffic accidents is expressed as:

$$P(r)=(e^{\wedge}(-22.3)\ (\ [22.3]\ ^{\wedge}r))/r!$$

5. References

[1] Caliendo, C., Guida, M. and Parisi, A. (2007). *A Crash-Prediction Model for Multilane Roads. Accident Analysis and Prevention*, 39, 657-670

[2] Zha, L., Lord, D., Zou, Y. (2014*). The Poisson Inverse Gaussian (PIG) Generalized Linear Regression Model for Analyzing Motor Vehicle Crash Data. Journal of Transportation Safety & Security*, (), 00–00. doi:10.1080/19439962.2014.977502

[3] Ma, L., Quiao, W., Yan, X. (2014). *A Quasi-Poisson Approach on Modeling Accident Hazard Index for Urban Road Segments*. Discrete Dynamics in Nature and Society, 1-8.

[4] Ünlü, H. K., Young, D. S., Yiğiter, A., & Hilal, Ö. L. (2020). *A Mixture Model with Poisson and Zero-truncated Poisson Components to Analyze Road Traffic Accidents in Turkey.* Journal of Applied Statistics, (), 1–15. doi:10.1080/02664763.2020.1843610

[5] Ab. Ghani, A. R., Sanik, M. E., & Mohd Mokhtar, R. A. (2019). *Comparison of Accident Prediction Model between ANN and MLR Models.*

[6] Geurts, K., Wets, G., Brijs, T., & Vanhoof, K. (n.d.). *Identification and Ranking of Black Spots.*

[7] Rahim, S. A. S. M., Marjan, J. M., & Voon, W. S. (2013). *Guideline on Accident- Prone Area Identification for Automated Enforcement System (AES)*. Malaysian Institute of Road Safety Research (MIROS). https://doi.org/ 10.4236/jtts.2017.72015

[8] Singh, R. K., & Suman, S. K. (2012). *Accident Analysis and Prediction of Model on National Highways.* International Journal of Advanced Technology in Civil Engineering.
[9] Omar, N., Prasetijo, J., Daniel, B. D., & Abdullah, M. A. E. (2017). *Accident Analysis and Highway Safety.* EDP Sciences. https://doi.org/10.1051/matecconf/201710308002
[10] Sapuan, M. S., Razali, A. M., Zamzuri, Z. H., Ibrahim, K. (2016). *THE 2016 UKM FST POSTGRADUATE COLLOQUIUM: Proc. of the Universiti Kebangsaan Malaysia Simulation on Poisson and Negative Binomial Models of Count Road Accident Modeling,* 1784(), (Faculty of Science and Technology 2016 Postgraduate Colloquium - Selangor, Malaysia)
[11] Department of Public Works and Highways (2004). *Accident Blackspot Investigation Handbook Project*
[12] Zhang, J., Wang, J., & Fang, S. (2019). *IEEE 2019 5th Int. Conf. on Transportation Information and Safety (ICTIS) - Prediction of Urban Expressway Total Traffic Accident Duration Based on Multiple Linear Regression and Artificial Neural Network,* (), 503–510

IOP Conf. Series: Earth and Environmental Science 1184 (2023) 012023 doi:10.1088/1755-1315/1184/1/012023

Integrating Fine Ceramic Tile Aggregates in Concrete Hollow Blocks

J Ramirez[1], C Remolisan, MJ Coscos and HA Canseco-Tuñacao

[1]Cebu Technological University-Main Campus, Cebu, Philippines

Email: juvelyn.ramirez@ctu.edu.ph, cristine.remolisan@ctu.edu.ph, markjason.coscos@ctu.edu.ph, hanaastrid.tunacao@ctu.edu.ph

Abstract. This study investigates the effect of fine ceramic tile aggregates (CTA), specifically floor and wall tiles, as a replacement for fine aggregates in non-load bearing Concrete Hollow Block's (CHB) fresh and hardened properties. The experimental design of this study performed was in line with the American Society of Testing and Materials (ASTM) standard procedure. Results for the workability of CHB with CTA showed comparable behavior to the control mix, which ranged between low to medium slump values (49 mm-59 mm). Specimens with CTA are characterized by higher water absorption and lesser density and moisture content. All the CHB with CTA replacement yielded higher compressive strength than the control mix. The 30% CTA substitution gained the highest compressive strength of 3.20 MPa (462.12 psi) for 28-days of curing. Overall, the physical and mechanical qualities of the samples with CTA percentage performed better than the control mix.

Keywords: Ceramic tile, aggregate substitute, CTA, CHB, ASTM, compressive strength

1. Introduction

Ceramic is a general term used to refer to ceramic products (Subedi, Wagle, & Basnet, 2020). Most buildings utilize fundamental construction materials, most of which are ceramic items. Most commonly manufactured ceramics include wall tiles, floor tiles, sanitary ware, household and technical ceramics. The development of the Philippines ceramic tiles industry is directly related to the growth of the construction industry. With the increase of residential, commercial, and government projects throughout the country, the demand for ceramic tiles as flooring and walls during construction also boosted. In fact, the construction of residential buildings in 2019 grew by 10%, amounting to 69.2% of the total number (PSA, 2021). The growing improvement of residential infrastructures in the country drives the demand for ceramic tiles. The Figure 1 showed the Philippine ceramic tile subsectors by application.

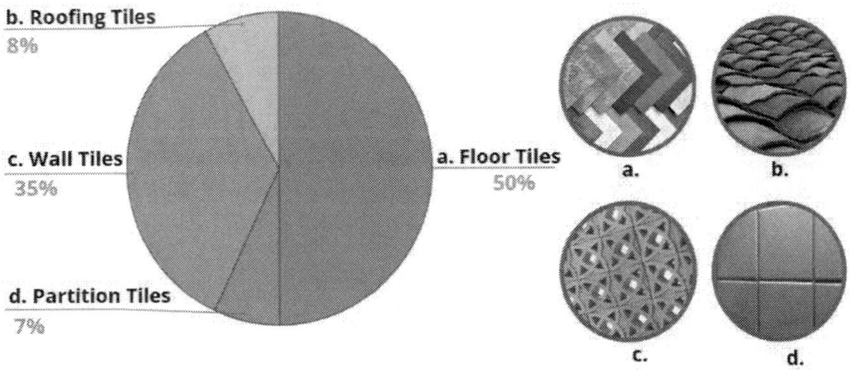

Figure 1 – Philippine Ceramic Tiles Market Share by Application (2018)

However, despite the enhancing benefits of ceramics, the increase in the accumulated quantity of ceramic waste has become one of the industrial sector's major environmental issues. Ceramic waste is one of the most significant contributors to the Construction and Demolition Waste (CDW) under the stony fraction with about 54% (Raval et al., 2013; El-Dieb et al., 2018; Juan et al., 2010). It can be generated during dressing, polishing, and manufacturing of ceramics. Moreover, in the study by Senthamarai & Devadas-Moharan (2005), about 30% of the daily manufacturing volume in the ceramic industry goes to waste.

Ceramic waste can be categorized into two types in accordance with the source of its raw materials in line with the report of Pacheco-Torgal & Jalali (2010). The first category is formed through generated fired ceramic by structural ceramic factories that utilized only red pastes for a product such as brick, blocks, and roof tiles manufacture. Meanwhile, the second is composed of fired ceramic wastes, which are produced in stoneware ceramic (wall, floor tiles, and sanitary ware). In each category, the fired ceramic waste was classified according to the production process differentiating them by using red or white ceramic pastes. However, the red paste is more frequent and much higher in volume.

Therefore, this authenticates the need to explore innovative ways of re-using ceramic waste. Aggregates constitute about 70%-80% of the total constituents in concrete production. This huge demand for sand is one of the reasons for sand mining in the country, which is the second most-consumed natural resource (Beiser, 2019; Peduzzi, 2014). The exploitation of huge quantities of natural resources always brings significant environmental impacts (Sonak et al., 2006; Kondolf, 1994). Among these adverse ecological effects are loss of habitat, loss of floodplain (Koehnken & Rintoul, 2018), the threat of water supply and dropping of groundwater table (Villioth, 2014), and damage to ecosystems (Torres et al., 2021).

Numerous researchers, such as Senthamarai & Devadas-Manoharan (2005) and Suzuki et al. (2009), identified that ceramics have the potential to replace natural aggregates. Furthermore, Cristiano (2014) confirmed that ceramic waste could be used safely in the production of concrete due to its favorable properties. It is hard and highly resistant to biological, chemical, and physical degradation. Some important works have already been carried out with this type of waste in construction materials (Binici, 2007; Braga, Brito, & Veiga, 2012; El-Dieb et al., 2018; Subedi, Wagle, & Basnet, 2020; Torkittikul & Chaipanich, 2010). In the study of Awoyera et al. (2016), the influence of ceramic tile wastes on the structural properties of concrete that was made using laterite was recently investigated. It was reported that ceramic-based lateralized concrete performed considerably well when compared to conventional concrete. Additionally, Ravindra et al. (2015) concluded their study that ceramic tile waste can be effectively used as a natural fine aggregate with 10% replacement as recommended.

Ceramic waste utilization can solve aggregate shortages in various construction sites (Awoyera et al., 2016). However, according to Medina et al. (2011), ceramic waste as an aggregate in manufacturing concrete has not been researched as thoroughly as other construction and demolition waste (CDW).

2. Research Significance

The study of replacing aggregates with CDW was a trend for the past years. Ceramic waste under the CDW was utilized as in concrete, asphalt concrete and solid concrete blocks. It can lessen environmental issues associated to aggregate mining and waste disposal. However, there was still no study of ceramic tile aggregate (CTA), especially wall and floor tiles, as a partial substitute for sand in non-load bearing concrete hollow blocks (CHB). This means that using (CTA) in CHB needs further research. Hence, the study investigates the fresh and hardened properties of CHB partially substituted with CTA.

3. Experimental Investigation

The experimental design of this study performed was in line with the American Society of Testing and Materials (ASTM) standard procedure. The research experiment was conducted by following Figure 2. First was the preparation of the materials to be used. The ceramic tile wastes were manually hammered into fine aggregates and sieved CTA underwent saturated surface-dry (SSD) condition. It was soaked for 24 hours prior to proportioning and mixing.

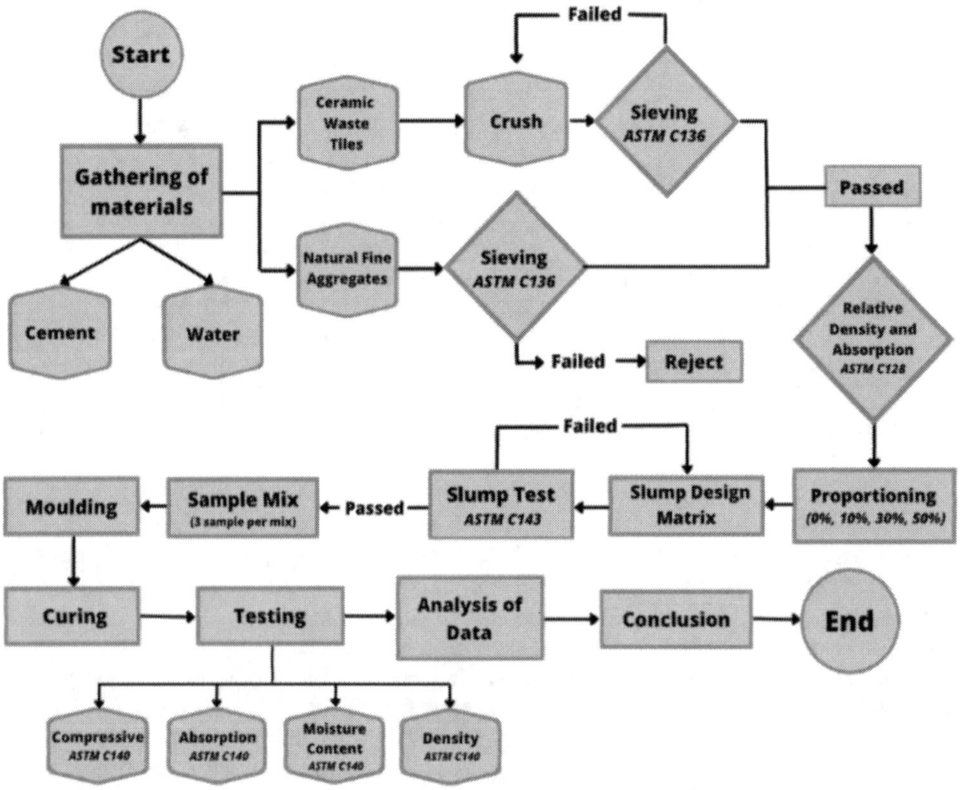

Figure 2 – Methodology Flow

A ratio of 1:7 cement-aggregates was adopted for the fabrication of CHB. The batches applied by weight were made according to percent replacement (0%, 10%, 30%, and 50%) of CTA. Mix proportion ratios by weight summarized in Table 1. A water to cement ratio of 0.5 was the basis in the study from the result of the slump design matrix. However, the 0.5 w/c ratio in the mix was adjusted by deducting the moisture content present in the NFA. Hence, a varying w/c ratio is reflected in Table 1.

Table 1 – Design Mix

Mix Proportion by Weight	Materials (kg)			
	Water	Cement	NFA	CTA
0%	5.25	19.29	135.00	-
10%	5.69	19.29	121.50	13.50
30%	6.57	19.29	94.50	40.50
50%	7.45	19.29	67.50	67.50
Total	24.97	77.16	418.50	121.50

*NOTE: *CTA passing Sieve No. 4; **for 60 CHBs*

Meanwhile, a slump design matrix was used to determine if the mix design could pass the slump test. A slump test was conducted under ASTM C143 to investigate the workability of the mixes. The mixing of the cast-in-place CHB was done manually with the use of a shovel on a clean base. First, the measured quantity of aggregate was spread on the surface then cement was added on top. Then, the dry components were mixed thoroughly for about 2 to 3 minutes with the shovels until the mixture attained a homogeneous color. A ditch was made at the center of the dry mixture. Finally, the measured quantity of water was added to the center and mixed for 5 minutes until uniform color and consistency of concrete was obtained. The CHB were then formed and placed in an undisturbed area for at least 48 hours. The samples were then cured by fully submerging in water at room temperature until the test age (7, 14, and 28 days).

Consequently, in line with ASTM C140, the hardened properties of the CHB are evaluated. These specimens are those trial mix designs that qualified for the fresh property (slump) specification. The samples were tested for compressive strength at Department of Public Works and Highways (DPWH) Region VII testing laboratory and absorption, moisture content, and specific gravity at Mega Testing Center Inc.

3.1. Materials

The materials utilized in this study were water, Ordinary Portland Cement (OPC), NFA, and CTA. The cement used for the experiment was an Ordinary Portland Cement (OPC) type 1 that serves as the binding material for the CHB according to ASTM 150. Both the CTW and NFA employed in this study were bought from local ceramic shops in Cebu. The waste tiles were manually crushed into fine aggregates with varying particle sizes less than or equal to the fraction of 4.75 mm (Sieve No.4) based on ASTM C136. As shown in Figure 3a, the samples that exceed the fraction of 4.75 mm underwent another hammer crushing to ensure that it would not exceed the maximum sample size. Meanwhile, the laboratory tests the aggregates include sieve analysis, water absorption, and specific gravity and are summarized in Table 2.

Figure 3 – Material and Specimens

a. Crushed Ceramic Tile Aggregates b. Concrete Hollow Block

Table 2 – Aggregate Properties

Properties	Natural Fine Aggregates (NFA)	Ceramic Tile Aggregates (CTA)	ASTM Standards
Specific Gravity	2.71	2.55	-
Water Absorption	6.16%	8.46%	-
Fineness Modulus	2.82	3.38	2.3 - 3.1
Moisture Content	3.58%	13.21%	-

3.2. Specimens
A total of 48 non-load bearing CHB specimens with a length of 0.4 m (16 in), thickness of 0.1 m (4 in), and height of 0.2 m (8 in) were fabricated. Figure 4 showed the guide geometry of the CHB.

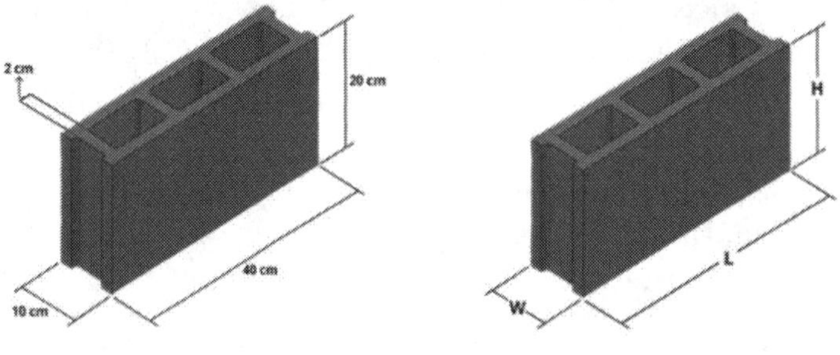

Figure 4 – CHB Geometry

3.3. Item of Investigation
The CHB specimens were air dried for at least 24 hours after it was removed from the curing pond after the curing periods of 7, 14, and 28 days. The actual dimensions (length, width, and height) were measured and recorded before testing. The average compressive strength of 3 specimens were recorded and evaluated based on the minimum compressive strength under the ASTM specification. Moreover, one specimen was used for the determination of absorption, moisture content, and specific gravity.

4. Analytical Investigation
ASTM C129 covers the standards for hollow and solid non-load bearing masonry. Masonry units are required to have a minimum net area compressive strength of 3.45 MPa (500 psi) for an individual unit, and three masonry units are required to have a minimum net area compressive strength of 4.14 MPa (600 psi). Compressive strength value can be obtained from equation 1.

IOP Conf. Series: Earth and Environmental Science 1184 (2023) 012023 doi:10.1088/1755-1315/1184/1/012023

$$\text{Compressive Strength} = \frac{P_{max}}{A_n} \text{ (MPa)} \tag{1}$$

The results were analyzed using one-way Analysis of Variance (ANOVA) to determine if the CTA replacements have a significant effect on the compressive strength of CHB. The significance level was specified to 0.05 and with a confidence level of 0.95 for this work.

5. Experimental Results And Discussion

5.1. Sieve Analysis

The sieve analysis performed in the experiment was in line with ASTM C136. Both natural and ceramic tile aggregates passed the particle size range of 0.075 mm (No. 200) to 4.75 mm (No. 4). Figure 5 displayed the difference in gradation patterns. The natural aggregates formed a continuous granulometric curve with a good gradation pattern. While for CTA, the analysis showed a range between 0.355 mm (No. 50) to 0.6 mm (No. 30) in particle size diameter with a very good gradation pattern. The nominal maximum particle size diameter was 4.75 mm (No. 4) falling within the ASTM acceptable range. The same pattern of gradation for CTA fine aggregate is observed in the study of Adekunle et al. (2017).

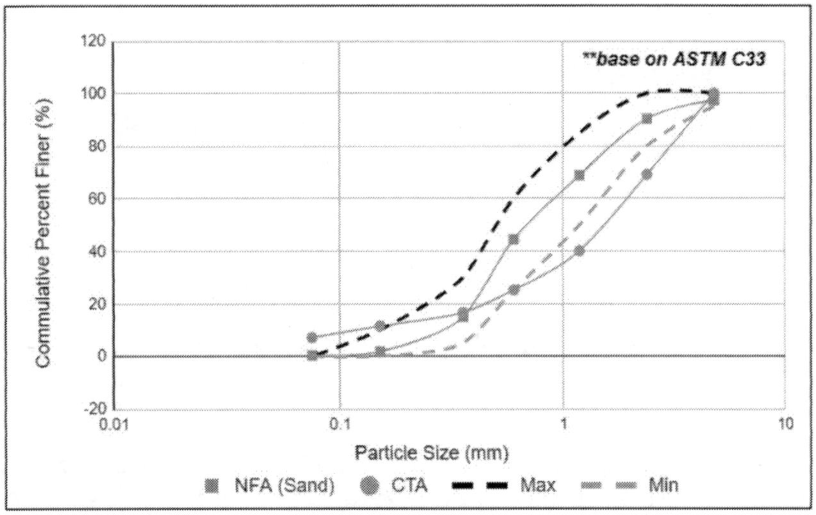

Figure 5 – Aggregate Particle Size Distribution

By performing sieve analysis with standard sieve, the fineness modulus was computed. OBE Ravindra et al. (2018) stated in his publication the significance of FM, that a very coarse or very fine sand will result in poor concrete mixes. He further stated that at FM 3.0, the best workability and compressive strength of concrete can be attained. The result of the FM in the aggregate is in Table 2 with CTA and NFA having 3.38 and 2.8, respectively. Based on this, NFA is finer in average than CTA. The higher the value of FM, the coarser the aggregate. Generally, a lower FM results in more paste, making the concrete more workable. This behavior can be seen in the slump test result in Figure 6 in which the control mix garnered the highest workability.

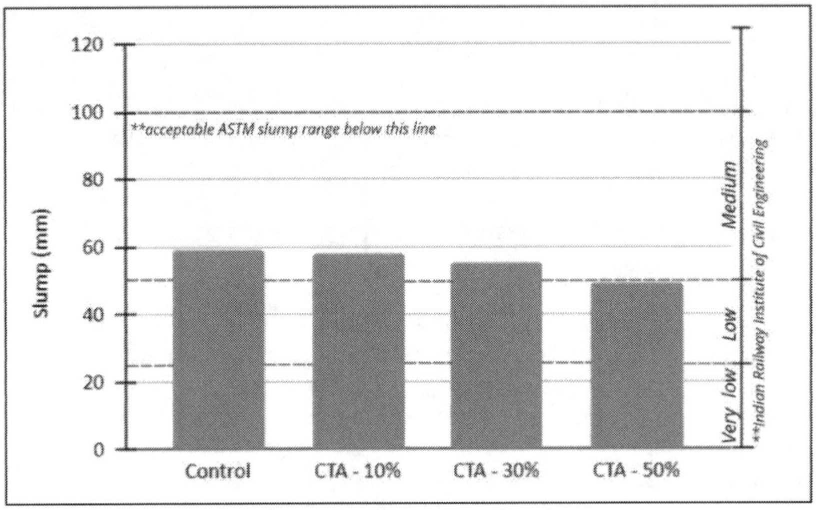

Figure 6 – Slump Test Result

5.2. Fresh Property Of Chb

5.2.1. Slump The workability of each fresh mix with constant w/c of 0.5 was measured through slump tests according to ASTM C143 and is presented in Figure 6. The slump of the mixes decreased as the amount of CTA replacement increased. The results indicate low to medium utility for the mixes, with control mix having the highest slump of 59 mm (2.36 in) and the CTA 50% mix with a slump of 49 mm (1.96 in). The outcomes showed that all the mixes passed the target slump of 100 mm (4 in). This slump values could be affected by the glazy surface of the ceramics (Awoyera et al., 2015) increasing its workability however, the angular shaped CTA with its sharp edges resulted to better bonding with other material resulting to low slump (Mashitah et al., 2008; Daniyal & Ahmad, 2015). The workability of the mixes in the experiment resulted to decreasing slump values which can be attributed to the governing properties of CTA which is its angular and sharp particles based on the results in Figure 6.

5.3. Hardened Properties of CHB

5.3.1. Water Absorption The water absorption results in Figure 7 showed similar trends for the 7, 14, and 28-day curing periods. The absorption values of CHB increased as the amount of CTA replacement increased. The high absorption values of CHB with CTA replacement were related to the high absorption of the aggregate. The CTA has absorption value of 8.46% while the NFA has 6.16% as reflected in Table 2. Similar results were found in the study of Falek et al. (2017) and Ravindra et al. (2015) where recycled aggregates are characterized by a strong ability to absorb water. After 60 minutes, the RA has absorption value of up to 90% higher than NA and porosity rate of 17% (Falek et al., 2017). Additionally, CTA has absorption of up to 17% higher than NFA in the study of Ravindra et al. (2015).

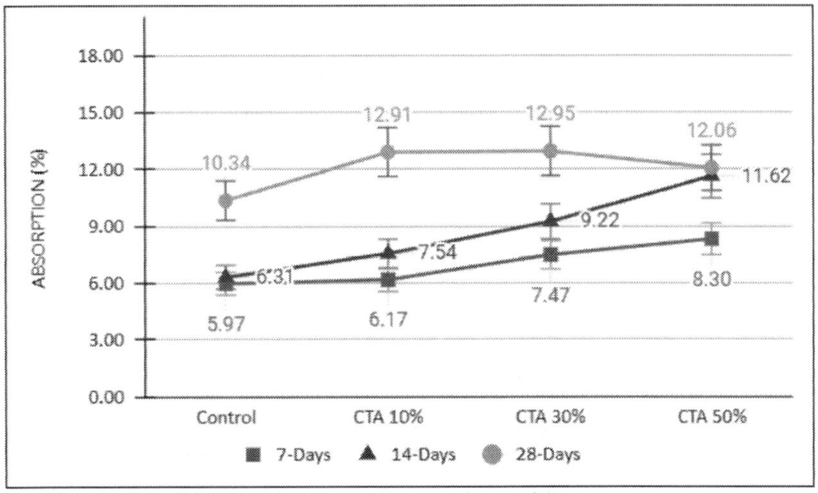

Figure 7 – Water Absorption

In the study of Medina et al. (2013) regarding the properties of recycled ceramic aggregates on concrete, it was observed that as the CTA replacement increases, the total porosity and absorption values also increases. In their research, the absorption value of CTA is 2.4 times higher than the NFA due to the high porosity property of CTA (28.13% higher than NFA). Moreover, there is higher presence of micro-pores on the surface of the CTA than NFA which leads to an increase in the water absorption level of the specimen with replacement (Anufrik et al., 2016). The rising values of absorption in CHB with replacement in Figure 7 can be linked to the high porosity and micro-pores of CTA.

Furthermore, the high absorption of CTA can be associated with the slump loss of the slump loss of the fresh concrete. Also, NFA was round and smooth while CTA was angular and elongated which makes it more workable. According to Awoyera et al. (2018), since CTA has rough surface it has better bonding characteristics than NFA, but it requires more cement paste to be more workable. In figure 6, it is displayed that as the CTA increases, the slump decreases due to its high absorption, shape and texture of the aggregate.

5.3.2. Specific Gravity/Relative Density The specific gravity of the CHB units in all curing periods declined as the CTA replacement inclined. The decrease in specific gravity can be explained by Law of Mixtures. According to Dolores et al. (2017), the properties and composites are influenced by various factors, one of which is the proportions of the matrix and reinforcement making the composite. When the Law of Mixtures is applied to this research, the addition of the CTA, which is lighter than NFA by 5.9% shown in Table 2, naturally made the resulting hollow blocks lighter. This result also conformed to the findings of previous study (Ravindra et al., 2015), albeit the CTA is used as partial replacement for concrete blocks and not CHB.

Additionally, the comparison for the actual versus the theoretical values of specific gravity is tabulated in Table 3. It was found that there is a discrepancy in the values this can be due to varying moisture content in the specimen. The decrease in the density of the blocks would correspond to a reduced dead load if CHB was used in a structure. The values for the specific gravity were summarized in Figure 8.

Table 3 – Specific Gravity: Actual vs. Theoretical

IOP Conf. Series: Earth and Environmental Science 1184 (2023) 012023 doi:10.1088/1755-1315/1184/1/012023

Sample Mix	7- Days Curing		14- Days Curing		28- Days Curing	
	Actual	**Theoretical**	**Actual**	**Theoretical**	**Actual**	**Theoretical**
Control	2.20	2.20	2.10	2.10	2.13	2.13
CTA 10%	2.19	2.19	2.08	2.09	2.09	2.12
CTA 30%	2.11	2.16	2.04	2.06	2.08	2.09
CTA 50%	2.09	2.14	2.01	2.04	2.02	2.07

** *considering moisture content obtained for control mix*

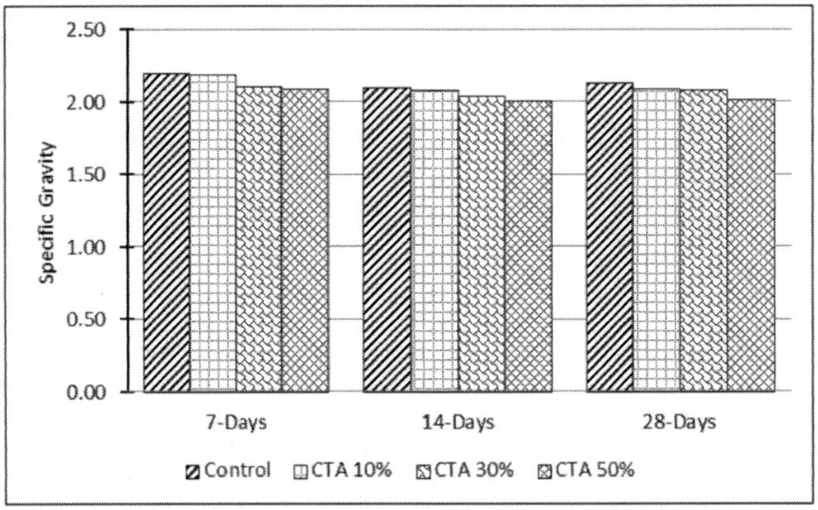

Figure 8 – Specific Gravity/Relative Density

Masonry units are classified to three according to weight, lightweight (< 1680 kg/m3 [105 lb/ft3]), medium weight (1680 kg/m3 [105 lb/ft3] < ≤ 2000 kg/m3 [125 lb/ft3]), and normal weight (> 2000 kg/m3 [125 lb/ft3]). The density of the specimens presented in Table 5 ranges from 1800 kg/m3 (112.5 lb/ft3) to 1976 kg/m3 (123.5 lb/ft3). Following the ASTM specifications, the blocks were classified as medium weight.

Table 4 – Density of CBH Specimens (kg/m^3)

Curing Age	7 Days	14 Days	28 Days
Control	1952	1976	1928
CTA 10%	1891	1934	1847
CTA 30%	1853	1864	1842
CTA 50	1835	1800	1801

** *based on 1 specimen*

5.3.3. Moisture Content The moisture content of the samples reduced as the CTA replacement increased as displayed in Figure 9. The CTA underwent SSD condition and its moisture content is recorded as 13.21% while NFA is in in-situ condition with moisture content of 3.58% shown in Table 2. According to the result of Zaharieva et al. (2003), recycled aggregates have high absorption capacity and moisture, however the water content in their specimen with replacement released the moisture quickly than the control after air drying. Figure 10 showed the state of the specimen after 48 hours of air drying before testing.

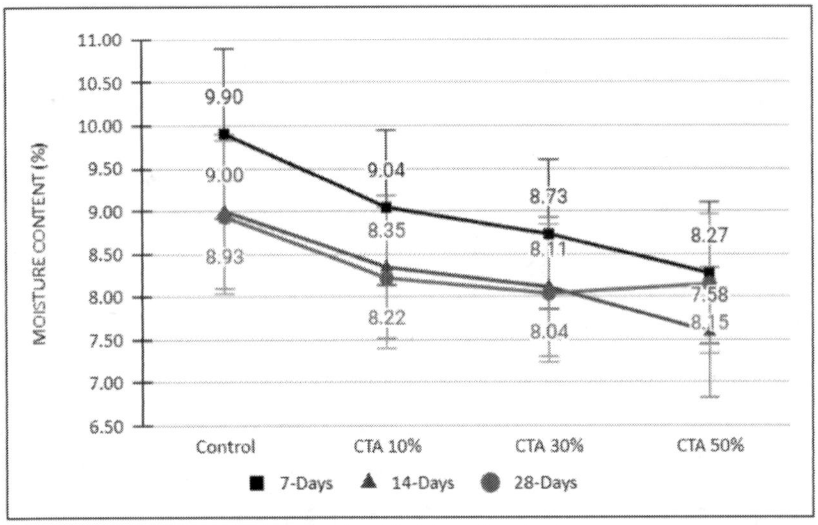

Figure 9 – Moisture Content

Figure 10 –State of the CHB's after 48 Hours of Air Dry

5.3.4. Compressive Strength The compressive strengths of CHB at curing ages 7, 14, and 28 were showed in Figure 11. It revealed that specimens with CTA replacements yield higher compressive strengths than the control mix. Also, the increase in strength was also observed as the curing age increased. In study of Ravindra et al., (2015) and Awoyera et al., (2015), concrete blocks with CTA have higher strength compared to the control. CTA 30% specimens have notably the highest compressive strength throughout the testing. It yielded 2.33 MPa (337.94 psi), 2.51 MPa (364.05 psi), and 3.20 MPa (464.12 psi) for the 7, 14 and 28-day curing respectively. An average gain of 64.80% compressive strength is observed in 30% replacement higher than the control mix. According to Silva et al. (2014), various studies were conducted on Recycled Masonry Aggregates (RMA) specifically ceramic bricks, roofing and wall tiles with substitution rates from 10-100%. It showed that the mechanical strength compressive or tensile strength of the specimens were acceptable until a replacement rate of 40%.

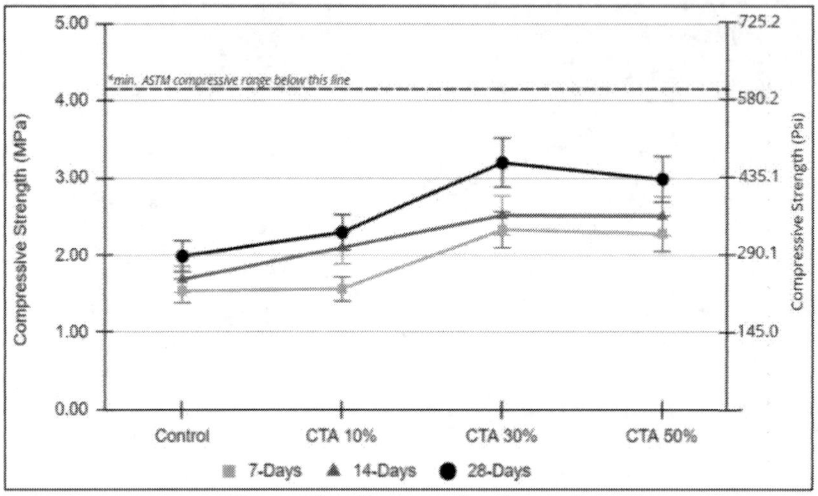

Figure 11 – Compressive Strength

Kallak et al. (2009) further explained that the decrease on compressive strength as the replacement exceed 40% is because of the following; (1) the increasing amount of CTA failed to develop proper/adequate bond with concrete and cement matrix due to the increased in glazed particles, (2) due to the high porosity of the crushed waste ceramic aggregates, it requires more cement paste to be in a workable mixture which can affect the strength of the sample due to the weaker bonding of the materials. Additionally, Liu et al. (2015) worked on concrete with ceramic waste and sand as fine aggregates and published that ceramic waste as an effective replacement of sand and could be replaced by 26.9–47.6%.

Based on the result of the experiment, beyond the 30% replacement of CTA the compressive strength started to decline due to increase of CTA. The increase in the amount of CTA resulted to the increase of glazed particles in the specimen. The glaze surface of CTA leads to a lower cohesion and engagement between the crushed ceramic tiles and cement paste affecting the bonding of the materials (Wadie et al., 2017; Daniyal & Ahmad, 2015). The results of the study in Figure 11 showed that the compressive strength at CHB with CTA 50% started to decline, it can be connected to the increasing amount of glazed particles in the specimen affecting the bonding of materials.

Additionally, Table 2 showed that the specific gravity of the CTA is lower than the NFA which is 2.55 and 2.71, respectively. The low specific gravity of the CTA contributed to less dense specimens with replacements. However, the intrinsic properties of the aggregates do not generally dictate the performance of the specimens (Burland, 1990; Kim et al., 2018). In fabricating masonry units, the performance (compressive strength) is merely related to the bonding capacity between the aggregates and cement paste (Corinaldesi, 2009).It was observed that CTA 30% was the optimum percent replacement which yielded the highest compressive strength showed in Figure 11, though it has lower workability than the control. This is due to the water added using the 0.5 w/c ratio is enough to lubricate the particles to aid the compaction of the mix, replacing the water in the voids with mortar and aggregates.

Meanwhile, at CTA 50% replacement the compressive strength started to decrease as displayed in Figure 11. Since CTA has a high water absorption (Ravindra et al., 2015) the water added in the mix was not enough to cover the voids due to the high CTA amount, thus the water was insufficient to hydrate the cement and develop. Also, this made the fresh mix of CTA 50% replacement less workable and harder to compact.

Using one-way ANOVA, there was a significant difference in the compressive strength of CHB in all specimens with CTA (14-days curing) and 30% CTA replacement (28-days curing). The results are presented in Table 4.

Table 5 – Analysis of Variance on Compressive Strength

Samples	P –Values		
	7-Day	**14-Day**	**28-Day**
Control vs. CTA 10%	0.952	0.003	0.338
Control vs. CTA 30%	0.075	0.001	0.021
Control vs. CTA 50%	0.053	0.011	0.153

*** significant if p-value < 0.05, otherwise not significant*

On the other hand, the CHB specimens tested did not pass the minimum required compressive strength of the ASTM standards of 3.45 MPa (500 psi) for an individual unit, and 4.14 MPa (600 psi) for an average of three units. Furthermore, it was observed that even though the tested specimens were dry on the surface, the inside of the specimens was still moist, which affected the compressive strength as showed in Figure 3b and Figures 8a, 8b and 8c. According to Shen & Xu (2019), as the moisture content in a sample increases, the compressive strength decreases. Moreover, it is observed that the failure initiated due to cracks development along the length and width of the CHB unit due to the smaller contact area, and crushing of CHB units at the bottom end since the inside is still wet. Figure 12 showed the failed CHB after the test.

Control CTA 10% CTA 30% CTA 50%

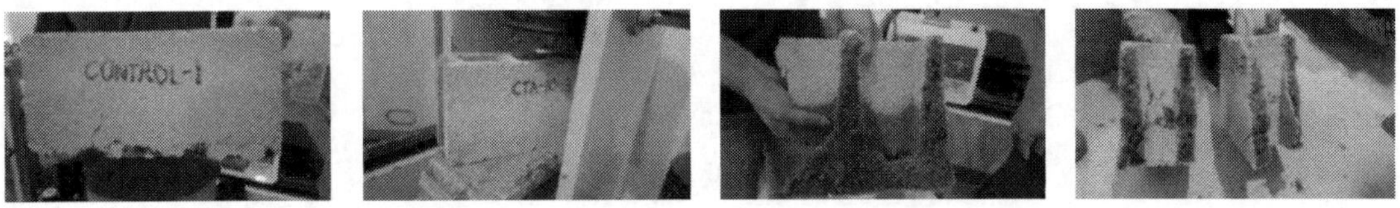

Crushing started at the bottom | Cracking started at the bottom | Cracks along the length | Cracks along the width

Figure 12 –Types of Failure after Testing

6. Further Research and Conclusion

Improving the quality of concrete masonry in the Philippines and finding new materials as a substitute for the natural aggregates needs effort and experimentation. This research will contribute new knowledge to engineers, enlightening the users on the potential uses of ceramic waste in a new light. However, this study has some gaps in the knowledge of research that follows the findings; hence the researchers have devised the following:

1. Use other curing methods. The future researcher can use other curing processes, such as continually spraying water on the specimen to determine if it affects the strength of samples compared to totally submerging it. However, further study and investigation are needed because of the lack of data on the amount of water needed to be sprayed in the curing of CHB.
2. Dry the samples thoroughly. It was advisable to air dry the specimen for more than 48 hours to ensure that it was completely dry before testing as it affects the strength of the CHB.
3. Utilize mechanical equipment. The use of mechanical equipment plays a vital role in the success of your study. A jaw crusher will significantly help the production of CTA as it takes a long time to crush it manually. Moreover, the researchers did manual cast-in place mixing of CHB. Hence, an automatic CHB machine is recommended to have consistent and uniform samples.
4. Apply alternative capping method. If there is no sulphur available for capping, gypsum cement can be utilized instead. Refer to ASTM C1552 for the procedure of using this material.

This paper was conducted to investigate the effects of utilizing CTA as a replacement for the NFA in manufacturing masonry concrete hollow blocks. Within the limited scope of the experiments carried out in this experimentation to determine the fresh and hardened properties of CHB, the following conclusions can be drawn:

1. The workability of CHB with CTA was comparable to the control mix, which ranged between low to medium workability. However, the CTA 50% mix was an exception due to its low workability of 49 mm (1.96 in) slump.

2. The CHB partially substituted with CTA yielded higher compressive strengths than the control mix (with 0 % replacement). The CTA 30% gained the highest value with an average of 64.80% higher than the control mix.

3. Using a CTA substitute, CHB can have higher water absorption and lesser density and moisture content. This will minimize the dead load of a building.

References

[1] Adekunle, A. A., Abimbola, K. R., & Familusi, A. O. (2017). Utilization of construction waste tiles as a replacement for fine aggregates in concrete. Engineering, Technology & Applied Science Research, 7(5), 1930-1933.

[2] Anufrik, S. S., Kurian, N. N., Zhukova, I. I., Znosko, K. F., & Belkov, M. V. (2016). Chemical composition of ceramic tile glazes. Journal of Applied Spectroscopy, 83(5), 764-770.

[3] ASTM C128. (2015). Standard Test Method for Relative Density (Specific Gravity) and Absorption of Fine Aggregate. ASTM International, West Conshohocken, 15, 1–6. https://doi.org/10.1520/C0128-15.

[4] ASTM C129. (2015). Standard Specification for Non Loadbearing Concrete Masonry Units, ASTM International, West Conshohocken, PA. Retrieved from: www.astm.org

[5] ASTM C136. (2017). Standard Test Method for Sieve Analysis of Fine and Coarse Aggregates. ASTM International, West Conshohocken, 16a, 1–5. https://doi.org/10.1520/C0090-16A.2

[6] ASTM C140. (2020). Standard Test Methods for Sampling and Testing Concrete Masonry Units and Related Units, ASTM International, West Conshohocken, PA. Retrieved from www.astm.org

[7] ASTM C143. (2020). Standard Test Method for Slump of Hydraulic-Cement Concrete, ASTM International, West Conshohocken, PA. Retrieved from: www.astm.org

[8] ASTM C150. (2020). Standard Specification for Portland Cement, ASTM International, West Conshohocken, PA. Retrieved from: www.astm.org

[9] ASTM C1552. (2016). Standard Practice for Capping Concrete Masonry Units, Related Units and Masonry Prisms for Compression Testing, ASTM International, West Conshohocken, PA. Retrieved from: www.astm.org

[10] Awoyera, P. O., Akinmusuru, J. O., Ndumbuki, J. M. (2016). Green concrete production with ceramic wastes and laterite. Construction and Building. Materials, Vol. 117, pp.29-36.https://doi.org/10.1016/j.conbuildmat.2016.04.108.

[11] Awoyera, P. O., Ndambuki, J. M., Akinmusuru, J. O., & Omole, D. O. (2018). Characterization of ceramic waste aggregate concrete. HBRC journal, 14(3), 282-287.

[12] Beiser, V. (2019, November 18). Why the world is running out of sand. Retrieved from BBC: https://www.bbc.com/future/article/20191108-why-the-world-is-running-out-of-sand

[13] Binici, H. (2007). Effect of crushed ceramic and basaltic pumice as fine aggregates on concrete mortars properties. Construction and Building Materials, 1191-1197.

[14] Braga, M., Brito, J., & Veiga, R. (2012). Incorporation of fine concrete aggregates in mortars. Construction and Building Materials, 960-968.

[15] Burland, J. B. (1990). On the compressibility and shear strength of natural clays. Géotechnique, 40(3), 329-378

[16] Corinaldesi, V., & Moriconi, G. (2009). Behaviour of cementitious mortars containing different kinds of recycled aggregate. Construction and building materials, 23(1), 289-294.

[17] Cristiano, M. (2014). The use of ceramic waste aggregates in concrete: a literary review. 10.13140/2.1.2349.1846.
[18] Daniyal, M., & Ahmad, S. (2015). Application of waste ceramic tile aggregates in concrete. International Journal of Innovative Research in Science, Engineering and Technology, 4(12), 12808-12815.
[19] Dolores, A. J. S., Lasco, J. D., Bertiz, T. M., & Lamar, K. M. (2020). Compressive strength and bulk density of concrete hollow blocks (CHB) infused with low-density polyethylene (LDPE) pellets. Civil Engineering Journal, 6(10), 1932-1943.
[20] El-Dieb, A., Taha, M., & Abu-Eishah, S. (2018). The Use of Ceramic Waste Powder (CWP) in Making Eco-Friendly Concretes
[21] Falek, K., Aoudjane, K., Kadri, E. H., & Kaoua, F. (2017). Influence of recycled aggregates on the mechanical and tribological behavior of concrete. Energy Procedia, 139, 456-461.
[22] Gavriletea, M. D. (2017, May 17). Environmental impacts of sand exploitation. Analysis of sand market. Sustainability, 9(7), 1118. : https://www.mdpi.com/2071-1050/9/7/1118/htm
[23] Juan, A., Medina, C., Guerra, I., Moran, J., Aguado, P., Sanchez, I., Rodriquez, O. (2010). Re-use of ceramic wastes in construction. Ceramic Materials, 197-214.
[24] Kallak, F. S. (2009). Use of crushed bricks as coarse aggregate in concrete. Tikrit Journal of Engineering Sciences, 16(3), 64-69.
[25] Kim, Y., Hanif, A., Kazmi, S. M., Munir, M. J., & Park, C. (2018). Properties enhancement of recycled aggregate concrete through pretreatment of coarse aggregates–Comparative assessment of assorted techniques. Journal of cleaner production, 191, 339-349.
[26] Koehnken, L., & Rintoul, M. (2018). Impacts of sand mining on ecosystem structure, process and biodiversity in rivers. World Wildlife Fund International. : Retrieved from https://d2ouvy59p0dg6k.cloudfront.net/downloads/sand_mining_impacts_on_world_rivers__final_.pdf
[27] Kondolf, G. M. (1994). Geomorphic and environmental effects of instream gravel mining. Landscape and Urban planning, 28(2-3), 225-243. Retrieved from https://doi.org/10.1016/0169-2046(94)90010-8
[28] Liu, F., Liu, J., Ma, B., Huang, J., & Li, H. (2015). Basic properties of concrete incorporating recycled ceramic aggregate and ultra-fine sand. Journal of Wuhan University of Technology-Mater. Sci. Ed., 30(2), 352-360.
[29] Mashitah, M. D., Kin, C. C., & Badorul, A. H. (2008, September). Recycling of homogenous ceramic tiles for the production of concrete block. In International symposium on environmental management: hazardous-environmental management toward sustainability (pp. 22-23).
[30] Medina, C., Juan, A., Frías, M., Sánchez de Rojas, M.I., Morán, J.M., Guerra, M.I. (2011). Characterization of Concrete made with Recycled Aggregate from Ceramic Sanitary Ware. Mater. Construction, 61 [304], 533–546. http://dx.doi.org/10.3989/mc.2011.59710
[31] OBE, R. K. D., Lye, C. Q., Lynn, C. J., & Elgalhud, A. A. (2018, November). Sustainable construction materials for concrete: A question of responsible use. In IOP Conference Series: Materials Science and Engineering (Vol. 442, No. 1, p. 012001). IOP Publishing.
[32] Pacheco-Torgal, F., Jalali, S. (2010) Reusing ceramic wastes in concrete, Construction and Building Materials. 832-838. ISSN 0950-0618. Retrieved from http://dx.doi.org/10.1016/j.conbuildmat.2009.10.023
[33] Peduzzi, P. (2014). Sand, rarer than one thinks. Environmental Development, 11, 208-218. Retrieved from https://archive-ouverte.unige.ch/unige:75919
[34] Philippine Statistics Authority (PSA). (2021). Construction Statistics from Approved Building Permits Second Quarter 2021. Retrieved from https://psa.gov.ph/content/construction-statistics-approved-building-permits-second-quarter-2021
[35] Raval, A., Patel, I., & Pitroda, J. (2013). Ceramic Waste: Effective Replacement of Cement For Establishing Sustainable Concrete. International Journal of Engineering Trends and Technology (IJETT), 2324-2329.

[36] Ravindra R., Anand-Kumar BG., Santhosh N. (2015). Behavior of concrete masonry blocks with ceramic waste. Recent Trends in Civil Engineering & Technology.2 (3): 35–40p.

[37] Senthamarai, R.M., Devadas-Manhoharan, P. (2005) Concrete with ceramic waste aggregate. Cem Concr Comp, 27 [9–10], 910–913. http://dx.doi.org/10.1016/j.cemconcomp.2005.04.003 Figure4.Acoustic absorption coefficient of CW composite products.0.90.80.70.60.5Sound absorption coefficient0.40.30.20.101001000OPSCW-CFrequency (Hz)

[38] Shen, J., & Xu, Q. (2019). Effect of moisture content and porosity on compressive strength of concrete during drying at 105 C. Construction and Building Materials, 195, 19-27.

[39] Silva, R. V., De Brito, J., & Dhir, R. K. (2014). Properties and composition of recycled aggregates from construction and demolition waste suitable for concrete production. Construction and Building Materials, 65, 201-217.

[40] Sonak, S., Pangam, P., Sonak, M., & Mayekar, D. (2006). Impact of sand mining on local ecology. Multiple dimensions of global environmental change. Teri Press, New Delhi, 101-121.

[41] Subedi, B., Wagle, D., & Basnet, K. (2020). Utilization of Crushed Ceramic Tile Wastes as Partial Replacement of Coarse Aggregate in Concrete Production. International Journal of Engineering Research & Technology (IJERT), 1572-1584.

[42] Suzuki M., Meddah-Mohammed S., Sato R. (2009). Use of porous ceramic waste aggregates for internal curing of high-performance concrete. Cement and Concrete Research. 39:373–81p.

[43] Torkittikul, P., & Chaipanich, A. (2010). Utilization of ceramic waste as fine aggregate within Portland cement and fly ash concrete. Cement & Concrete Composites, 440–449.

[44] Torres, A., Simoni, M., Keiding, J., Müller, D., zu Ermgassen, S., Lui, J., Jaeger, J., winter, M., Lambin, E. (2021, May 21). Sustainability of the global sand system in the Anthropocene. One Earth 4, no. 5 (2021): 639-650. : https://www.sciencedirect.com/science/article/pii/S259033222100230X

[45] Villioth, J. (2014, August 5). Building an economy on quicksand. Retrieved from Environmental Justice Organisations, Liabilities and Trade: http://www.ejolt.org/2014/08/building-an-economy-on-quicksand/

[46] Wadie, E. H., Sadek, E. F., & Abdel Wahad, M. M. (2017). The use of ceramic tile waste as aggregates in concrete. Int. J. Sci. Eng. Res, 8, 1461-1468.

[47] Zaharieva, R., Buyle-Bodin, F., Skoczylas, F., & Wirquin, E. (2003). Assessment of the surface permeation properties of recycled aggregate concrete. Cement and Concrete Composites, 25(2), 223-232.

Coconut Shell Ash as Partial Substitute to Cement in Concrete Hollow Blocks

J S Lumbab[1,2], J Bongo[1], R M Moralde[1] and G I Villaceran[1]

[1]Civil Engineering Department, Cebu Technological University Corner M.J. Cuenco Ave. & R. Palma St., Cebu City, 6000, Philippines

[2]E-mail: jeonel.lumbab@ctu.edu.ph

Abstract. Due to heightened demand for construction materials and environmental crisis, this study presents an alternative to cement using upcycled agricultural waste to help the construction industry and address environmental issues. The study focuses on substituting cement in concrete hollow blocks with coconut shell ash at 0, 10, 20, and 30 percent replacement. Four-inch width specimens are fabricated after slump test and underwent compressive strength and water absorption tests at 7, 14, and 28 curing periods. These are some of the insights of the study: (a) slump height of CHB with 30% substitution is the lowest at 4 mm, (b) the 10% CSA performed well among CHB with CSA gaining 3.7454 MPa, and (c) CSA decreased the water absorption to up to 4.6% at 20% replacement and moisture content to a value of 18.22% at 30% substitution after 28 days curing.

1. Introduction

The volume of production of coconut in the Philippines amounted to 13.8 million tons in 2016, for which around 2 million tons of its shells are left as a residue that adds to the agricultural waste of the Philippines and contributes negatively to the environment [1].

However, according to [2], recycling the coconut shell wastes into coconut shell ash as a partial replacement for cement is an effective way to not only lessen environmental pollution and burden but also reduce the cost of concrete production.

In 2021, the overall volume of cement production worldwide reached up to 4.4 billion tons, which is over thrice the statistics 26 years ago, indicating the growth of the construction industry [3]. Cement manufacturing begins with the extraction of raw materials, which include clay and limestone. Then the raw meal is produced by grinding these ingredients into a fine powder known as the raw meal. To break down chemical bonds and arrange them into new compounds, the raw meal is heated in a cement kiln to a sintering temperature of up to 1450°C. This process generates a significant amount of carbon dioxide (CO2), reaching up to 1.6 billion tons per year [4]. Cement production has a noticeable impact on global warming, acidification, and photochemical oxidation. The largest environmental emissions, especially GHG emissions, are generated in the calcination stage from calcium carbonate decomposition and coal consumption. Therefore, finding raw materials and fuel substitutions can reduce environmental emissions [5].

On the other hand, coconut is a non-fossilized, biodegradable organic material that may replace fossil fuels and supplement other renewable energy sources [6]—burning either fossil fuels or biomass releases CO2. However, according to a study conducted by [7], coconut shell-derived energy can save 18.3% of GHGs emission that causes global warming, 64.1% of the same emission that results in ozone depletion, and 71.5% of non-renewable energy consumption.

Coconut Shell Ash comprises 45.02% of SiO2 or silicon dioxide, commonly known as silica [8]. Silica in fine-grained form holds a high pozzolanic activity; this makes silica a valuable component of the binder when producing concrete. It imparts strength to the concrete [9]. Coconut Shell particles are found to have high lignin and lower cellulose. The coconut shell has 46% lignin. Lignin adds to the workability of concrete as it is more weather resistant. The cellulose content is 26.6%. Low cellulose indicates less moisture absorption [8]. Based on [10], CSA can be classified as a Class N pozzolan, which, by definition, is a raw or calcined natural pozzolan that meets specified requirements. It contains some di, reaching, opaline cherts and shales, tufts and volcanic ashes or pumicites, and various materials requiring calcination, such as clay and shale. The maximum loss of ignition for a Class N pozzolan is 10%, and it is noted that CS has a low LoI of 4.28%, which means that the amount of unburned carbon in CSA is minimal. This is beneficial for the concrete since a large percentage of losses during combustion results in a decreased strength of concrete because it reduces the pozzolanic activity [11].

The chemical properties of CSA indicate that it is a potential alternative to cement. Thus, this research study utilizes coconut shell ash as a partial substitute to Portland cement. The characterization of nonloadbearing concrete hollow block with coconut shell ash as a partial replacement to cement is to be determined. Specifically, the objectives of this research work are to: (1) investigate the fresh properties of the CHB mixture slump incorporated with coconut shell ash (CSA) at 10%, 20%, and 30% replacement of Portland cement, and (2) investigate the hardened properties of CHB incorporated with coconut shell ash (CSA) at 10%, 20% and 30% replacement of Portland cement.

2. Materials and Methods

The masonry units were composed of natural fine aggregates (NFA) and coconut shell ash (CSA). The NFA was obtained from a local supplier in Cebu City and has passed the sieve analysis. It has a specific gravity, water absorption, and fineness modulus of 2.77, 2.9%, and 2.45, respectively.

On the other hand, the coconut shell ash was secured by burning coconut shells gathered around Cebu Province. The shells were first scraped off coconut meat and husk and sun-dried for 48 hours before being subjected to open-air burning for around three hours and left to cool for twelve hours. The material passed through a 75 μm (No. 200) sieve was collected as CSA. Data from a study by [12] about the percentage of oxide composition of CSA compared to cement was referenced as the chemical properties of the CSA. The large amount of silica in the CSA indicates that it is a cementitious material and can be used as a substitute for OPC [13]. Meanwhile, the information regarding the comparison of the physical properties of cement and CSA was taken from a study by [14], where CSA has a lower specific gravity with 1.33 g/cc but has higher water absorption and fineness modulus, which is 25% and 8%, respectively. Due to limited financial resources, the chemical and physical properties of CSA were not checked and tested.

a.) collected CS b.) cleaned CS c.) crushed CS

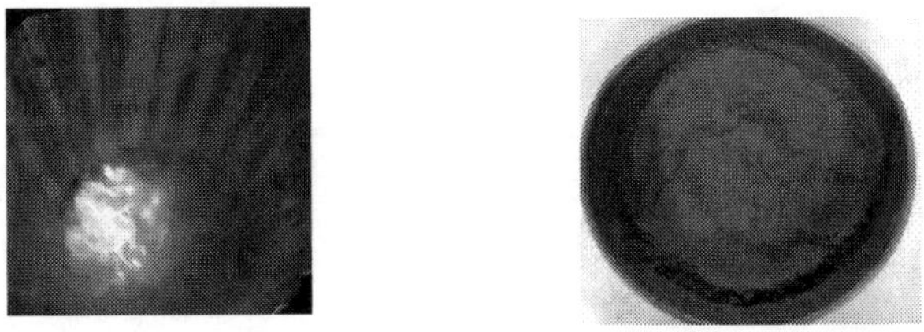

d.) burning CS e.) sieved CSA

Figure 1. Process of collecting CSA.

2.1 Experimental Investigation

Concrete hollow block units, four inches in width and incorporated with coconut shell ash at 0, 10, 20, and 30-percent replacement to ordinary Portland cement, were investigated in the study. The materials were prepared and selected accordingly to ensure conformance with the ASTM standard for assurance that the characterization of the CHB was solely influenced by the varying CSA percentage involved in the mix. The design ratios used in the experiment were established following the cement-to-aggregate ratio and water-to-cement ratio of 1:7 and 1:2, respectively. The masonry units were fabricated using a manual hollow block moulder and were cured for 7, 14, and 28 days. Both the fabrication and curing of the CHBs took place at the Cebu Technological University – Main Campus. For each curing period, the specimens were tested for compressive strength at the Department of Public Works and Highways (DPWH) testing center and density and water absorption in accordance with [15] at the Megatesting Center, Inc.

2.2 Statistical Treatment

One-way ANOVA was used to compare the compressive strength of the samples using manual computation and checking through MS excel. The percent replacement of CSA was the independent variable while the compressive strength was the dependent variable. Also, the 0% replacement was the control setup while the rest were the experimental setups. The following were the hypotheses drawn:

H_o: There is no significant difference in the compressive strengths between the control and experimental setups.

H_i: There is a significant difference in the compressive strengths between the control and experimental setups.

3. Results and Discussion

3.1 Fine Aggregates

Based on the sieve analysis conducted, the particle size distribution of the NFA is illustrated in Figure 1. The fineness modulus of the aggregate resulted in 2.45 and using the gravimetric procedure, the density, specific gravity, and absorption of the fine aggregates were 2400kg/m^3, 2.77, and 2.9%, respectively.

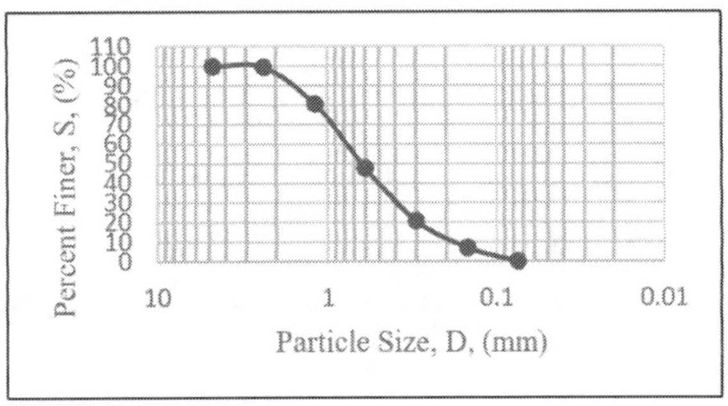

Figure 2. Particle size distribution curve of NFA.

3.2 Slump Results

The slump test is conducted in conformance with [16]. Table 1 shows the slump results in millimeters of each of the Mix IDs. The range of slump is around 4-10 mm. As observed in the result, the slump height decreases as the CSA percent replacement increases. This indicates that concrete mix with CSA results to lower workability, but still passes the required slump behaviour for concrete masonry units as per ASTM standards since the mix did not fall away or shear off.

Table 1. Slump Height per MIX ID

MIX ID	Slump Height (mm)
CC	10
CSA1	7
CSA2	5
CSA3	4

3.3 Net Area Compressive Strength

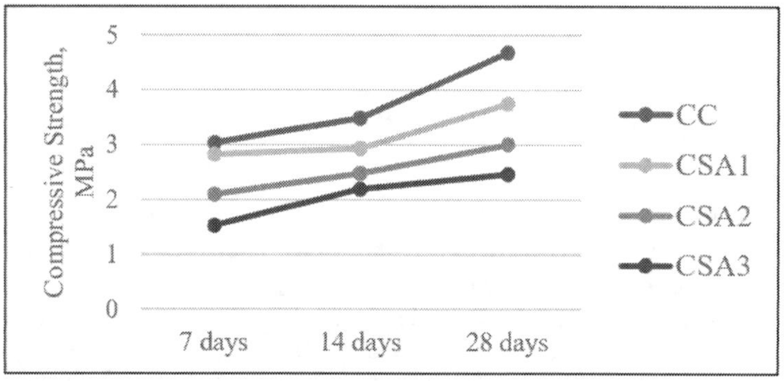

Figure 3. Consolidated compressive strength of units

Figure 3 show the effect of CSA percentage replacement on the compressive strength of concrete hollow blocks. The control sample (CC) achieves the highest compressive strength among all mix ID in every curing period; 3.0318 for 7th day, 3.4825 for 14th day, and 4.6830 for 28th day. This is then followed by CSA1 of 10% replacement with strength of 3.7454 MPa, CSA2 with 3.0017 MPa, and CSA3 gaining 2.4627 MPa, at 28 days curing age. It is observed that the compressive strength of CHB increases with age. However, only the control specimen has reached the required compressive strength for nonloadbearing concrete masonry which is 4.14 MPa at 28 days curing period. Also, it is important to note that there is a difference in the equivalent water to cement ratio (w/c) per curing age of the units considering the moisture content of the fine aggregates on the day of fabrication. This variation is presented in Table 2, where the equivalent w/c of samples for the 7-day curing is 0.5288 that is 0.8% higher than that of the 14th and 28th day samples which is only 0.5246. Nevertheless, the results in compressive strength of each of the mix ID can be freely compared between the same curing periods.

Table 2. Equivalent water to cement ratio

Sample Fabrication for	In Situ Weight (g)	Oven-dry Weight (g)	Moisture Content (%)	$w/c_{equivalent}$
7th Day	500	486	2.8807	0.5288
14th Day	500	488	2.4590	0.5246
28th Day	500	488	2.4590	0.5246

Meanwhile, the compressive strength of the CHB incorporated with 30% CSA decreased by 47.4% relative to the control at 28 days. Likewise, a 35.9% decrease is observed with CSA2 and 20% for CSA1. This indicates that compressive strength decreases as CSA percent replacement increases. This outcome can be attributed to the chemical composition of CSA. CSA has higher silicon dioxide (silica) which is 66.32% compared to cement that is only 15.99%. The unreacted silica remains in the matrix as a stable material and will not stimulate any chemical reaction. In contrary, it possesses lower calcium oxide (CaO) of 6.25% in contrast with cement that has 58.86% CaO. This CaO in the binder reacts with water in the mix and produces calcium hydroxide, $Ca(OH)_2$. In turn, this $Ca(OH)_2$ reacts with carbon dioxide in the air in a process called carbonation. Subsequently, this process increases the mechanical strength of concrete. Hence, CHB with CSA, which possess significantly lesser CaO that is a crucial substance for the carbonation process, results in lower compressive strength.

Furthermore, one-way ANOVA was used to compare the compressive strengths of each mix ID. Since the F-computed value of 41.4395 is greater than the F – tabular value of 4.066 at 0.05 level of significance with 3 and 8 degrees of freedom, the null hypothesis is rejected. For 95% degree of confidence, the P-value was computed to be 0.000032 which is less than 5% or 0.05, which means that there is a significant difference in the compressive strength of four design mixes.

3.4 Water Absorption and Moisture Content

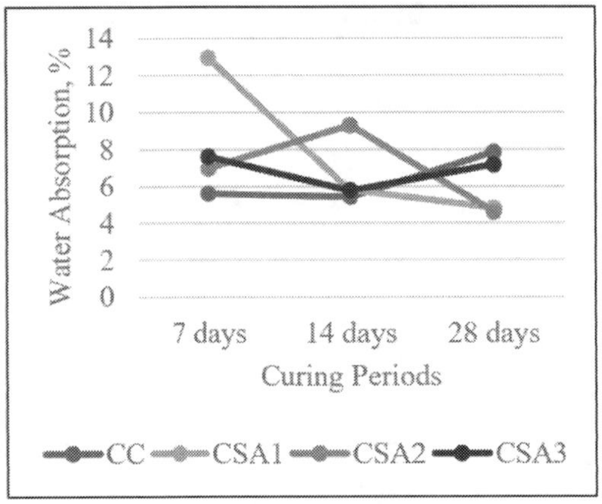

Figure 4. Water absorption of units.

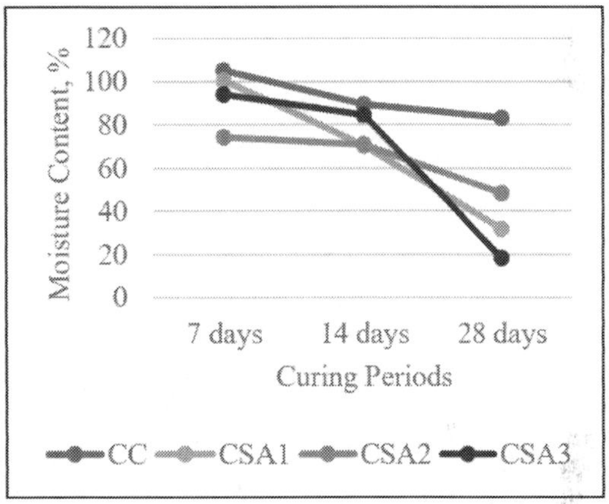

Figure 5. Moisture content of units.

One sample per mix ID per curing batch was tested for water absorption and moisture content in the Megatesting Center. Based on the results in Figure 4, at seven days, CSA1 has the optimum water absorption percentage of 12.94%, and the CC has the lowest water absorption with 5.6%. While at 14 days, CSA2 has the optimum water absorption with 9.29 %, and the CC still has the lowest absorption with 5.42%. Thus, at seven and 14-day, the control obtained the lowest water absorption, with 5.6 and 5.42, respectively, which indicates that CHB with CSA replacements has higher water absorption than the control. However, during a 28-day curing period, CC has the optimum water absorption with 7.85%, and CSA2 has the lowest absorption with 4.6%. Denser samples appears to have exhibited lower water absorption. This is due to the fewer amount of void that occupy the CHB when the unit is denser, hence, less water can penetrate.

Meanwhile, CC gained the highest moisture content of 83.2% on the 28th day curing while CSA3 exhibited the lowest which amounted to 18.22% as seen in Figure 5. It can be observed from the data that moisture content of CHB decreases with curing age. This happens because as hydration continually occurs, the pores in the matrix are gradually consumed and filled with calcium silicate hydrate. Thus, less water now exists in the masonry unit.

Meanwhile, it can also be observed that moisture content decreases as the percent replacement of CSA increases. This is a good indication for CHB as high moisture content affects the compressive strength of concrete negatively.

3.5 Density

Density is a measure of the ratio of the mass of a substance to its volume. It can be noticed in Figure 6 that the density of the control sample decreases with age from 2015 kg/m3 at 7 days curing to 1988 kg/m3 at 14 days, and 1903 kg/m3 at 28 days curing. Moreover, it is also observed that at an early age, at seven and 14 days, the control has the highest density among all units, which is 2015 and 1988, respectively. However, at 28 days, the control sample now has the lowest density with 1903 and the CSA1, CSA2 and CSA3 exhibits higher density with 2013, 2022, and 1935 kg/m3, respectively. Although the control sample has the lowest density, passing as medium weight CHB, it still exhibits the highest compressive strength among all the samples. This entails that density is not the sole factor that affects the mechanical strength of CHB; a masonry unit can be lighter in weight but still perform well. The amount of cement and the curing still plays a crucial role in the compressive strength of CHB.

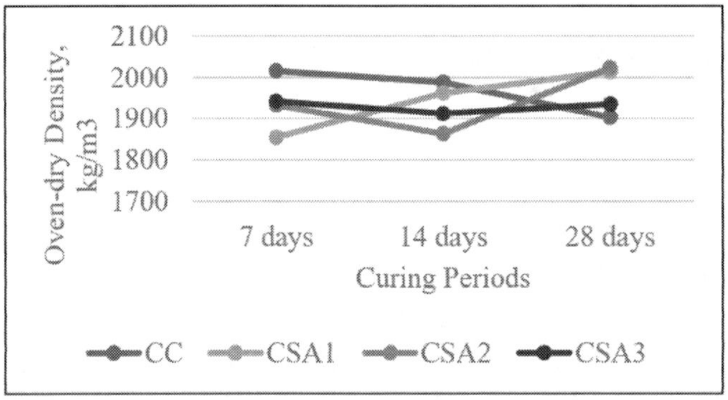

Figure 6. Oven-dry density of units.

4. Conclusion

Important conclusions are summarized as follows.

1. Partial replacement of CSA resulted in decreased slump height of concrete mixture, indicating very low workability. This decrease in workability is due to higher water absorption of CSA. However, the resulting slump height ranges from 4-10mm, which is still within the range required for CHB units, which must be a maximum of 100mm.
2. Compressive strength of all specimens increases relative to curing age. The control specimen obtained the optimum compressive strength in 7, 14 and 28 days curing periods; 3.0318 MPa, 3.4825 MPa, and 4.6830 MPa, respectively. Thereafter, only the control specimen reached the minimum requirement for nonloadbearing CHB at 28 days curing age. Thus, indicates that CSA is not suitable as a substitute to cement in CHB at 10 percent replacement and higher and that studies with lower CSA replacement are recommended using 0.5 w/c.
3. Compressive strength of the specimens decreases relative to the increase of CSA replacements. Nonetheless, 10 percent CSA replacement performed well among other percent replacements with CSA. It resulted to an average decrease in compressive strength of 14.23% and obtained 3.7454 MPa compressive strength, while 20 and 30 percent substitution obtained 3.0017 and 2.4627 MPa, respectively, at 28 days curing period.
4. CHB with CSA have higher water absorption at 7 and 14 days compared to the control specimen. However, during the 28-day curing period, the control specimens have optimum water absorption.
5. Partial replacement of CSA has reduced the moisture content of the CHB units. In addition, the moisture content of all specimens decreases relative to the curing period.

6. Partial replacement of CSA makes the CHB units less dense compared to the control specimens at seven and 14-day curing periods. However, at 28-day, the control is less dense compared to the 10, 20, and 30 CSA percent replacements. Also, the density of the control specimen decreases with age.

7. With the density results, it can be concluded that CC and CSA3 passed as medium weight concrete hollow blocks with density less than 2000 kg/m3.

5. References

[1] Caparino, O. (2018, February 22-23). Status of agricultural waste and utilization in the Philippines [Conference presentation]. 2018 International Forum on Sustainable Application of Waste-to-Energy in Asia Region, Busan, Korea. https://sustainabledevelopment.un.org/content/unosd/documents/37656.Philippines-Power%20Point%20Presentation-2-21-18.pdf

[2] Desai B.M. et al. (2017). Experimental study on concrete by partial replacement of cement with coconut shell ash incorporating steel fibres: A review. International Journal for Research in Applied Science & Engineering Technology, 5(3), pp. 161-164. http://dx.doi.org/10.22214/ijraset.2017.3029

[3] Garside, M. (2022). Global cement production 1995-2021. Mining, Metals & Minerals. https://www.statista.com/statistics/1087115/global-cement-production-volume/

[4] Leetham, D. (2015). The cement manufacturing process. Retrieved from https://www.thermofisher.com/blog/mining/the-cement-manufacturing-process/#:~:text=Cement%20manufacturing%20is%20a%20complex,C%20in%20a%20cement%20kiln

[5] Song, D. et al. (2015).Life-cycle environmental impact analysis of a typical cement production chain. Appl Energy. http://dx.doi.org/10.1016/j.apenergy.2015.09.003

[6] Espina, R. et al. (2022). The optimal high heating value of the torrefied coconut shells. Engineering, Technology & Applied Science Research, 12(3). http://dx.doi.org/10.48084/etasr.4931

[7] Sreejith, C. C. (2013). Life cycle assessment of producer gas derived from coconut shell and its comparison with coal gas: an Indian perspective. International Journal of Energy and Environmental Engineering, 4(8). http://dx.doi.org/10.1186/2251-6832-4-8

[8] Adeala, A. J., Olaoye, J. O., & Adeniji, A. A. (2020, January). Potential of coconut shell ash as partial replacement of ordinary Portland cement in concrete production. International Journal of Engineering Science Invention, 9(1), pp. 47-53.

[9] Pokorny et al. (2016, July). Coagulated silica - a-SiO2 admixture in cement paste [Conference paper]. THERMOPHYSICS 2016: 21st International Meeting. http://dx.doi.org/10.1063/1.4955254

[10] ASTM C618 (2015). Standard Specification for Coal Fly Ash and Raw or Calcined Natural Pozzolan for Use in Concrete. ASTM International, West Conshohocken, PA.

[11] Nagarajan, V. (2014). Experimental study on partial replacement of cement with coconut shell ash in concrete. *International Journal of Science and Research, 3(3).* https://www.ijsr.net/archive/v3i3/MDIwMTMxMjYw.pdf

[12] Joshua et al. (2018). Data on the pozzolanic activity in coconut shell ash (CSA) for use in sustainable construction. Data in Brief, 18, pp. 1142-1145. https://doi.org/10.1016/j.dib.2018.03.125

[13] Bheel, N., & Mangi, S. (2021, February). Coconut shell ash as cementitious material in concrete: A review. https://doi.org/10.17576/jkukm-2020-33(1)-03

[14] Gummadi, C. & Srikanth, G. (2012). Experimental investigation on partial replacement of cement with coconut shell ash and silica fume in concrete. *Education 3(6),* pp. 7-15

[15] ASTM C140 (2022).Standard Test Methods for Sampling and Testing Concrete Masonry Units and Related Units. ASTM International, West Conshohocken, PA.
[16] ASTM C143 (2015). Slump of Hydraulic Cement Concrete. *ASTM International*, West Conshohocken, PA.

IOP Conf. Series: Earth and Environmental Science 1184 (2023) 012025 doi:10.1088/1755-1315/1184/1/012025

The Effect of Natrium Hydroxide Molarity Variation and Alkali Ratio on the Compressive Strength of Geopolymer Paste and Mortar

Mariamah[1*], Nursiah Chairunnisa[1], Ratni Nurwidayati[1]
[1]Civil Engineering Study Program, Lambung Mangkurat University, Banjarmasin, Indonesia, 70123
*Corresponding author: mariamah.5599@gmail.com

Abstract. Geopolymers have great potential to replace conventional cementitious materials and are classified as green materials. Their main constituents are silica (Si) and aluminum (Al), which can be found in the fly ash material from industrial processes of the PLTU. The parameters that affect the strength of the concrete are influenced by the molarity of natrium hydroxide and the ratio of alkalis ($NaOH/Na_2SiO_3$). This study examined the characteristics of the geopolymer paste and mortar specimens. Furthermore, mixture composition was carried out using the weight ratio method. The mortar produced consists of 65% fine aggregate and 35% paste, while the pasta contains 60% fly ash and 40% alkaline solution. The molarities of NaOH used were 8M, 10M, 12M, and 14M with a base ratio of 1.5, 2, and 2.5 $Na_2SiO_3/NaOH$. The curing of specimens was carried out with the humid temperature method. The results showed that the highest compressive strength of geopolymer paste and mortar was obtained at 12M NaOH molarity with an alkaline ratio of 2.5, namely 12.59 MPa and 21.75 MPa at 28 days, respectively. NaOH of more than 12M can cause a decrease, while the base ratio of 2.5 has the highest strength. The results also showed that the higher the molarity of NaOH and the greater the alkaline ratios, the longer the setting time.

1. Introduction

Over the years, there has been a tremendous increase in infrastructure development, which also led to increased cement production. Previous studies revealed that the cement and cement-based building material industries contributed approximately 8-10% of the amount of carbondioxide emissions in the world. This percentage is likely to continue increasing along with the rapid development of infrastructures [1]. One of the efforts to reduce the use of cement is the production of geopolymer concrete, which is also known as eco-green construction. This new type of concrete can replace the cement by 100% as a binder, and its major constituents include silicon dioxide (Si) and aluminum (Al). These materials can be found in newly produced fly ash from the combustion of PLTU coal, which contains chemical compounds of pozzolanic ash. PLTU Asam-Asam is located in South Kalimantan as a coal producer for electricity demand and fly ash waste. Furthermore, the waste can be used as a binder for geopolymer concrete. Some of the advantages of the concrete include good mechanical properties, high initial strength, low chloride ion diffusion rate, as well as stable heat and fire resistance [2]. In the mixture, the silicate source and aluminum are reacted with an alkaline solution as an activator to create a polymerization reaction. This then leads to the formation of geopolymer concrete without the use of

IOP Conf. Series: Earth and Environmental Science 1184 (2023) 012025 doi:10.1088/1755-1315/1184/1/012025

cement. To carry out the polymerization reaction, it is necessary to use sodium hydroxide or NaOH, which has strong alkaline properties as the reactant, as well as sodium silicate or Na_2SiO_3 as a catalyst. The molarity of NaOH and Na_2SiO_3/NaOH ratio (alkaline) are important parameters affecting the strength of the geopolymer concrete produced [3].

Herwani et al. [4] revealed that the molarity of NaOH solution as an activator has an effect on the compressive strength of the product. The basic material used for the production process is fly ash obtained from PLTU Suralaya. The NaOH molarities used include 10M, 12M, and 14M. The test results showed that the optimal compressive strength was obtained at a concentration of 12M. Further study was carried out by Farizka and Darma et al. [5] on the influence of different molarities, namely 6M, 8M, and 10M with fly ash from PJB Tanjung Jati B. Based on the test results, the 8M NaOH solution had the highest compressive strength compared to the 6M and 10M solutions. Sengkey et al. [6] explored the operation of the alkaline activator by testing the strength at the age of seven days using a geopolymer mortar based on class C fly ash obtained from Amurang PLTU, North Sulawesi. The mortar was made with variations of 8M, 10M, and 12M NaOH, as well as a Na_2SiO_3/NaOH ratio of 1.5, 2, and 2.5. The test results showed that the highest compressive strength was obtained with a molarity of 10M, and a ratio of 2.5. The Na_2SiO_3/NaOH ratios used in the study of Joseph and Matthew [7] were 1.5, 2, 2.5, 3, and 3.5. As the alkaline solution composition increased from 1.5 to 2.5, the compressive strength of the geopolymer concrete also increased, and it decreased sharply from 3 to 3.5. The decrease above a ratio of 2.5 was because the amount of sodium hydroxide became insufficient for the dissolution process.

Based on previous studies, there is a gap analysis in terms of the optimum NaOH molarity to the compressive strength value. This indicates that further study is still required on the use of fly ash as a raw constituent material of geopolymer. A mixture of the concrete, produced from PLTU Asam-asam, South Kalimantan with different NaOH molarities of 8M, 10M, 12M, and 14M, as well as alkaline ratio of 1.5, 2, and 2.5 was tested at 3, 7, 14, and 28 days. Therefore, this study aims to determine the effect of NaOH molarity and the alkali ratio on compressive strength.

2. Research Methodology

2.1. Materials

Fly ash used in this study was class F processed from PLTU Asam-Asam, and its chemical compositions along with the specifications are presented in Table 1. Based on ASTM C-618, the sample was classified as type F based on the amount of SiO_2, Al_2O_3, and Fe_2O_3 at 74.51%, as shown in Table 1. The activators used in this study include sodium hydroxide (NaOH) and sodium silicate (Na_2SiO_3). The sodium hydroxide used was in the form of crystals (flakes) with a purity of 98%, and it was dissolved in water to obtain a NaOH solution. The Na_2SiO_3 was a thick liquid, and its composition is presented in Table 2.

Table 1. Chemical composition PLTU Asam-Asam fly ash.

	Parameter	Result (% weight)	Methods
Ash Analysis (Dry Basis)	Silicon Dioxide as SiO_2	40.92	ASTM D 6349-13
	Aluminium Trioxide as Al_2O_3	10.08	
	Iron Oxide as Fe_2O_3	23.51	
	Calcium Oxide as CaO	12.86	
	Magnesium Oxide as MgO	8.97	
	Sodium Dioxide as Na_2O	0.26	
	Potassium Oxide as K_2O	0.74	
	Manganase Oxide as MnO_2	0.49	

Titanium Dioxide as TIO_2	0.62
Phosphorus Pentoxide as P_2O_5	0.20
Sulphur Trioxide as SO_3	0.82
Undetermined	0.53

Table 2. Compotitions of Na_2SiO_3.

Parameters	Units	Results	Methods
SiO_2 Content	Mass fraction, %	30.6200	
Na_2O Content	Mass fraction, %	9.4200	
Ratio SiO_2/Na_2O		3.2500	SNI 0127:2015
Spesific Gravity at 20 °C		1.4216	
Baume at 20 °C	°Be	43.0000	

The aquadesh water was used in the study with a normal pH of 7, and it met the water standards for the production of pastes and mortars. The fine aggregate was obtained from Barito River sand for mortar specimens.

2.2. Methods
Geopolymer paste and mortar specimens were prepared using the similar method as for conventional Geopolymer paste and mortar specimens were prepared with a similar method used for the production of conventional OPC paste and mortar. Furthermore, the basic ingredients of the mixture include fly ash of PLTU Asam-Asam, water, natrium silicate (Na_2SiO_3), and natrium hydroxide (NaOH), as shown in Figure 1. The proportion by weight of the geopolymer paste mixture was 65% fly ash and 40% alkali activator (binder proportion). The Molarity variation consists of 8M, 10M, 12M, and 14M, while the Na_2SiO_3/NaOH ratios were 1.5, 2.0, and 2.5. The weight proportion of the geopolymer mortar specimen used was 65% fine aggregate and 35% paste.

Geopolymer specimens were produced by blending the constituent materials into a mixer. The paste was then placed in a cylindrical mold with a size of 28 x 56 mm, while the mortar was transferred to a cube mold with a size of 50mm x 50mm x 50mm.

The test results were analyzed and compared for each setting time and compressive strength of the paste and mortar. A total of 24 cylinder specimens were cast for each set of parameters, six each for the compressive strength at 3, 7, 14, and 28 days using the ASTM C109 procedure. The test results with load cells are presented in Figure 2. A total of 288 cylinder specimens were cast for this experimental test. The variations and nomenclature of test objects consist of 24 different sets of paste and 3 sets of mortar specimens, as shown in Table 3 and Table 4. Furthermore, 27 cube samples were cast for each set of parameters, three each for compressive strength at 7, 14, and 28 days.

The setting time test was then carried out using Vicat apparatus with the procedure from ASTM C191. The initial setting time (hours/minutes) was reached when the geopolymer paste lost its plastic properties or began to stiffen, and the measurement was taken when the penetration was 25 mm from the surface. Meanwhile, the final time was obtained when the paste began to harden, and it was measured when the needle could not penetrate further or at zero value.

Table 3. Cylinder specimens series of geopolymer paste.

IOP Conf. Series: Earth and Environmental Science 1184 (2023) 012025 doi:10.1088/1755-1315/1184/1/012025

No	Series	Variable		Total sample
		Molarity (Mol)	Alkaline ratio	
1	GP-8A	8.0	1.5	24
2	GP-8B	8.0	2.0	24
3	GP-8C	8.0	2.5	24
4	GP-10A	10.0	1.5	24
5	GP-10B	10.0	2.0	24
6	GP-10C	10.0	2.5	24
7	GP-12A	12.0	1.5	24
8	GP-12B	12.0	2.0	24
9	GP-12C	12.0	2.5	24
10	GP-14A	14.0	1.5	24
11	GP-14B	14.0	2.0	24
12	GP-14C	14.0	2.5	24
		Total samples		288

Table 4. Cubes specimens series of geopolymer motar.

Series	Variable		Total sample
	Molarity (Mol)	Alkaline ratio	
GM-8	8.0	2.5	24
GM-10	10.0	2.5	24
GM-12	12.0	2.5	24
Total samples			72

3. Result and Discussion

3.1. Setting Time Check

The setting time of the geopolymer was analyzed by replacing ordinary cement with a mixture of fly ash, NaOH molarity, and an alkaline solution. This test was carried out to determine the effect of NaOH molarity and alkali ratio on the setting time of geopolymer paste, and the results are presented in Table 5.

Table 5. Results of initial and final setting times.

Code of specimens	Molarity (Mol)	Alkaline solution ratio	Initial setting time (Minutes)	Final setting time (Minutes)
GP-8A	8.0	1.5	600	1010
GP-8B	8.0	2.0	630	1040
GP-8C	8.0	2.5	650	1060
GP-10A	10.0	1.5	855	1245
GP-10B	10.0	2.0	885	1290
GP-10C	10.0	2.5	915	1320
GP-12A	12.0	1.5	935	1350
GP-12B	12.0	2.0	955	1410
GP-12C	12.0	2.5	1080	1470
GP-14A	14.0	1.5	1090	1810
GP-14B	14.0	2.0	1130	1850
GP-14C	14.0	2.5	1185	1890

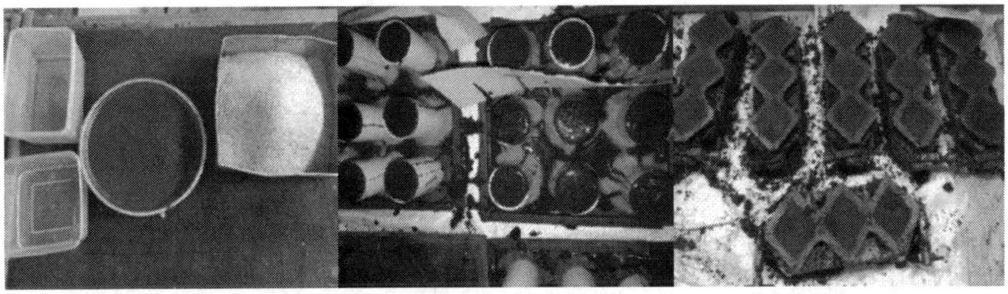

Figure 1. Basic ingredients, paste and mortar specimen.

Figure 2. Compressive strength with load cell.

Furthermore, the monitoring results were plotted in Figure 3 based on the relationship between NaOH molarity and the initial and final setting times for each alkaline ratio. Figure 4 showed the relationship between the alkaline ratio and the initial and final setting times for each molarity.

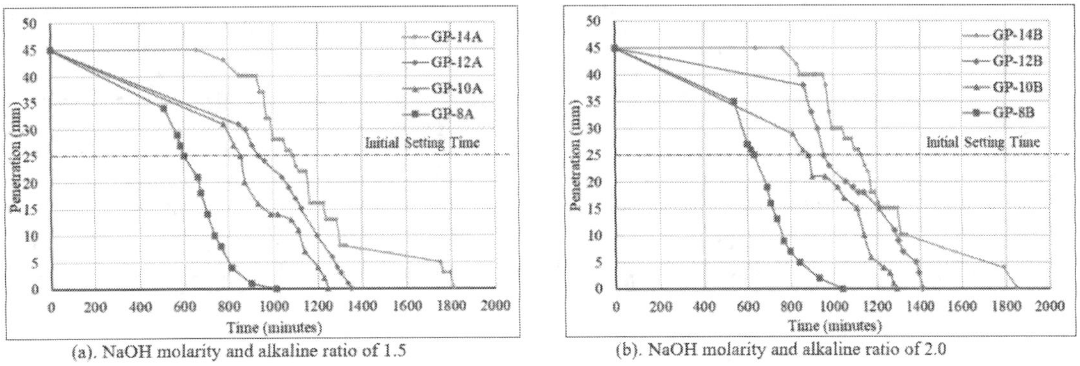

(a). NaOH molarity and alkaline ratio of 1.5

(b). NaOH molarity and alkaline ratio of 2.0

IOP Conf. Series: Earth and Environmental Science 1184 (2023) 012025 doi:10.1088/1755-1315/1184/1/012025

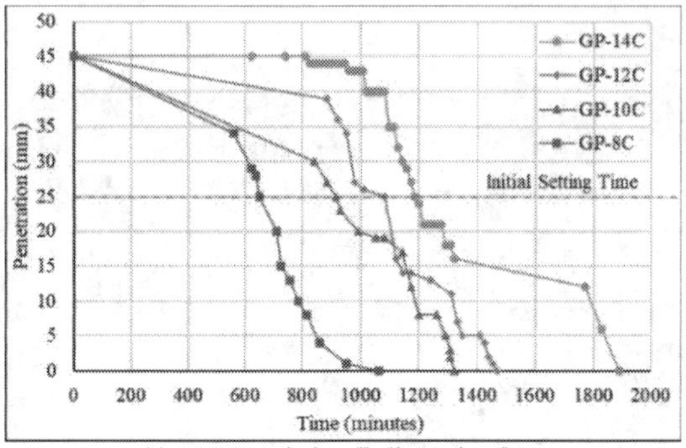

(c). NaOH molarity alkaline ratio of 2.5

Figure 3. Relationship between NaOH molarity and setting time at an alkaline ratio.

Figures 3a to 3c show that the setting time increases along with the molarity of NaOH. These results are in line with Cornelis et al. [8] and Qomaruddin et al. [9], where the higher the molarity of NaOH used, the shorter the time.

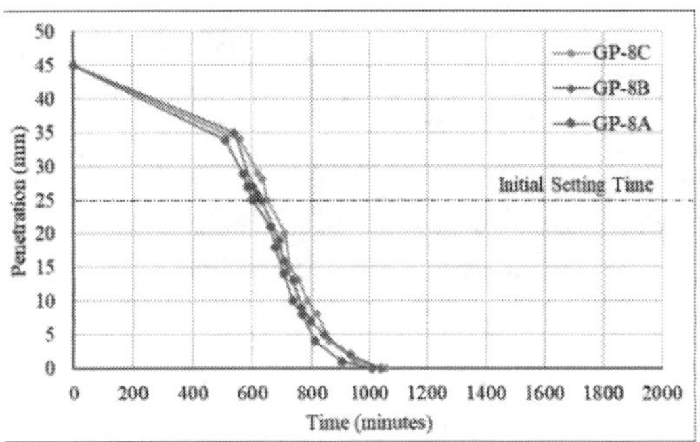

(a). Ratio of alkali and molarity of 8 NaOH

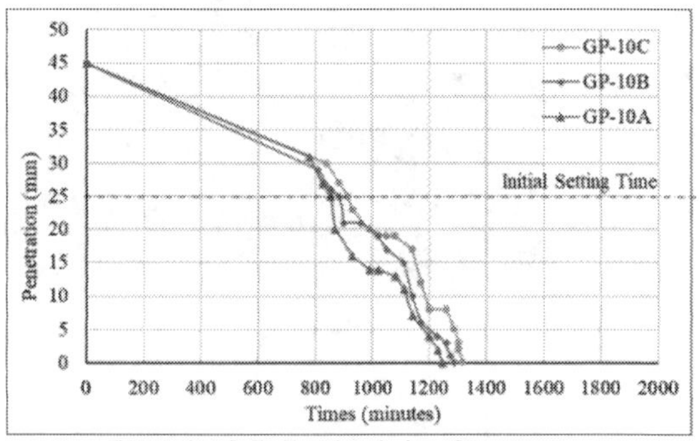

(b). Ratio of alkali and molarity of 10 NaOH

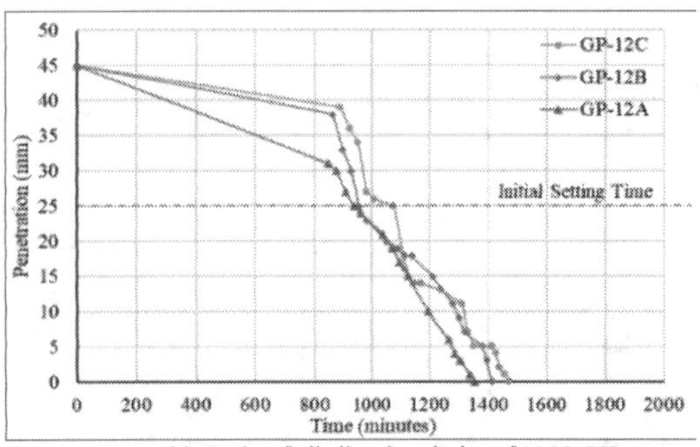

(c). Ratio of alkali and molarity of 12 NaOH

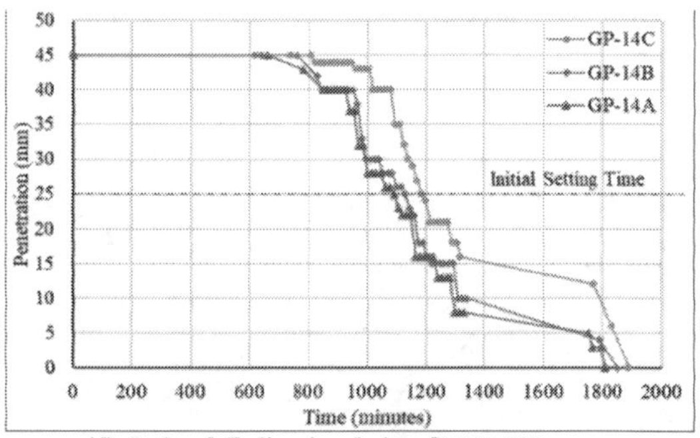

(d). Ratio of alkali and molarity of 14 NaOH

Figure 4. Relationship between alkaline ratio and NaOH molarity.

Figure 4 shows that the increase in the alkali ratio can retard the setting time of geopolymer. However, compared to the influence of NaOH molar concentration, there was no significant change. These findings are consistent with Satria et al. [10] that the setting time became longer as the alkaline ratio increased. The results also showed that the value obtained for geopolymer paste was higher than that of ordinary concrete. Qomaruddin et al. [9] revealed that adding high CaO-containing materials, such as waste carbide, gypsum, and lime to geopolymer paste can accelerate the setting time movement.

3.2. The effect of NaOH molarity and alkali ratio on geopolymer paste compressive strength

Figure 5 shows the compressive strength of pastes at 3, 7, 14, and 28 days. At 8M to 12M, there was an increase in the strength, but it varied inversely with 14M of NaOH. These findings are consistent with Abdullah et al. [11], and Herwani et al. [4] that 12M NaOH molarity gave the optimum result.

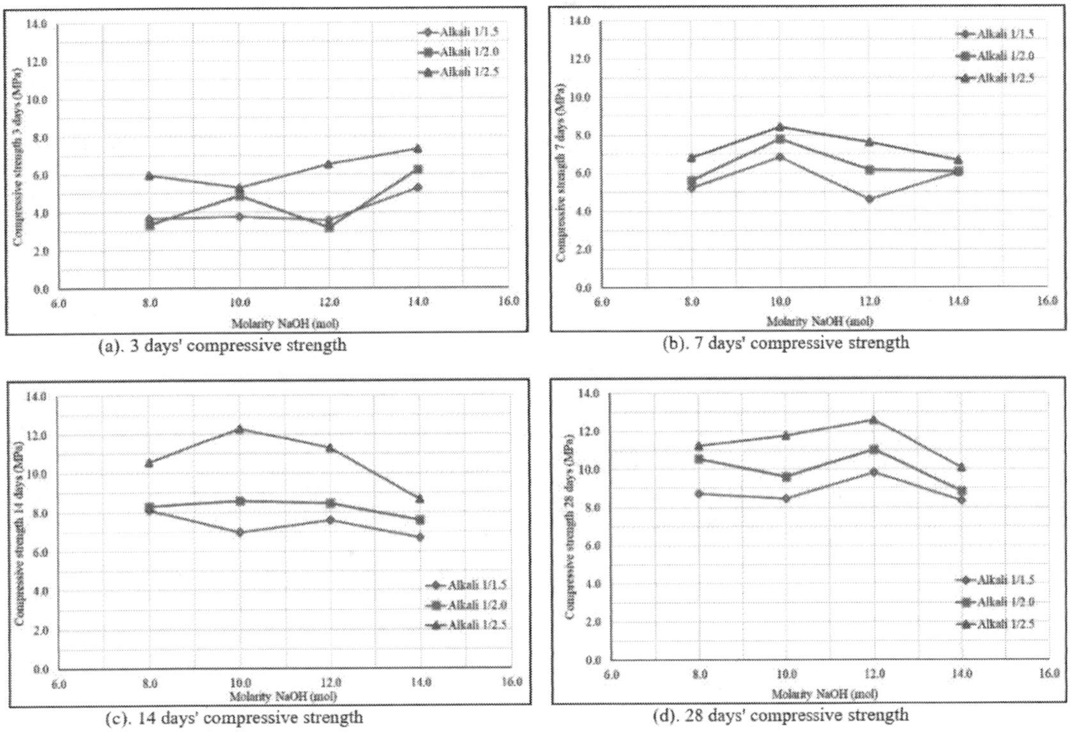

Figure 5. The graph of NaOH's molarity relationship to 3, 7, 14 and 28 days' compressive strength.

The influence of the alkaline ratio on the compressive strength of the paste is presented in Figure 6.

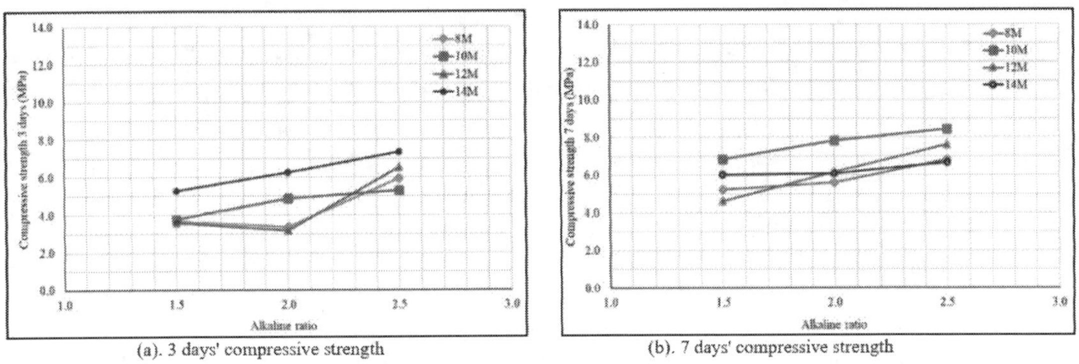

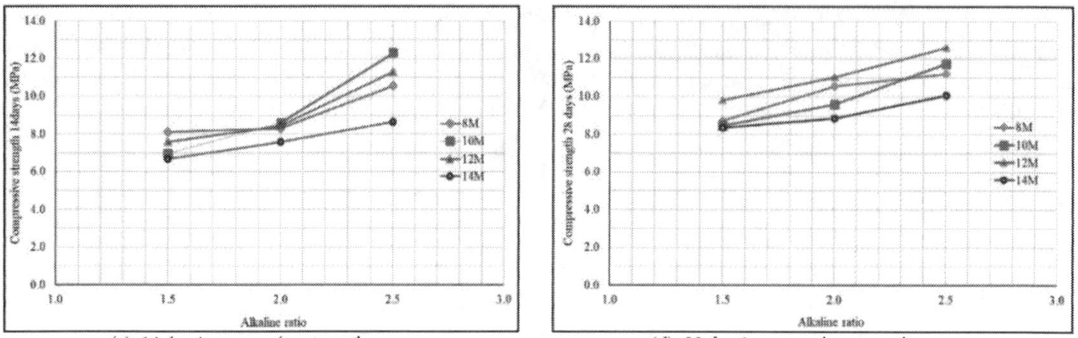

(c). 14 days' compressive strength

(d). 28 days' compressive strength

Figure 6. Relationship between alkaline and compressive strength of pasta at 3, 7, 14 and 28 days.

Figures 6a to 6d show that an alkaline ratio of 2.5 gave the maximum compressive strength for all variables of specimens. These findings are in line with Sengkey et al. [6] that the maximum value was obtained from a ratio of 2.5. This indicates that the greater the amount of Na_2SiO_3 in the paste, the greater the influence of the amount of SiO2 on the compressive strength [12].
The test results of paste specimens at the age of 28 days were tabulated in Figure 7. The three maximum compressive strength values of the geopolymer paste were obtained at NaOH of 8M, 10M, and 12M with a 2.5 alkaline ratio, namely 11.22 MPa, 11.76 MPa. and 12.59 MPa respectively. Based on the results, the increase in age gave a higher strength, and this is consistent with Nurwidayati, et al. [13].

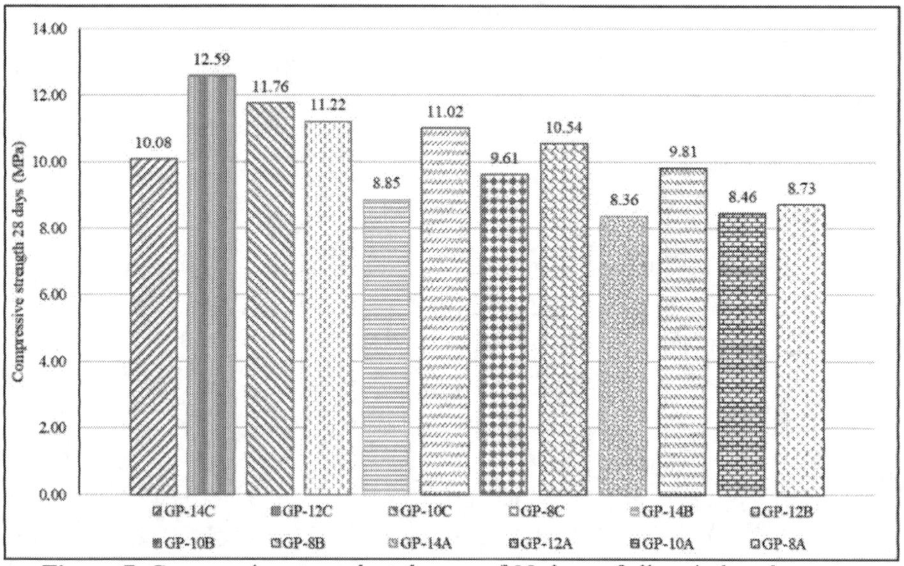

Figure 7. Compressive strength at the age of 28 days of all variations in paste.

3.3. Compressive Strenght Relationship Paste and Mortar

The fly ash to alkaline solution ratio used was 60/40.5 with NaOH concentrations of 8M, 10M, and 12M as well as alkaline ratio of 2.5. Using a mixture of 65/35 fine aggregate and paste, the geopolymer mortar was prepared. The measurement results of the compressive strength are presented in Table 8.

Table 8. Compressive strength of paste and mortar at 7, 14, and 28 days.

IOP Conf. Series: Earth and Environmental Science 1184 (2023) 012025 doi:10.1088/1755-1315/1184/1/012025

Molarity	Paste (MPa)			Mortar (MPa)		
	7 days	14 days	28 days	7 days	14 days	28 days
8	6,81	10,55	11,22	6,20	12,62	17,09
10	8,43	12,29	11,76	8,00	15,05	19,63
12	7,61	11,29	12,59	7,69	13,69	21,75

The compressive strength of the paste and geopolymer mortar at 7, 14, and 28 days is presented in Figure 6. This indicates that there is a difference between the value obtained for the paste and geopolymer mortar. Figure 8 showed that the compressive strength at 7 days increased along with the molarity from 8M to 10M, and then decreased at 12M. The strength of the paste and mortar at 10M increased by 23.79% and 29.03% respectively. At 14 days, the overall improvement was 16.49% and 19.25% at 10M molarity for geopolymer paste and mortar. At 12M, there was a decrease in the value recorded for two sample types. Meanwhile, the overall rise in 12M molarity for geopolymer paste and mortar was 12.21% and 27.26% at 28 days. The same pattern of the increase occurred at the age of 7, 14, and 28 days, but the compressive strength of the paste did not experience a significant increase compared to mortar.

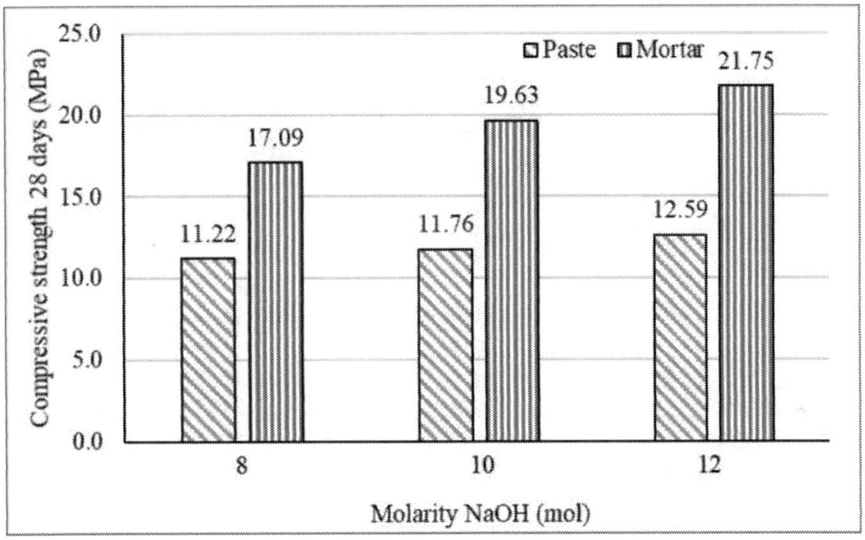

Figure 8. Compressive strength at 28 days of geopolymer paste and mortar.

The mortar strength value at 8M was 1.52 times that of the paste, while it is 1.67 times higher at 10M. Furthermore, at 12M, the compressive strength of the paste was 1.73 times that of the paste. Compared with geopolymer paste samples, geopolymer mortar provided greater compressive strength, and this is in line with Samantasinghar et al. [14].

4. Conclusion

Several findings can be drawn from the results of this study:

1. The higher the molarity of NaOH and the greater the ratio of alkali (Na_2SiO_3/NaOH), the lower the changes in setting time.
2. The optimum compressive strength of the geopolymer paste can be obtained with a molarity of 12M NaOH. Increasing the molarity of NaOH to >12M did not significantly affect the increase in compressive strength.

3. The compressive strength of the geopolymer paste was also influenced by the alkaline ratio. In this study, the maximum value was recorded at 2.5 alkaline ratio.
4. The increase of age at 7, 14, and 28 days provided the highest compressive for the paste.

References

[1] Hardjasaputra, Harianto; Ekawati, Esteriana. 1997. "Research Design of Geopolimer Concrete Based on Pltu Suralaya- Banten Fly Ash on Compressive and Flexible Strength," 24–33.

[2] Fang, Yuan, and Obada Kayali. 2013. "The Fate of Water in Fly Ash-Based Geopolymers." *In Construction and Building Materials*.

[3] Panjaitan, Putrilyan Ezra; Herlina, Liana. 2020. "Review Faktor - Faktor Yang Mempengaruhi Karakteristik Kuat Tekan Beton Geopolimer" 03 (02).

[4] Herwani, Ivindra Pane, Iswandi Imran, and Bambang Budiono. 2018. "Compressive Strength of Fly Ash-Based Geopolymer Concrete with a Variable of Sodium Hydroxide (NaOH) Solution Molarity." *MATEC Web of Conferences* 147: 1–5.

[5] Darma, Adi S; Farizka, Rahman N; Han, Ay Lie; Purwanto. 2018. "Studi Experimental Pengaruh Perbedaan Molaritas."Jurnal Karya Teknik Sipil, 7:89–98.

[6] Sengkey, Sandri Linna., et all. (2020). "Design Engineering Study of Compressive Strength and Durability of Fly Ash and Trass Based Geopolymer Mortar"

[7] Joseph, Benny, and George Mathew. 2012. "Influence of Aggregate Content on the Behavior of Fly Ash Based Geopolymer Concrete." *Scientia Iranica*.

[8] Cornelis, Remigildus, Henricus Priyosulistyo, Iman Satyarno, and Rochmadi. 2018. "The Investigation on Setting Time and Strength of High Calcium Fly Ash Based Geopolymer." *Applied Mechanics and Materials* 881 (May): 158–64. https://doi.org/10.4028/www.scientific.net/amm.881.158.

[9] Qomaruddin, Mochammad, Khotibul Umam, Istianah Istianah, Yayan Adi Saputro, and Purwanto Purwanto. 2019. "Pengaruh Bahan Kalsium Oksida Pada Waktu Pengikatan Pasta Beton Geopolimer Dan Konvensional." *EKSAKTA: Jurnal Ilmu-Ilmu MIPA* 19: 182–92. https://doi.org/10.20885/eksakta.vol19.iss2.art8.

[10] Satria, Juan; Sugiarto, Agung; Antoni; Hardjito, Djwantoro. 2015. "Karakteristik Beton Geopolimer Berdasarkan Variasi Waktu Pengambilan Fly Ash" 2 (2): 1–8.

[11] Abdullah, M. M.A., et all. 2011. "The Relationship of NaOH Molarity, Na2SiO3/NaOH Ratio, Fly Ash/Alkaline Activator Ratio, and Curing Temperature to the Strength of Fly Ash-Based Geopolymer." *Advanced Materials Research* 328–330 (September 2011): 1475–1482.

[12] Degirmenci, F. N. 2016, "Utilization of Natural and Waste Pozzolans As an Alternative Resource of Geopolymer Mortar." *Int. J. Civ. Eng*.

[13] R. Nurwidayati, M. B. Ulum, J. J. Ekaputri, Triwulan, P. Suprobo, "Material Science Forum", 841, 118-125 (2016)

[14] Samantasinghar, Subhashree, and Suresh Prasad Singh. 2019. "Fresh and Hardened Properties of Fly Ash–Slag Blended Geopolymer Paste and Mortar." *International Journal of Concrete Structures and Materials* 13 (1): 1–12. https://doi.org/10.1186/s40069-019-0360-1.

The compressive strength and the setting time of partial replacement of fly ash by palm oil fuel ash on geopolymer paste and mortar

R Nurwidayati[1,2*], A R Saidy[1], H Wardhana[2], N Chairunnisa[2] and A Anshar[2]

[1]Doctoral Department of Agricultural Science, University of Lambung Mangkurat, Indonesia
[2]Civil Engineering Program, Engineering Faculty, Lambung Mangkurat University, Indonesia
*Corresponding author: ratninurwidayat@ulm.ac.id

Abstract. This study replaced fly ash with palm oil fuel ash (POFA) in geopolymer paste and mortar. The alkaline activator was a mixture of sodium silicate solution and sodium hydroxide solution with a ratio of 2.5. Three variables were evaluated: the raw material combination (100% fly ash and 25% POFA substituted to fly ash), the mass ratio of raw material to alkali activator (60/40, 65/35, and 70/30), and NaOH molarity (8 M and 10 M). The geopolymer mortar was utilized based on the highest compressive strength of geopolymer paste. The ratio of fine aggregate to binder was 65 to 35. The result showed that the compressive strength of geopolymer paste increased as NaOH molarity increased. 25% substitution of POFA for fly ash increased the compressive strength of geopolymer paste from 12.0% to 27% at 28 days. The raw material/ alkali activator ratio of 65/35 indicated the highest compressive strength. The POFA replacement for fly ash had a longer setting time than 100% fly ash. The compressive strength ratio of mortar to the geopolymer paste was 0.66 and 0.59 for the 65/35 and 70/30 ratios of raw material to alkali activator, respectively.

Keywords: palm oil fuel ash, geopolymer, compressive strength, sodium silicate, sodium hydroxide

1. Introduction

As a coal producing province, South Kalimantan has a power plant that serves the electricity needs of both South and Central Kalimantan Provinces, named the Asam-Asam power plant, located in Tanah Laut Regency. In its manufacturing process, the Asam-Asam power plant, with a capacity of 465 megawatts, generates 160 tons of bottom and fly ash daily [1]. The number of industries that use coal as an energy source, which is continually growing, is directly proportional to the abundance of coal. In the short and long term, fly ash and bottom ash harm ecosystems, the environment, and human health.

According to Government Regulation (PP) Number 22 of 2021, fly ash and bottom ash are residual wastes from power plant combustion that become non-hazardous and toxic materials waste [2]. Prior to the regulation, the use of fly ash and bottom ash in construction was still restricted because it was categorized as hazardous and toxic waste. After the regulation is published, fly ash and bottom ash may be optimized as raw materials for development and infrastructure, presenting a significant opportunity to use the fly ash and bottom ash. However, given that the amount of waste generated each year is

IOP Conf. Series: Earth and Environmental Science 1184 (2023) 012026 doi:10.1088/1755-1315/1184/1/012026

increasing, the use of fly ash and bottom ash, which is only about 5%-10% of the total volume of fly ash and bottom ash, is deemed insufficient [1].

The palm oil waste problem is also occurring in South Kalimantan. The palm oil industry is one of the essential agro-industries [3]. In South Kalimantan, the area of oil palm plantations is 106,000 hectares, with 45 palm oil mills, with crude palm oil (CPO) production reaching 1,561,147 tons per year [4]. Palm oil waste ash is often called palm oil fuel ash (POFA). Like fly ash and bottom ash, POFA also requires a large area for disposal. POFA continues to increase yearly, and it is still underutilized, so that it can cause environmental damage. POFA has high pozzolanic characteristics. It is highly reactive and exhibits satisfactory micro-charging capabilities and unique properties.

Cement production, on the other hand, contributes to the greenhouse effect because one ton of CO_2 is released into the atmosphere per ton of cement produced [5]. One of the efforts to reduce the use of cement and increase industrial waste, which is high in Si and Al, is a geopolymer. Geopolymer concrete is also known as green concrete because it does not contain cement. The advantages of geopolymer include high strength, good bonding with steel bars, and high durability compared to normal concrete [6,7]. However, the characteristic of geopolymer depends on the raw material [8]. Hence, local waste industries in South Kalimantan are required to evaluate raw materials for geopolymer paste and mortar.

2. Material and method

2.1. Materials

In this study, the raw material blend was conducted between fly ash and POFA and 100% fly ash as an aluminosilicate source to evaluate the effect of substituting POFA for fly ash. Fly ash was collected from the Asam-Asam power plant, as seen in figure 1 (a). POFA was a regional waste from the palm oil industry in South Kalimantan, Indonesia, as seen in figure 1 (b). Fly ash and POFA were sieved using a 45 μm mesh sieve to ensure the fineness material was less than 34% (fineness requirements of ASTM C618). The chemical compositions of fly ash and POFA were determined using X-ray Fluorescence (XRF), as shown in table 1. Barito River sand with a specific gravity of 2.64 was used as the fine aggregate. The physical properties of fly ash, POFA, and fine aggregate are summarized in table 2.

Table 1. Chemical composition of fly ash and POFA (% weight).

Oxide	SiO_2	Al_2O_3	Fe_2O_3	CaO	MgO	Na_2O	K_2O	MnO_2	TiO_2	P_2O_5	SO_3
Fly ash	40.92	10.08	23.51	12.86	8.97	0.26	0.74	0.49	0.62	0.2	0.82
POFA	55.39	5.74	4.17	9.24	9.44	0.59	7.99	0.2	0.32	5.86	0.53

Table 2. Physical properties of materials.

Properties	Fly ash	POFA	Fine aggregate
specific gravity	3.09	2.35	2.64
volume weight (gr/cm^3)	1.23	0.63	1.5
water absorption (%)	-	-	0.28

Figure 1. (a) fly ash, (b) POFA.

Alkali activators combined sodium silicate and sodium hydroxide with an alkali solution ratio (sodium silicate to sodium hydroxide in mass) of 2.5. The characteristics of sodium silicate are indicated in table 3. The sodium hydroxide solution was prepared one day before casting the specimens.

Table 3. Sodium silicate specifications.

Parameter	
SiO_2 (%)	30.62
Na_2O (%)	9.42
Ratio (SiO_2/Na_2O)	3.25
Specific Gravity at 20°C	1.42
Baume at 20°C (°Be)	43.00

2.2. Methods

The geopolymer paste setting time and compressive strength were evaluated in the first phase of this study. Two types of molarity of sodium hydroxide were 8 M and 10 M, and two blended raw materials were 100% fly ash and 25% fly ash, replaced by POFA. There were three compositions of raw material to alkali activator ratio by weight, which was 60/40, 65/35, and 70/30. Sodium silicate was mixed with sodium hydroxide one hour before casting. The raw material was put in a bowl mixer, and the alkali activator was added slowly and blended for two minutes. The fresh geopolymer paste was then cast in a 28mm diameter cylinder with a height of 56mm. The paste mold was opened after 24 hours, as seen in figure 2 (a). Wrapped the specimens in a moist towel was selected as a curing system in this research until the days of testing, as seen in figure 2 (b). A total of 144 specimens were evaluated the compressive strength at ages 3, 7, 14, and 28 days. The specimens are listed in table 4. Specimen G1 indicated 100% fly ash as raw material, while G2 denoted 75% fly ash and 25% POFA. The number eight means 8 M NaOH, and the number ten means 10 M NaOH. The raw material to alkali activator ratio variable represented labels A, B, and C for 60/40, 65/35, and 70/30, respectively.

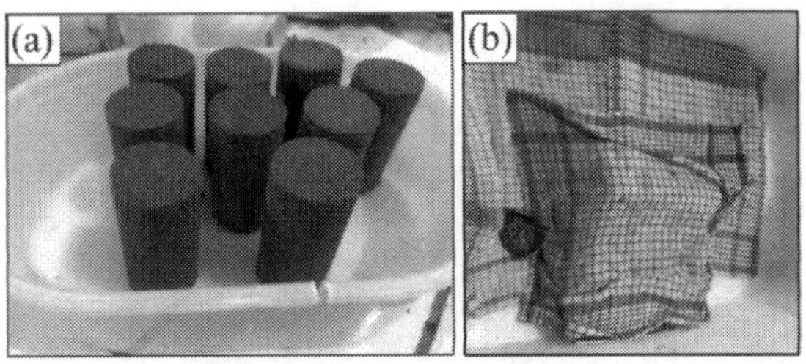

Figure 2. (a). Geopolymer paste specimens (b). Cure proses.

In the second stage, the compressive strength of the geopolymer mortar was evaluated at 7, 14, and 28 days. Geopolymer mortar composition was based on the highest compressive strength of geopolymer paste. Pour the fine aggregate into the fresh geopolymer paste using the paste mixing method described previously. Mix for an additional 2 minutes. The mortar composition was 65% fine aggregate and 35% geopolymer paste. Next, pour fresh geopolymer mortar into the 50-mm cube molds, as figured in figure 3. As with paste-curing systems, the mortar was covered with a damp cloth.

Figure 3. Geopolymer mortar specimens.

IOP Conf. Series: Earth and Environmental Science 1184 (2023) 012026 doi:10.1088/1755-1315/1184/1/012026

Table 4. Geopolymer paste specimens.

	FA/POFA (%)	NaOH Molarity	Raw material/alkali solution ratio
G1-8A	100/0	8	60/40
G1-8B	100/0	8	65/35
G1-8C	100/0	8	70/30
G2-8A	75/25	8	60/40
G2-8B	75/25	8	65/35
G2-8C	75/25	8	70/30
G1-10A	100/0	10	60/40
G1-10B	100/0	10	65/35
G1-10C	100/0	10	70/30
G2-10A	75/25	10	60/40
G2-10B	75/25	10	65/35
G2-10C	75/25	10	70/30

3. Results and discussion

3.1. Setting time

A setting time evaluation was conducted, as shown in figure 4. The initial setting of all mixtures varied from 9.3 to 20.2 hours, and the final setting was between 16 and 27 hours. The setting time reported in this study was longer than in previous studies [9–11]. Replacing 25% of fly ash with POFA, which has a higher SiO_2 content and lower CaO compares to fly ash, significantly prolongs the initial and final setting time. The existence of POFA indicated a longer setting time. These results agreed well with the finding by Huseien et al. [3]. Raw materials with low CaO tend to provide a longer setting time [12–14]. The high SiO_2 content of POFA allows for pozzolanic reactions and delays the setting time when CaO is added to produce additional C-S-H gels [10]. SiO_2 content in POFA in this research was 55.39% higher than fly ash (40.92%). Reducing 25% of the fly ash and replacing it with POFA increases the SiO_2 content of the raw material.

Higher molarities of NaOH significantly increased the setting time for 100% fly ash. However, when fly ash was replaced with 25% POFA, the increase in setting time was insignificant. The ratio of raw material to alkali activator also affects the setting time. The higher the raw material ratio to the alkali solution, the shorter the setting time. These findings were consistent with Kamhangrittirong et al. [11,15]. The higher the ratio of raw material to alkali solution, the less solution in the mix, the faster the dehydration of water, and the faster the rate of geopolymerization. As a result, the hardening time is reduced [9].

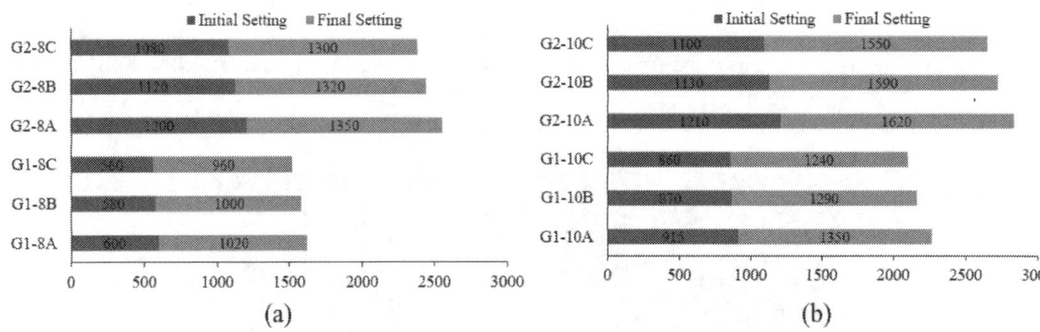

Figure 4. Setting time of geopolymer paste. (a) 8 M NaOH (b) 10 M NaOH.

3.2. Compressive Strength of Geopolymer Paste

The compressive strength test on the geopolymer paste cylinder is captured in figure 5. The compressive strength of geopolymer paste based on the raw material of 100% fly ash and 25% POFA substitution with various ratios of raw material to alkaline solution and the two molarities of NaOH is presented in figure 6. At 3, 7, 14, and 28 days, the average compressive strength of three specimens was evaluated.

The compressive strength of specimens tends to increase with age of the specimens. G2-10B had the highest compressive strength of 16.02 MPa. The effect of POFA replacements on fly ash compressive strength at early ages, at 3 and 7 days, for 10 M NaOH and raw material to alkali solution 70/30, demonstrated a decrease from 6.51 to 5.45 MPa and from 10.48 to 8.41 MPa, respectively. Ranjbar et al. [16] reported that increasing the ratio of silica/alumina (SiO_2/Al_2O_3) reduced the early compressive strength and postponed the geopolymerization process. Fly ash and POFA had SiO_2/Al_2O_3 ratios of 4.06 and 9.65, respectively. The incomplete geopolymerization process, where Al_2O_3 showed a higher dissolution rate during the early stages of geopolymerization, was blamed for the decreased compressive strength [3]. When 25% POFA was used to replace fly ash, the compressive strength was higher than when 100% fly ash was used. At 8 M NaOH, increasing compressive strength at 28 days of age was 20.96%, 16.29%, and 10.73%, whereas, at 10 M NaOH, the increase was 21.24%, 15.66%, and 18.14%, respectively, for raw material to alkali activator ratios of 60/40, 65/35, and 70/ 30. These results indicated that the 25% substitution of POFA improved the mechanical properties of the geopolymer paste.

The results in figure 6 also demonstrate the effect of NaOH concentration on the compressive strength of geopolymer paste. The compressive strength increased as the NaOH concentration increased from 8 to 10 M. This same behavior occurred in two raw material compositions and three ratios of raw material to alkali solution. Fly ash particles dissolved more readily at greater NaOH concentrations, which improved geopolymerization [17]. The ratio of raw material to alkali solution affected the compressive strength of geopolymer paste. The compressive strength of the 65/35 ratio was higher than the 60/40 ratio. However, it was lower than the 70/30 ratio. It should be noted that the optimum raw material to alkali solution ratio for the geopolymer paste obtained was 65/35.

Figure 5. The compressive strength test of geopolymer paste and mortar.

3.3. Compressive Strength of Geopolymer Mortar

The geopolymer mortar was manufactured based on the highest compressive strength of paste geopolymer. Figure 7 illustrates the compressive strength of geopolymer mortar at 7, 14, and 28 days. Two ratios of raw material to alkali solution were conducted at 25% fly ash replacement by POFA and 10 M NaOH. Identical to geopolymer paste, the ratio of raw material to alkali solution of 65/35 exhibits higher compressive strength than 70/30. The increasing ratio from 65/35 to 70/30 decreased the compressive strength by 33.4, 10.4, and 10.4%, respectively, at 7, 14, and 28 days. The ratios between the compressive strength of mortar to paste, respectively, for 65/35 and 70/30, were 0.66 and 0.59.

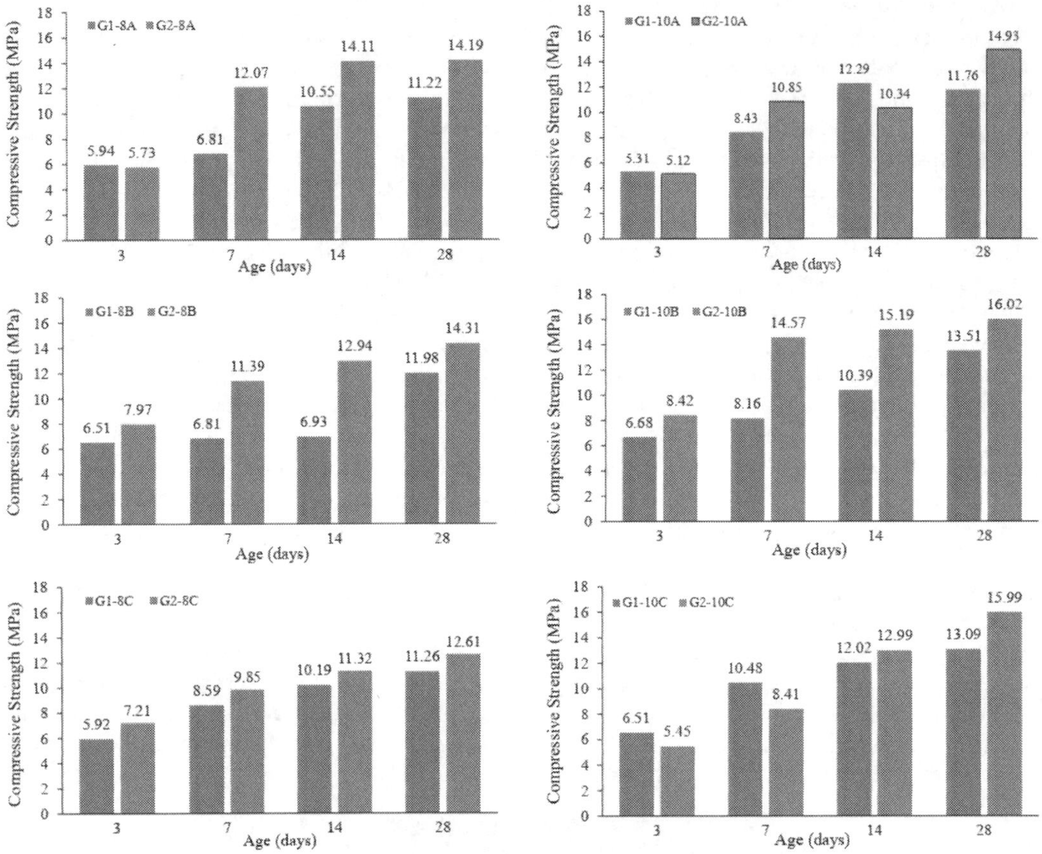

Figure 6. The compressive strength of geopolymer paste.

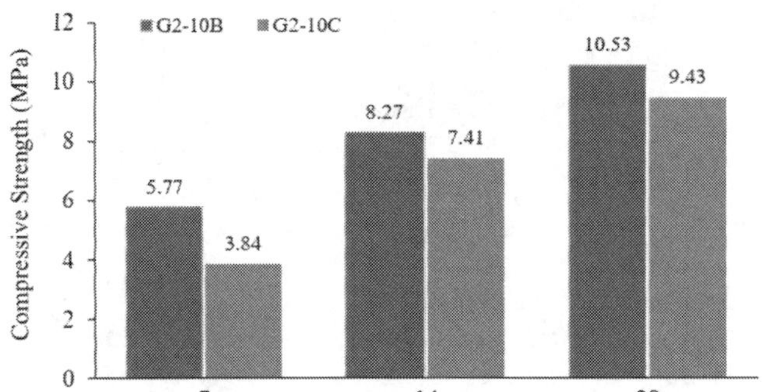

Figure 7. Effect of raw material to alkali solution ratio to the compressive strength of mortar geopolymer.

4. Conclusion

Based on the description and discussion above, the following conclusions are drawn:

1. The 25% POFA replacement for fly ash had a longer setting time than 100% fly ash.
2. Higher NaOH concentration caused prolonged setting time.

3. Higher raw material to alkali solution ratio insignificant lengthens the setting time.
4. A 25% substitution of POFA for fly ash increased the compressive strength of geopolymer paste.
5. The compressive strength of geopolymer paste increased as NaOH molarity increased.
6. The raw material/alkali activator ratio of 65/35 indicated the highest compressive strength.
7. The compressive strength ratio of mortar to the geopolymer paste was 0.66 and 0.59 for the 65/35 and 70/30 ratios of raw material to alkali activator, respectively.

References

[1] Trianto G A 2022 PLN Optimalkan Limbah Batu Bara Untuk Bangun Infrastruktur di Kalsel *https://web.pln.co.id/*
[2] Pemerintah Indonesia 2021 *Peraturan Pemerintah Republik Indonesia Nomor 22 Tahun 2021 tentang Pedoman Perlindungan dan Pengelolaan Lingkungan Hidup*
[3] Huseien G F, Tahir M M, Mirza J, Ismail M, Shah K W and Asaad M A 2018 Effects of POFA replaced with FA on durability properties of GBFS included alkali activated mortars *Constr. Build. Mater.* **175** 174–86
[4] Yasinta 2022 Pemprov Kalsel Prioritaskan Perkebunan Kelapa Sawit Berkelanjutan Melalui RAD-KSB
[5] Davidovits J 1994 Properties of Geopolymer Cements *First Int. Conf. Alkaline Cem. Concr.* 131–49
[6] Nurwidayati R, Ekaputri J J, Triwulan and Suprobo P 2020 Effect of embedment length on bond strength of geopolymer concrete *AIP Conf. Proc.* **2291**
[7] Nurwidayati R, Ekaputri J, Triwulan and Suprobo P 2019 Bond Behaviour Between Reinforcing Bars and Geopolymer Concrete By Using Pull-out Test *MATEC Web Conf.* **280** 04008
[8] Nurwidayati R, Ulum M B, Ekaputri J J, Triwulan and Suprobo P 2016 Characterization of fly ash on geopolymer paste *Materials Science Forum* vol 841 pp 118–25
[9] Yaseri S, Hajiaghaei G, Mohammadi F, Mahdikhani M and Farokhzad R 2017 The role of synthesis parameters on the workability, setting and strength properties of binary binder based geopolymer paste *Constr. Build. Mater.* **157** 534–45
[10] Karakoç M B, Türkmen I, Maras M M, Kantarci F, Demirbog˘a R and Toprak M U 2014 Mechanical properties and setting time of ferrochrome slag based geopolymer paste and mortar *Constr. Build. Mater.* **72** 283–92
[11] Kamhangrittirong P, Suwanvitaya P and Chindaprasirt P 2011 Synthesis and Properties of High Calcium Fly Ash Based Geopolymer for Concrete Applications *36th Conference on Our World in Concrete & Structures* (Singapore)
[12] Nath P and Sarker P K 2014 Effect of GGBFS on setting, workability and early strength properties of fly ash geopolymer concrete cured in ambient condition *Constr. Build. Mater.* **66** 163–71
[13] Diaz-Loya E I, Allouche E N and Vaidya S 2011 Mechanical Properties of Fly-Ash-Based Geopolymer Concrete *ACI Mater. J.* **108** 300–7
[14] Diaz E I, Allouche E N and Eklund S 2010 Factors affecting the suitability of fly ash as source material for geopolymers *Fuel* **89** 992–6
[15] Nath P and Sarker P K 2015 Use of OPC to improve setting and early strength properties of low calcium fly ash geopolymer concrete cured at room temperature *Cem. Concr. Compos.* **55** 205–14
[16] Ranjbar N, Mehrali M, Behnia A, Alengaram U J and Jumaat M Z 2014 Compressive strength and microstructural analysis of fly ash/palm oil fuel ash based geopolymer mortar *Mater. Des.* **59** 532–9
[17] Somna K, Jaturapitakkul C, Kajitvichyanukul P and Chindaprasirt P 2011 NaOH-activated ground fly ash geopolymer cured at ambient temperature *Fuel* **90** 2118–24

IOP Conf. Series: Earth and Environmental Science 1184 (2023) 012027 doi:10.1088/1755-1315/1184/1/012027

Viability of Napier Grass Ash as Partial Substitute of Cement in CHB

H A R Canseco-Tuñacao[1,2], J M A Nogra[1], M F M Rosell[1] and J C L Alquilita[1]

[1]Civil Engineering Department, Cebu Technological University, Cebu, Philippines
[2]Corresponding author: hanaastrid.tunacao@ctu.edu.ph

Abstract. The effects on the physical and mechanical properties of Napier grass ash (NGA) as substitute for ordinary Portland cement (OPC) on concrete hollow block (CHB) were investigated. Mix ratio of 1:2:4 with water-cement ratio of 0.60, and 10%, 15%, and 20% NGA replacements were adapted. Experimental procedures and testing conformed to American Society of Testing and Materials (ASTM) standards. Slump values, in comparison to control, showed a decline of 16.67%, 25%, and 58.33% for 10%, 15%, and 20% NGA replacements, respectively, which correlates to water absorption result. The denser the CHB, the lower the moisture content. Compressive strength results revealed that all NGA substitutes performed better than the control sample. NGA-20% had the highest compressive strength of 4.2 MPa (individual unit) and 3.9 MPa (average of 3 units) after 28 days of curing. Overall, NGA replacement was not viable for load-bearing but ideal for nonload-bearing CHB using the mix design.

1. Introduction

One of the biggest contributors to progress, lensed through infrastructures, is the substance called cement. Cement is a combination of raw materials such as limestone, clay, and marl and is usually grey. It acts as an adhesive to bind sand, gravel, and particulate matter when mixed with water. According to the Portland Cement Association and Rubenstone [1], an average of a hundred million metric tons of cement per year is consumed worldwide since 2000. This is due to the increasing demands for construction and the ease of use of cement.

With the increase in demand comes a shortage in supply. In the Philippines, Mogato [2] reported that the local cement industry cannot meet the annual demand due to the Build, Build, Build Program of the government. Despite the Philippine Mineral Resources Accounts report last 1994 [3] that the country has 50 billion metric tons of non-metallic mineral reserves and 39 percent of this is limestone, this resource is non-renewable. The lime, made from calcined limestone, that composed cement cannot be readily replaced by natural means to keep up with the fast-paced consumption. Thus, it becomes a challenge to researchers to explore alternative and sustainable partial replacements for cement.

Napier grass (Pennisetum purpureum) is a popular specie for pasture in the Philippines [4]. It is an adaptable specie being able to breed in both dry or wet conditions and even in limited spaces. It is a popular as a cut-and-carry forage in the tropics. Napier grass has more narrow leaves and taller stems than sugarcane. Napier grass grows well after being planted and can survive more than three years. Figure 1 shows the image of a Napier grass.

Figure 1. Napier Grass

The ash coming from Napier grass leaves is used as a partial replacement for cement in making concrete hollow blocks. According to Cordeiro & Sales [5], at 20% Napier grass ash replacement, no difference in compressive strength for the 35 MPa concrete was observed at 28 days of testing. Moreover, the elephant grass Napier variant (EGN) shows high pozzolanic activity after 7 days of curing with at least 85% lime making it supplementary cementing material [6]. In addition, results at 28 days showed that 15% Pennisetum purpureum ash (PPA) or Napier Grass Ash replacement in mortar to be optimum [7]. Furthermore, silica is the main oxide in Napier grass ash, reaching 55.7% by weight, making it a good candidate to be a pozzolanic mineral addition to OPC mixtures [8]. Lastly, at 20% cement replacement of raw NGA and NGA further burned at 400 degrees Celsius, showed mechanical strength results of 77.48 MPa and 80.7 MPa, respectively. Control mortar was at 83.5 MPa. This indicated that NGA can meet the requirement of being a pozzolan regardless if NGA undergone further burning. Table 1 is the comparison between the components present in Napier grass ash, fly ash, and OPC according existing studies.

Table 1. Percentage by Weight of the Chemical Composition of NGA vs. OPC and Fly Ash

Component	Ordinary Portland Cement [9]	Fly Ash [9]	Napier Grass Ash [5]	Napier Grass Ash [6]	Napier Grass Ash [8]
SiO_2	21.25	59.94	56.2	80.00	55.70
Al_2O_3	4.33	22.87	22.1	0.44	2.68
Fe_2O_3	1.85	4.67	6.1	0.77	1.61
TiO_2	0.13	0.94	No Data	No Data	No Data
CaO	64.30	3.08	No Data	1.85	8.44
MgO	1.81	1.55	No Data	1.11	19.78
SO_3	3.70	0.35	2.3	0.65	1.25
K_2O	0.71	2.19	7.4	7.05	9.94
Na_2O	0.17	0.62	No Data	No Data	0.00
Cl	No Data	No Data	No Data	0.95	0.22
P_2O_5	No Data	No Data	No Data	0.56	0.40
LOI	1.50	3.34	4.4	6.31	No Data

Cordeiro & Sales (2015) [5], Nakanishi et al. (2014) [6], and Rosello et al. (2018) [8] classified NGA under Class C of ASTM C618-19 [11], Standard Specification for Coal Fly Ash and Raw or Calcined Natural Pozzolan for Use in Concrete, with the chemical requirements of a minimum of 50% for the combination of silicon dioxide (SiO_2), aluminum oxide (Al_2O_3), and iron oxide (Fe_2O_3), and maximum of 5% for the Sulfur Trioxide (SO_3), 3% for moisture content, 6% for loss on ignition (LOI). Silicon dioxide, a pozzolanic substance in cementitious binders, are known to improve concrete density and affects the degree of hydration of cementitious materials thus affecting the rate of development of concrete properties [12][13]. Aluminum oxide added at a range of 10-15% by weight of cement reduced free chloride content [14]. Meanwhile, sulfur trioxide at excessive amounts causes an increase in the drying shrinkage of cement [15]. Based on the tabulated related studies, all three Napier grass ash have chemical composition in accordance to ASTM C618-19.

In the Philippines, concrete hollow blocks CHBs plays an important role in the partition of structures as it has a relatively low cost and ease in installation [16]. It is characterized as a block made of concrete with hollows or spaces with shell thickness and web thickness conforming to the requirements from ASTM C90 [17]. CHBs are categorized as load-bearing and nonload-bearing blocks. The thickness ranges from 15 to 20 centimeters and 7 to 10 centimeters for load-bearing and nonload-bearing blocks, respectively. Load-bearing blocks carry loads, such as live loads and dead loads including its own weight. Meanwhile, nonload-bearing blocks serves as partitions and dividers, walls and fences [18].

Existing studies adapted partial cement substitutes in concrete materials using Napier grass ash but there are no published studies that has utilized NGA inn concrete hollow blocks in particular. Thus, the study is pursued to impart useful information in the field of concrete-making.

2. Materials and Methods

2.1 Material Preparation

The leaves of raw Napier grass were sourced from Alcoy, Cebu, Philippines. The samples obtained were screened and cleaned with water to remove unwanted material. Cleaned samples were sun-dried for a minimum period of 5 days until the consistent weight of the dried sample was achieved. Dried samples were cut into small strips and milled in a blender. A thick pot was utilized placed above the flame from the gas stove to obtain ashes from the dried Napier grass leaves. Ashes obtained were sieved using a 90-micron (No.170) sieve as per ASTM C184 [19]. In addition, it was subjected to density test based on ASTM 311.

Sieving of the sand, as per ASTM D 2487-17 [20], was conducted to separate the fine and coarse aggregates used in the mix. Sieve analysis, water absorption, and specific gravity tests for the aggregates followed ASTM C33 [21]. Type 1 ordinary Portland cement conformed to ASTM C150 [22].

2.2 Material Proportion

The mix ratio of 1:2:4 (cement : fine sand : coarse sand) was adapted from Cabahug et al. (2016) [23] study for load-bearing CHB. The target individual unit compressive strength is 12.4 MPa (1800 psi) for average of three units as per ASTM C90 [17]. Water-cement ratio was 0.60. Ordinary Portland Cement was replaced with Napier grass ash at 0%, 10%, 15% and 20%.

2.3 Fabrication and Curing of CHB

Figure 2 shows the CHB dimensions which conform to ASTM C90. In this study, the fabrication of CHB was done manually. One CHB was fabricated per mix at an average duration of five (5) minutes from mixing to molding. The demolded CHB was air-dried for 48 hours before fully-submerged in the curing pond. A total of 48 specimens were made. Twelve specimens, 3 for control and per replacements, were sent for absorption test after 28-days curing period. Thirty-six samples, 3 for control and per replacements at 7, 14, and 28 days curing periods, were used for compressive strength test as per ASTM C140 [24].

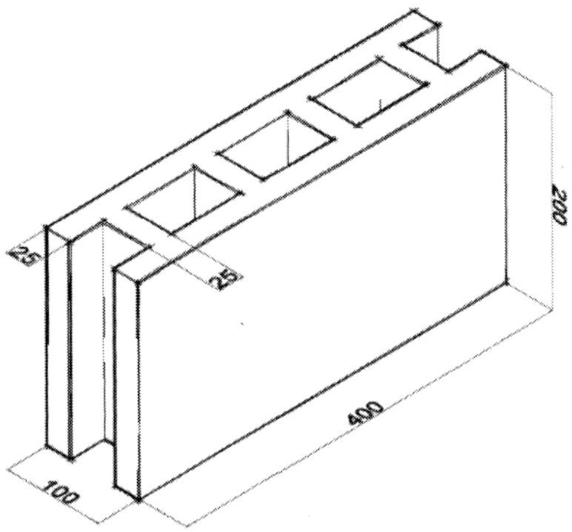

Figure 2. CHB Dimensions (in mm)

2.4 Analytical Investigation

2.4.1 Slump and Physical Tests

ASTM C143 [25] was conducted to determine slump. Three batches for each percentage replacement were measured and the average was considered. The target slump was from low (50 mm) to 0. The received weight and dimensions were measured based on ASTM C140 [24].

2.4.2 Mechanical Tests

ASTM C140 [24] was followed to determine the mechanical properties of CHB which includes water absorption, moisture content, density, net volume and area, and compressive strength. Twelve specimens undergone water absorption test. In addition, 36 CHBs were sent for compressive strength test. The equations below were utilized.

$$\text{Absorption, \%} = \left[\frac{(W_s - W_d)}{(W_s - W_i)}\right] \times 100 \tag{1}$$

$$\text{Moisture content, \%} = \left[\frac{(W_{\gamma} - W_d)}{(W_s - W_d)}\right] \times 100 \tag{2}$$

$$\text{Density (D), } \frac{kg}{m^3} = \left[\frac{W_d}{(W_s - W_i)}\right] \times 1000 \tag{3}$$

$$\text{Volume } (V_n), m^3 = \left(\frac{W_d}{D}\right) \times 10^6 = (W_s - W_i) \times 1000 \tag{4}$$

$$\text{Net Area } (A_n), m^3 = \left(\frac{V_n}{H}\right) \times 10^3 \tag{5}$$

$$\text{Net Area Compressive Strength (MPa)} = \left(\frac{P_{max}}{A_n}\right) \tag{6}$$

IOP Conf. Series: Earth and Environmental Science 1184 (2023) 012027 doi:10.1088/1755-1315/1184/1/012027

3. Results and Discussion

3.1 Material Properties

Sieve analysis as per ASTM C33 [21] of natural fine and coarse sand was conducted in accordance to ASTM C136 [26]. Table 2 presents that both aggregates were in accordance to ASTM standards.

Table 2. Sieve Analysis of Aggregates

Natural Fine Sand (NFS)			**Natural Coarse Sand (NCS)**		
Sieve Designation	Mass Percent Passing	ASTM Standards	**Sieve Designation**	Mass Percent Passing	ASTM Standards
4.75mm (No.4)	98	95-100	**4.75mm (No.4)**	85	85-100
2.36 mm (No.8)	94	80-100	**2.36 mm (No.8)**	11	10-40
1.18 mm (No.16)	70	50-85	**1.18 mm (No.16)**	2	0-10
600-µm (No.30)	37	25-60	**600-µm (No.30)**	0	-
300-µm (No.50)	18	5-30	**300-µm (No.50)**	0	0-5
150-µm (No.100)	6	0-10	**150-µm (No.100)**	0	-

Table 3. Aggregate Properties

Properties	**Natural Fine Sand (NFS)**	**Natural Coarse Sand (NCS)**	**ASTM Standard Range**
Specific Gravity	2.71	2.53	-
Water Absorption	2.25	6.16	-
Fineness Modulus	2.77	3.02	2.3 – 3.1
Average Moisture Content	3.45	1.63	-

In addition, Table 3 summarizes the aggregates' specific gravity, water absorption, moisture content, and fineness modulus. Results showed that the fineness modulus of the aggregates are also within ASTM standards. From Table 3, the moisture content of natural fine sand (NFS) is twice compared to natural coarse sand (NCS). Density test for NGA yielded 2520 kg/m3 (Gs=2.52) which conforms to ASTM C618 [11] standard range of specific gravity, 2.2-2.8, for Type C fly ash.

3.2 Design Mixture Proportions

Based from gathered material properties, the design mixture proportions were calculated for and are shown in Table 4 below.

Table 4. Mix Proportions by Weight (kg)

Material	NGA Replacement			
	0% NGA0	10% NGA10	15% NGA15	20% NGA20
Water	1.179	1.179	1.179	1.179
NGA	0.000	0.540	0.810	1.080
OPC	6.664	5.998	5.665	5.331
NFS	12.015	12.017	12.017	12.017
NCS	22.038	22.038	22.038	22.038

3.3 Fresh Properties

Based on Figure 3, a very low slump was achieved for all mix. NGA10, NGA15 and NGA20 obtained slumps of 10 mm, 9 mm, and 5 mm, respectively, which translates to 16.67%, 25%, and 58.33% decline compared to the control NGA0 (12 mm). It shows that the higher NGA replacement to cement, the more the decrease in slump. The same decrease has been observed from the study of Lv et al. [10] where the flowability of mortars decreased to 175 mm for NGA-Raw and 185 mm for NGA-400 in comparison to the control specimen at 195 mm. The decrease in slump was accounted for due to the irregular and rough structure of NGA particles and due to its higher LOI attributing to increased carbon content. Carbon with very high specific surface area results to higher water absorption. From Table 1, it can be seen that the LOI for NGA and Fly Ash are more than double to the LOI of OPC. Higher LOI for NGA resulting to higher carbon content resulted in lower flowability as indicated by the results in Figure 3.

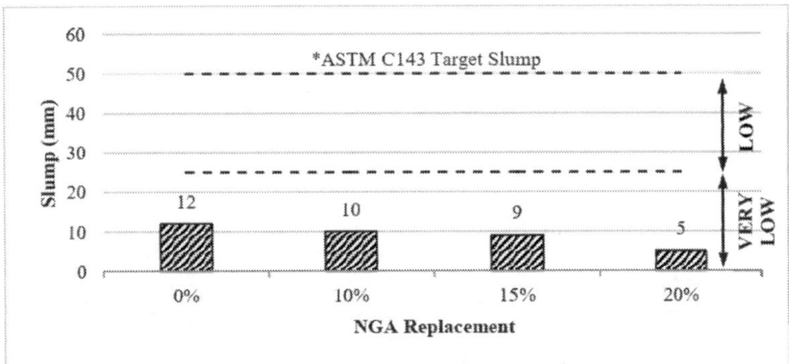

Figure 3. Slump Values

3.4 Mechanical Properties

Thirty-eight specimens were recorded to have either its length, width, and/or height smaller by 0.5 mm to 2.5 mm. This can be correlated to manual compaction and demolding. Nevertheless, all discrepancies are within the limits of ±3 mm as set forth by ASTM C90 [17].

Masonry units are classified to three according to weight as lightweight (< 1680 kg/m3 [105 lb/ft3]), medium weight (1680 kg/m3 [105 lb/ft3] < 2000 kg/m3 [125 lb/ft3]), or normal weight (> 2000 kg/m3 [125 lb/ft3]). Following the ASTM specifications, the blocks were classified as medium weight except for NGA-10% that reached normal weight as reflected in Table 5.

Table 5. Oven-dried Density of CHB Specimens in kg/m^3

Specimen	28 Days Curing Period	ASTM C90 Density Classification
NGA0	1932	Medium Weight (1680-2000)
NGA10	2011	Normal Weight (2000 or more)
NGA15	1997	Medium Weight (1680-2000)
NGA20	1930	Medium Weight (1680-2000)

Laboratory test results shown in Figure 4(a) suggests that with higher NGA replacements, the higher the water absorption of CHB. Compared to the control sample NGA0 (86.35 kg/m3), NGA10, NGA15, and NGA20 showed increased water absorption at 36.83%, 1.85%, and 52.80%, respectively. A similar trend was exhibited on mortar specimens of Wan Ibrahim et al. [7]. For mortars containing NGA, absorption rate was observed to be higher compared to the control at different testing age. In addition, based on ASTM C90 [17] standards for minimum water absorption, 288 kg/m3 for lightweight, 240 kg/m3 for medium weight, and 208 kg/m3 for normal weight, the water absorption of NGA0, NGA10, NGA15, and NGA20 are within standard limits.

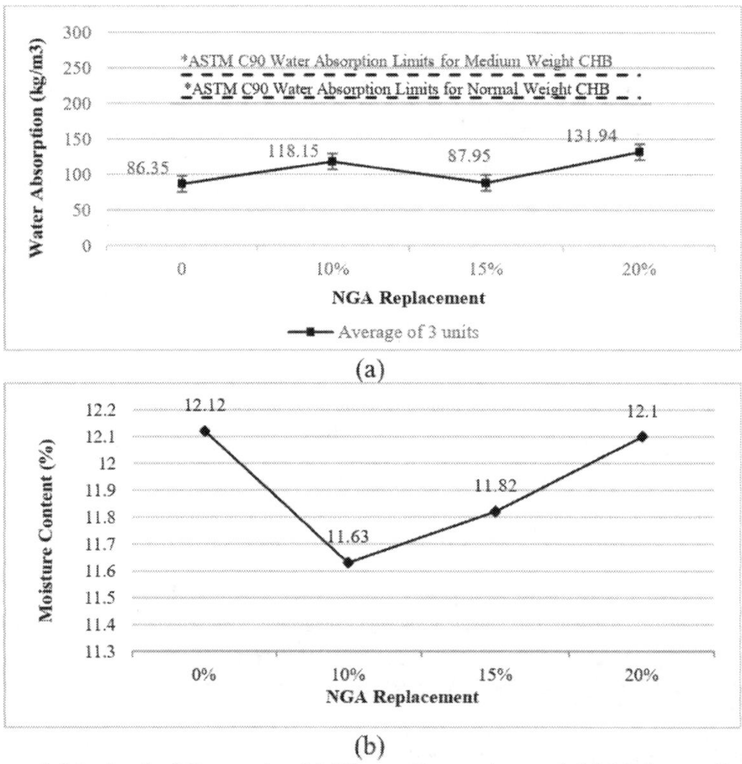

Figure 4. Mechanical Properties (a) Water Absorption and (b) Moisture Content

Figure 4(b) shows that NGA10 has the least moisture content. Moreover, an increasing moisture content is observed with increased NGA replacement. Meanwhile, recalling Table 5, it shows that oven-dried density of CHB peaked at 2011 kg/m3 for NGA10 then declines to 1997 kg/m3 and 1930 kg/m3 for NGA15 and NGA20, respectively. These data shows that the denser the specimen, the lower its moisture content. This can be attributed to the filler effect resulting to higher compaction of the components reducing the voids of the CHB. Thus, decreasing its ability to hold water. This signifies that among the specimen design, NGA10 has the desirable amount of NGA enough to fill in the voids.

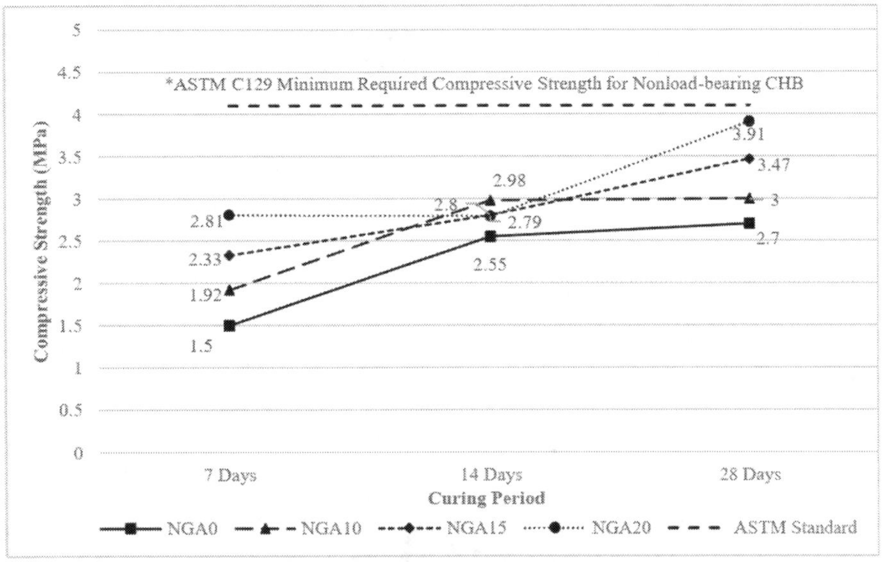

Figure 5. Compressive Strength Results

Using interquartile range rule, there were no outliers detected on compressive strength results. Figure 5 shows a unique increasing trend for the specimens with NGA replacements. NGA10 started with the lowest increase in strength of 1.92 MPa at 7 days and then increased to 3.0 MPa at 14 days then almost remained constant at 28 days. For NGA15, a continuous increasing trend was observed with strengths of 2.3 MPa, 2.8 MPa, and 3.5 MPa at 7, 14, and 28 days, respectively. NGA20% started with the highest strength of 2.8 MPa at 7 days, remained constant at 14 days, and then peaked at 3.9 MPa at 28 days. It was found that at 10% replacement, peak strength is achieved early while at 20%, strength remained constant up to 14 days then suddenly peaked at latter stage.

NG10, NG15, and NG20 performed better than the control NG0 at all curing periods with an increase of 28%, 55.33%, and 87.33% at 7 days, 16.86%, 9.80%, and 9.41% at 14 days, and 11.11%, 28.52%, and 44.81% at 28 days, respectively. The increase is attributed to the behavior of NGA. Lv et al. [10] highlighted that the pozzolanic reaction in mortar specimens counteracts the reduced cementitious effect due to NGA substitution. Moreover, NGA promoted C-S-H gel formation resulting to improved mechanical strength of the material. Wan Ibrahim et al. [7] stated that mortars with NGA showed increased compressive strength due to pore refinement as a result of the pozzolanic reaction of the material with cement particles. Moreover, since NGA particles are finer, there is a faster reaction with $Ca(OH)_2$ resulting to increased compressive strength in mortar specimens. It is, however, observed that in contrast to common notion of achieving increased compressive strength by increasing density by packing methods [27], results in this experimental program are otherwise. From Table 5, it indicates that NGA10 is the most dense specimen while NGA15 to NGA20 showed a decrease in density with increase in NGA. However, Figure 5 shows that the least dense specimen with NGA replacement, NGA20, acquired highest compressive strength in most of the testing days. This indicates that level of pozzolanic reaction increased with increase in NGA replacement. Shown in Table 6 is the strength activity index of the specimens. It is defined as the ratio of compressive strength of the specimens with NGA to the strength of the control. According to ASTM C618 [11], specimens with 20% of cement replacement by mass by another material should exhibit a strength activity index greater than 75% to be considered as a pozzolan during the 7 and/or 28 testing day. From the same table, it is shown that all specimens have indices greater than 75% at all testing days indicating the pozzolanic activity of NGA in the CHB specimens.

Table 6. Strength Activity Index (%)

Specimen	Testing age (days)		
	7	14	28
NGA0	100	100	100
NGA10	128	117	111
NGA15	155	109	129
NGA20	145	110	145

It is to note that 6 CHBs reached the minimum required compressive strength of 3.5 MPa (500 psi) for an individual unit while NGA-20% at 28-days curing almost achieved the 4.1 MPa (600 psi) for an average of three units for nonload-bearing CHB as per ASTM C129 [28]. Figure 4(b) and Figure 6(a) results and a photo of the specimens prior to testing, respectively, which indicate the factor resulting to lesser achieved compressive strength. The CHBs contain an average 12% moisture before being sent for testing as indicated in Figure 4(b). Moreover, this is observable in the photo shown in Figure 6(a).

(a)

(b)

Figure 6. CHB Specimen Photos (a) CHB Condition Prior to Compressive Strength Test and (b) After Testing

In the study conducted by Chen et al. [29], degree of saturation of concrete is inversely proportional to the compressive strength. When the degree of saturation is 52%, 87%, and 92%, the compressive strength decreases by 16.1%, 19.2%, and 38.4% respectively. Figure 6(b) shows the failure patterns of CHBs that were observed to start from the bottom where moisture is concentrated.

4. Conclusions and Recommendations

A study was conducted to investigate the effects of utilizing NGA as partial replacement for OPC in fabricating CHB. Efforts were made to perform physical and mechanical tests based on ASTM standards. Important conclusions are summarized as follows.

1. Increase in NGA replacement resulted to decrease in slump which correlated to an increase water absorption. Slump and water absorption results for 10%, 15% and 20% NGA replacement are within standard limits.

2. Density of CHB is inversely proportional to its water content. The most dense, NGA10, had the lowest moisture. Density decreased with increased while moisture content increased with NGA replacement. Among the specimens, CHBs with 10% NGA replacement was most compact due to the effective filler effect of NGA at this percentage replacement.

3. Compressive strength results show that all CHB with NGA substitutes performed better than control sample. Increase in compressive strength was observed with increase in NGA replacement with peak strength of 3.91 MPa at 28 days for NGA20. Increase in strength was observed with increase in NGA replacement regardless of the decrease in density, as is the common trend in concrete behavior indicating increased pozzolanic reaction with increase in NGA with NGA20 with a strength activity index of 145% at 28 days.

References

[1] Rubenstone, J 2016 PCA Forecast Growth in Cement at World of Concrete 2016. Retrieved from https://www.enr.com/articles/38747-pca-forecasts-growth-in-cement-consumption-at-world-of-concrete-2016
[2] Mogato, A 2019 Cement shortage seen to delay construction projects-Pronove Tai. Retrieved from https://www.rappler.com/business/227862-pronove-tai-says-cement-shortage-will-delay-construction-projects/
[3] Philippine Statistics Authority 2022 Retrieved from https://psa.gov.ph/content/mineral-resources
[4] Rosacia, W Z, Rimando, R M, Perino, E A, Avanzado M B, Polinag M A and Rueda, L B 2007 Environmental Requirement and Management of Some Important Forage Species in the Philippines *Research Information Series on Ecosystems* 19 (1-3) pp. 1-20
[5] Cordeiro, G C, Sales C P 2015 Pozzolanic activity of elephant grass ash and its influence on the mechanical properties of concrete *Cement and Concrete Composites* 55 pp. 331-336
[6] Nakanishi, E Y, Frias M, Martinez-Ramirez, S, Santos, S, Rodrigues, M, Rodriguez, O, Savastano, H 2014 Characterization and properties of elephant grass ashes as supplementary cementing material in pozzolan/Ca(OH)2 pastes *Construction and Building Materials* 73 pp. 391-398
[7] Wan Ibrahim, M H, Al-Fasih, M Y, NIK PA, N N A, Jaya, R & Setiawan, M I 2021 Characterization of Mortar with Pennisetum purpureum ashes as cement replacement material *IIUM Engineering Journal* **22**(2) pp. 83-97
[8] Rosello, J, Soriano, L, Svastano, H, Borrachero, M V, Santamarina, P and Paya, J 2018 Microscopic Chemical Characterization and Reactivity in Cementing Systems of Elephant Grass Leaf Ashes *Microscopy and Microanalysis* 24 (6) pp. 593-603
[9] Borsoi, A, Collepardi, S, Coppola, L, Troli, R and Collepardi, M 2000 Effect of Superplasticizer Type on the Performance of High-Volume Fly Ash Concrete *ACI Special Publications* 195 pp. 17-28
[10] Lv, Y, Ye, G and De Schutter, G 2019 Characterization of Cogeneration Generated Napier Grass Ash and its Potential Use as SCMs *Materials and Structures* 52 (87)
[11] ASTM C618-19 Standard Specification for Coal Fly Ash and Raw or Calcined Natural Pozzolan

for Use in Concrete
[12] Xu, X and Lu, Z 2007 Effect of silicon dioxide on hydration and hardening of Portland and cement *Journal of the Chinese Ceramic Society* 35 pp. 478-484
[13] Berra, M, Carassiti, F, Mangialardi, T, Paolini, A E and Sebastiani, M 2012 Effects of nanosilica addition on workability and compressive strength of Portland cement pastes *Construction and Building Materials* 35 pp. 666-675
[14] Ramesh, K 2014 An Experimental Study on the Effect of Aluminum Oxide on Strength and Corrosion Resistance of Concrete *International Journal of Engineering Research & Technology* 3 (11)
[15] Zayed, A M, Brown, K and Hanhan, A 2004 Effect of Sulfur Trioxide Content on Concrete Structures using Florida Materials *The National Academies of Sciences, Engineering, and Medicine* 0510754
[16] Humanitarian Shelter Working Group 2014 Concrete Hollow Blocks (CHB). Retrieved from https://sheltercluster.s3.eu-central-1.amazonaws.com/public/docs/Key%20Messages%20CHB%20V1.1.pdf
[17] ASTM C90-22 Standard Specification for Loadbearing Concrete Masonry Units
[18] Fajardo, M 2000 Simplified Construction Estimate 3rd Edition *5138 Merchandising* pp. 41-42
[19] ASTM C184-94e1 Standard Test Method for Fineness of Hydraulic Cement by 150-&#181m (No. 100) and 75-&#181m (No. 200) Sieves
[20] ASTM D2487-17 Standard Practice for Classification of Soils for Engineering Purposes (Unified Soil Classification System)
[21] ASTM C33/C33M-18 Standard Specification for Concrete Aggregates
[22] ASTM C150-07 Standard Specification for Portland Cement
[23] Cabahug, R, Bacol, J, Lagutin, P J, Luniza, L J, Mamon, G C and Pilapil, P N E 2018Crumb Rubber Tire as Partial Replacement for Fine Aggregates in Concrete Hollow Blocks *Mindanao Journal of Science and Technology* 14
[24] ASTM C140/C140M-22b Standard Test Methods for Sampling and Testing Concrete Masonry Units and Related Units
[25] ASTM C143/C143M-12 Standard Test Method for Slump of Hydraulic-Cement Concrete
[26] ASTM C136/C136M Standard Test Method for Sieve Analysis of Fine and Coarse Aggregates
[27] Canseco-Tuñacao, H A R, Remoto, K, Melendres, K and Deguzman I M 2022 Recycled Coarse Aggregate from Concrete Waste Using DMDA for Concrete
[28] ASTM C129-17 Standard Specifications for Nonloadbearing Concrete Masonry Units
[29] Chen, X, Wanshan, H and Zhou, J 2012 Effect of moisture content on compressive and split tensile strength of concrete *Indian Journal of Engineering and Materials Science* 19 pp.427-435

The Grow Impact of Electrical Stimulation and Grow Lights on Lettuce Plant (Lactuca Sativa) in Hydroponics System

Ferdinand F. Batayola, Ph.D.[1, 1], Wilen Melsedec O. Narvios, Ph.D. [1,2], Jasper C. Mendez[1,3], Ralph M. Casinto[1,4], and Carl Jade Villegas[1,5]

[1]Department of Electrical Engineering, Cebu Technological University-Main Campus

[1]ferdinand.batayola@ctu.edu.ph, [2]wilenmelsedec.narvios@ctu.edu.ph, [3]jasper.mendez@ctu.edu.ph, [4]ralph.casinto@ctu.edu.ph, [5]carljade.villegas@ctu.edu.ph,

Abstract. In today's agricultural production, hydroponics has become one of the most popular approaches to producing vegetables in a limited space. With its technological advancement, the concept of farming is well adapted. In this study, the researchers develop a process to improve the growth yield production of lettuce through electrical stimulation and the application of grow lights through an aquarium-type hydroponics system. Three comparative set-ups were made during the experiment, a lettuce plant in the soil for the outdoor set-up, and two sets of lettuce planting in an indoor set-up provided with pink and blue LED lights. Each aquarium set-up undergoes an electrical stimulation with different settings of voltage. Based on the results, it was observed that the plant grew quicker in the aquarium set-up with the pink LED lights and a voltage setting of 20V. Based on observation, it showed a substantial impact on the growth as well as their quality, weight, and appearance. Results showed that electricity boosted the yield of lettuce plants where it only takes 24 days for the lettuce plant to be ready for harvest from a normal of 65 days to mature. On the other hand, the outdoor lettuce plant also developed and produced a result of its quality, height of the plant, and appearance. However, the plant takes a long time to grow and per inspection, the leaves eventually dry out. This study manifested that the electrical stimulation and growth lights could trigger biological activities that could enhance the efficient growth yield of hydroponically grown plants.

Keywords: *electrical stimulation, grow lights, hydroponics system, growth yield, agricultural production*

1. Introduction

Hydroponics is the practice of growing plants using only water, nutrients, and a growing medium. The techniques for adapting the system have been used for a long time ago. Although the general theory behind hydroponics farming remains the same, where modern technology has enabled the growth of plants faster, stronger, and healthier. The study was conceptualized to have an indoor type of hydroponics system for a controlled environment using grow lights and the application of electrical stimulation. The objective of the study was to investigate the applicability of electrical stimulation for the growth yield of plants particularly in the hydroponics system. It relates to the stimulation of plant development through electrifying the environment around the plant. The idea was supported by Xin Zhang's study as he highlighted that leafy vegetable production under a controlled environment using artificial lighting has many advantages over conventional greenhouses and open field production [1]. This is also supported by Gaikwad's study, according to him, hydroponics technology is the most intensive and versatile method of growing crops production as it allows optimum utilization of nutrient solution, water, and space, as well as better control of climate and plant protection factors [2]. Furthermore, he reiterated that hydroponics production increases the quality of crops and their productivity, which results in higher competitiveness and economic income wherein he specified that hydroponics technology provides a variety of benefits when compared with conventional farming methods. As encouraged by Malik's study, the hydroponics production system has the potential to revolutionize agricultural production that ensures food security in the era of accelerating urbanization [3].

The concept of electrical stimulation in agriculture was already adopted and disputed whether the idea can help the plant growth rate. There are several studies proved the positive effect of electricity on the contribution of plant growth. In this study, the researcher showed that plants can also be applied electricity to increase the plant's viability and growth rate. Using an electric circuit in applying electric current into the plants. The data collected supports the claim that electricity makes a higher growth. The objective of the study was also supported based on Wang's study. According to him, electrical stimuli such as electric fields, currents, and discharge plasma can both enhance and inhibit plant growth [4]. The study summarizes various studies performed using different stimulation methods such as direct and indirect stimuli. From his study he summarized that proper control of applied electrical energy can cause genomic modifications, stimulate enzyme activity, enhance photosynthesis, control pH and bacterial levels in the cultivation environment, and increase nutrient uptake, resulting in increased yield [4]. The thought of electrical stimulation on plants was also suggested based on Smirnov's study. According to him, the technique in the application of electrical currents will result in yield optimization and quality improvement, thus, he iterated that electricity is an abiotic stress stimulant [5]. Takaki's research summarized that electrical stimuli such as electric fields can be used for the promotion of plant growth mode change from vegetative to reproductive [6].

For an indoor hydroponics system, sunlight is one of the factors that affect the growth rate of plants. Sunlight contains a spectrum of colors that can stimulate the photosynthetic process of hydroponic plants to grow. However, due to some environmental conditions that are not ideal, nutrients for hydroponic plants from lighting may be lacking. To solve the problem, Prasetia presented in his study a system that can be proposed where the sun's nutrients are replaced through the light-emitting diode (LED) grow lights that can produce a color spectrum like indoor sunlight. He stressed that LED growth lights are superior in terms of fresh weight, the number of leaves, and plant height respectively. The LED grow light has a better performance compared to sunlight [7].

2. Related Literature

Hydroponics is a new approach to growing plants in soilless culture. It uses the mineral nutrient solution for hydroponics growing to feed the plants in water, without soil. The traditional farming system is incapable of meeting present and future food demands. As a result, there is a pressing need to create a new agricultural technique that encourages plants to develop more quickly. The hydroponic system should be able to meet the rapidly increasing demand at a low cost and with minimal natural resource consumption [8]. According to Seungjun Lee and Jiyoung Lee, hydroponic systems have grown in popularity across the world and are increasingly being utilized for a variety of reasons in different parts of the world [9]. Due to its highly effective resource management and high-quality food production, hydroponic agriculture is gaining favor across the world [10]. By extending crop production into the vertical dimension, vertical farming systems have been proposed as an engineering approach to boost productivity per unit area of cultivated land [11]. Agriculture's future is bright, as it will produce higher-quality crops by making farms smarter in terms of sensing and manipulating their control parameters [12].

Sources of light that are generally used for plant cultivation such as fluorescent, metal-halide, high-pressure sodium, and incandescent lamps. These sources are responsible to increase photosynthetic photon flux levels but their disadvantage is that they contain unnecessary wavelengths that are located outside the photosynthetically active radiation spectrum, and the promoting growth has low quality [13]. With the presence of a light-emitting diode (LED), its unique features include the ability to control the spectral composition, a small mass and volume, durability, minimum heating, long operating lifetimes, wavelength specificity and narrow bandwidth, relatively cool emitting surfaces, and the electrical input current is linear to the photon output [13]. These lights are commonly used in growing plants in hydroponics. Instead of natural light, LED is used to illuminate the plant and its light can aid plant growth [14]. Apparently, the main causes of differences in indoor and outdoor plant growth are temperature and light accessibility. As a result, the light quality is another component that may be optimized to prevent abnormal plant performance while the use of light-emitting diodes (LED) in plant growth facilities grows [15]. LED offers a lot of potential as a supplemental light source for plants, a new and adaptable light source with a cool emitting surface, wavelength specificity, and low power consumption [16]. Plants cultivated under LED lamps have equivalent or superior chlorophyll content and photosynthetic capacity than those cultivated under fluorescent lamps [17].

Based on the study of [13], combining red, blue, and white LEDs has several favorable impacts on lettuce plant growth, development, nutrition, appearance, and edible quality. The regenerative and hydroponic culture systems may associate with commercial requirements for speedy, large-scale, and accurate management of plant production. According to the study of Wei Gao, where its hydroponic plant is spinach, LEDs with a red: blue ratio of 1.2 produced the highest spinach shoot yield (42 g per plant) and the lowest nitrate content (535 mg kg1), resulting in lower power consumption due to increased total fresh weight (1.7 kWh per 100 g) and higher energy yield (46.8 g). LED illumination with a R:B ratio of 1.2 also guaranteed a higher light energy use efficiency (0.025–0.026). (red, blue, and green light fractions of 31 %, 26 %, and 41 %, respectively). In hydroponic spinach, a larger proportion of red light inhibited oxalate accumulation whereas a higher fraction of blue light promoted vitamin C accumulation and oxalate reduction [18].

In the study conducted by [14], they used a 3:1 mix of LED red and blue light bulbs, resulting in a reddish-purple color. When they experimented with light vegetable salad in a confined setting for ten days, 16 hours a day, the results showed that the plants grew steadily under the LED light. The temperature and the brightness from the LED may be regulated at a consistent level in the test. Also, blue light (wavelength 430-460nm) has the advantage of allowing Chlorophyll samples to absorb more light waves. Chlorophyll allows plants to increase their light output even more. Accelerating stem development improves stem strength while decreasing stem elongation results in robust, fresh green leaves. The light waves from chlorophyll samples may be better absorbed by red light (wavelength 630-660nm), speeding up the blooms and nourishing the flowers and fruit to perfection. As well as assisting in maximizing productivity. Root development is accelerated, resulting in robust roots. Stem growth is also accelerated [19].

Enhancing the production rate and biological activities in plants in the process of electrical stimulation is vital. Short exposure time with low amplitude current injection, electric field, and pulsed electric field will enhance the growth rate and yield of plants in hydroponics such as tomatoes, strawberries, lettuce, and flowers [20]. The alternating current electric field affects the growth of the lettuce plant and also increases the uptake of metal ions. Electrophysiology functions primarily because plants consist of living cells and the plasma membrane of any plant root is charged negatively [21]. With lettuce cultivated in a soilless closed environment, the effect of potassium silicate is administered at different degrees of electrical conductivity of the nutrient solution. [22]. The growth of lettuce plants with electric current is much better than without current, and the effect of plant growth has an impact on the system [34]. Electrical currents are used for the improvement of horticulture plant cultivation. In support of [23], electricity use approaches classification is provided, highlighting the fact that electricity is an abiotic stress stimulator [5] and an effective way of faster plant growth in hydroponics.

3. Methodology

The proposed design as shown in figure 1, consists of an air stone and an air pump that will aid the plants in getting adequate oxygen to survive in an indoor environment and prevent the nutrient solution samples from forming lumps. Aside from this, sensors such as pH, temperature, and humidity sensors were also installed in the system's design for monitoring. A comparative set-up was made during the experiment, as two sets of lettuce planting provided with pink and blue LED lights were used as grow lights as shown in figure 2 and figure 3. The design is to replace natural sunshine, promoting photosynthesis and providing the proper shade range for the plant to grow and thrive.

Likewise, in conducting the electrical stimulations as shown in figure 4, three sets of the experiment setup were carried out. The first set-up is using 10 volts with a current of 0.18 ampere, then 20 volts with a current of 0.40 ampere, and lastly 30 volts setting with a current of 0.60 ampere.

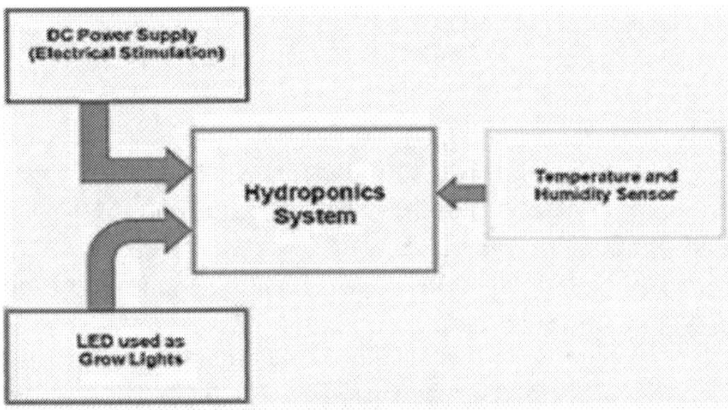

Figure 1. Hydroponics system design

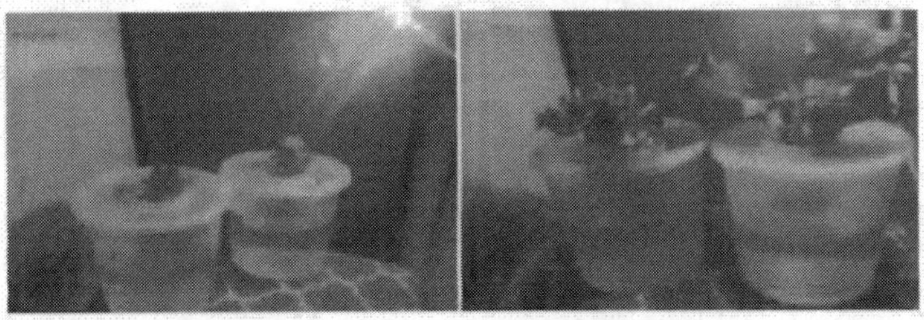
Figure 2. Hydroponics system set-up using the blue LED lights

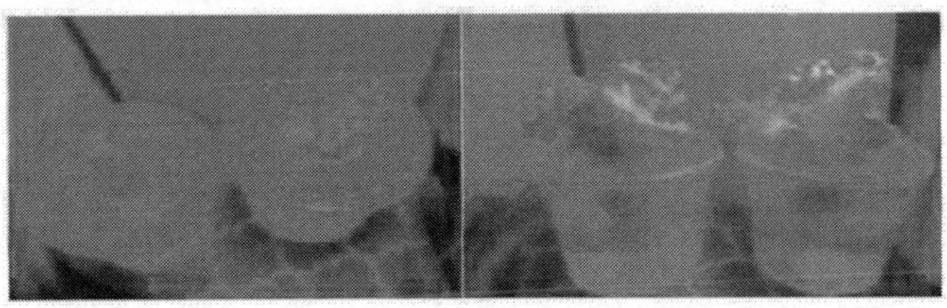
Figure 3. Hydroponics system set-up using the pink LED lights

IOP Conf. Series: Earth and Environmental Science 1184 (2023) 012028 doi:10.1088/1755-1315/1184/1/012028

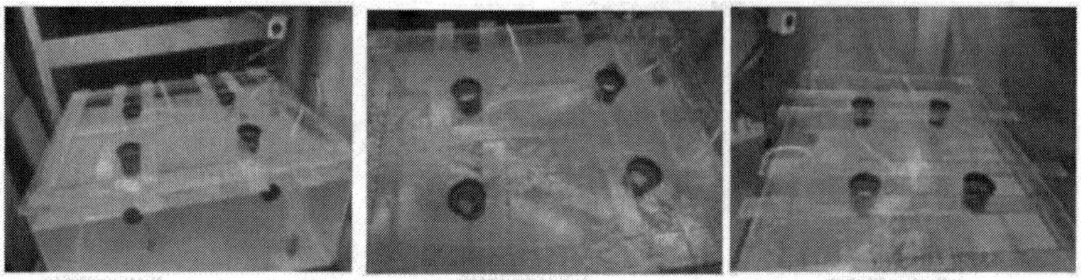

10Vset up 20V set up 30V set up

Figure 4. Electrical stimulation set-up

In conducting the experiment using the grow lights and electrical stimulation, the following parameters were used for data collection such as the appearance, thickness, and height of the plant.

4. Results and Discussion

Grow Lights:

Shown in figures 5 and 6 are the images of the appearance and thickness of the lettuce plant using pink and blue LED lights. Based on the results on the appearance of the lettuce plant, using pink LED light resulted in a better appearance with its light greenish and greenish color. With blue LED light, the lettuce appeared to dry out and its color turned to light brownish as shown in table 1.

Figure 5: Grow light pink LED bulb

Figure 6: Grow light pink LED bulb

Table 1: Summary of Data on the appearance of lettuce plant using LEDs a grow lights light pink LED bulb

No. Days	Using Pink LEDs	Using Blue LEDs
10 days	Light Greenish	Yellowish
20 days	Greenish	Light Yellowish
30 days	Light Greenish	Light Brownish

In terms of thickness as shown in table 2, the system with pink LED light used resulted in a good outcome because the plant is thicker compared to the system with blue LED light.

Table 2: Summary of Data on the thickness of lettuce plant using LEDs a grow lights light pink LED bulb

No. Days	Using Pink LEDs	Using Blue LEDs
10 days	Thin	Slightly Thin
20 days	Thick	Thin
30 days	Thick	Thin

In terms of height measurements, table 3, showed the comparative data between the pink and blue LEDs. Based on the results, using the pink LEDs has a favorable performance in terms of the height of the lettuce plant.

Table 3: Summary of Data on the height of lettuce plant using LEDs a grow lights light pink LED bulb

No. Days	Using Pink LEDs	Using Blue LEDs
10 days	70.63 cm	65.75 cm
20 days	105.56 cm	90.94 cm
30 days	120.36 cm	100.14 cm

Electrical Stimulation:

As shown in figure 7, the voltage set-up for the application of the electrical stimulation on plants using the 10V, 20 V and, 30V settings.

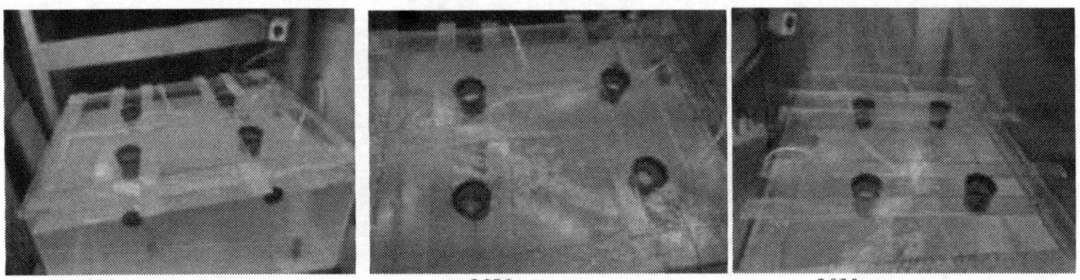

10Vset up 20V set up 30V set up

Figure 7. Electrical stimulation set-up

IOP Conf. Series: Earth and Environmental Science 1184 (2023) 012028 doi:10.1088/1755-1315/1184/1/012028

Shown in figures 8 and 9 are the images of the 10V and 20V set-up for monitoring the plant performance.

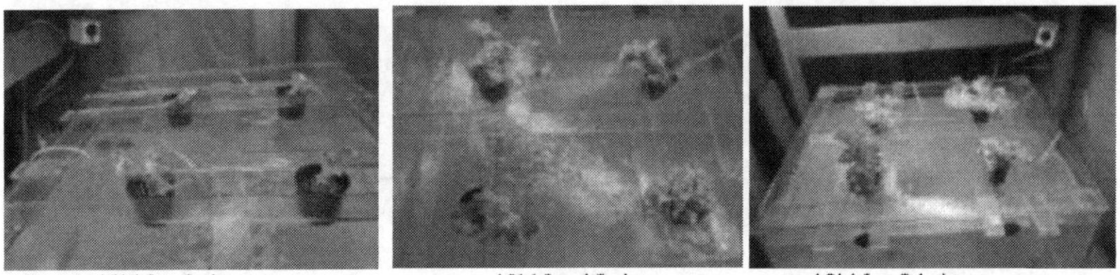

10V for 8 days 10V for 16 days 10V for 24 days

Figure 8. 10V electrical stimulation set-up monitoring

20V after 8 days 20V after 16 days 20V after 24 days

Figure 9. 20V electrical stimulation set-up monitoring

As shown in table 4, the comparative performance in terms of its appearance. Based on the results, using the 20V supply, the appearance of the lettuce plants is remarkable. It also showed that the plant in 30V supply resulted in an unseemly appearance.

Table 4: Summary of Data on the appearance of lettuce plants on the electrical stimulation set-up

No. Days	10Volts	20Volts	30Volts
8 days	Yellowish	Greenish	Yellowish
16 days	Light Greenish	Greenish	Pale Greenish
24 days	Greenish	Greenish	Light Brownish

For thickness monitoring using the different voltage settings as shown in table 5. Based on the results, showed that the thickness of the lettuce plant yielded a positive outcome in the 20V supply. The thickness of lettuce is less observed using the voltage application of a 30V supply.

Table 5: Summary of Data on the thickness of lettuce plants on the electrical stimulation set-up

No. Days	10Volts	20Volts	30Volts
8 days	Slightly Thick	Thick	Slightly Thick
16 days	Thick	Very Thick	Thick
24 days	Thick	Very Thick	Slightly Thick

Likewise, for the monitoring of the heights of the plant, table 6 showed the comparative performance of the different voltage settings. Based on the results, it showed that the 20V setting has a significant result in terms of the height of the plant.

Table 6: Summary of Data on the height of lettuce plants on the electrical stimulation set-up

No. Days	10Volts	20Volts	30Volts
8 days	85.26 cm	85.23 cm	82.37 cm
16 days	122.25 cm	123.14 cm	115.64 cm
24 days	132.56 cm	165.76 cm	110.33 cm

Table 7: Chart of the Comparative Data of Plant Height Monitoring Using Different Voltage Setting

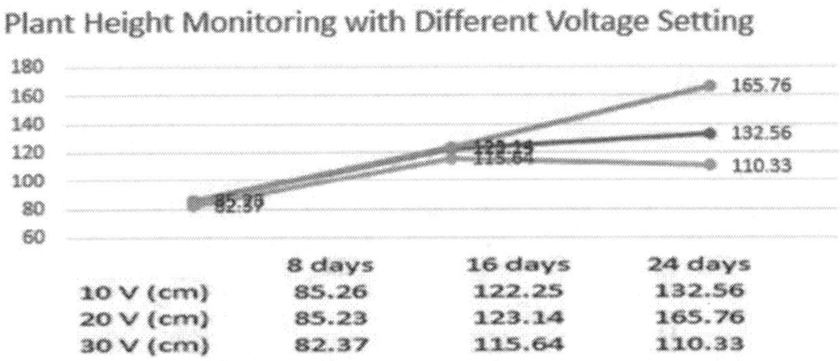

Based on findings, electrical stimulation has a substantial impact on the growth of lettuce plants as well as their quality, weight, and look. The plant grows quicker when the prototype system was used on the 20V inputs. Based on the settings specified on the system, the results from the 10 and 20-voltage inputs were similarly satisfactory in terms of their height and appearance. A higher voltage setup has caused the plant to develop slowly and for its roots to separate from the stem. Based on the study, it was observed that the plant grew quicker using pink LEDs and a voltage setting of 20 volts.

IOP Conf. Series: Earth and Environmental Science 1184 (2023) 012028 doi:10.1088/1755-1315/1184/1/012028

4. Conclusion

A controlled environment using artificial lighting and the application of electrical stimulation has advantages over a conventional hydroponics system. In this study, the growth rate of the lettuce plants has a significant result in terms of yield, length, weight, and appearance. From the study a pink LED light used as grow lights produced good results than blue LED lights. For electrical stimulation, the 20V setting also showed a better growth yield performance.

The combination of pink LED lights as grow lights and a voltage setting of 20V for electrical stimulation produced a remarkable performance on lettuce plants in the hydroponics system. A faster rate of harvest time was observed as it took only 24 days to harvest the lettuce plants from a traditional system of 40 days.

The electric stimulation can be used for the promotion of plant growth mode change from vegetative to reproductive. In the Philippines, agriculture fuels the economy. It is significant to study innovations like these to create more efficient ways to grow our crops.

5. References

[1] Xin Zhang, D.H. (2018). Effects of environment on the growth, photosynthesis, and quality of hydroponic lettuce in a plant factory. International Journal of Agriculture and Biological Engineering,33-40.

[2] Dinkar J. Gaikwad,S.M. (2020). Hydroponics Cultivation of Crops. In: Protected Cultivation and mart Agriculture,279-287.

[3] Arshad Mahmood Malik, K.M. (2019). Hydroponics Production Function, Returns to Scale and Domestic, Pakistan Journal of Agricultural Research,449-455.

[4] Wang,D.(2022).Improvement of Plant Growth and Control of Cultivation Environment Using Electrical Stimuli.Agritech: Innovative Agriculture Using Microwaves and Plasmas,227-246.

[5] Alexander Smirnov,Y.P.(2020).Technique of Plant Electrical Stimulation by Weak Electric Currents. Handbook of Research on Smart Technology Models for Business and Industry,21.

[6] Takaki, K. (2022). Promotion of Reproductive Growth of Mushroom Using Electrical Stimuli, Agritech: Innovative Agriculture Using Microwaves and Plasmas, 247-271.

[7] Yuda Prasetia,A.G. (2021). Evaluation of IoT- Based Grow Light Automation on Hydroponic Plant Growth. Jurnal Ilmiah Teknik Elektro Komputer dan Informatika (JITEKI),314-325.

[8] C.J.G Aliac and E. Maravillas,(2018). IoT Hydroponics Management System IEEE 10th International Conference on Humanoid, Nanotechnology, Information Technology, Communication and Control, Environment and Management (HNICEM),1-5

[9] S. Lee and J.Lee (2015).Beneficial Bacteria and Fungi in Hydroponic Systems: Type and Characteristics of Hydroponic Food Production Methods. Scientia Horticulturae, 206-215.

[10] N. Sharma.S. Acharya. (2018). Hydroponics as an Advanced Technique for Vegetable Production: An Overview. Journal of Soil and Water Conservation. Vol.17,364 -371.

[11] D.Touliatos, I.C. Dodd. (2016). Vertical Farming Increases Lettuce Yield Per Unit Area Compared to Conventional Horizontal Hydroponics. Food and Energy Security. Vol.5, No.3, 184-191.

[12] M.I. Alipio, A.E. dela Cruz (2017). A Smart Hydroponics Farming System Using Exact Inference in Bayesian Network. IEEE 6th Global Conference on Consumer Electronics (GCCE), 1-5.

[13] K.H.Lin, M.Y. Huang. (2013). The Effects of Red, Blue, and White Light-Emitting Diodes on the Growth, Development, and Edible Quality of Hydroponically Grown Lettuce (Lactuca Sativa). Scientia Horticulture, Vol. 150,86-91.

[14] L. Promratrak. (2017). The Effect of Using LED Lighting in the Growth of Crops Hydroponics. International Journal of Smart Grid and Clean Energy. Vol 6. No. 2, 133-140.

[15] C. Chiang, D. Bankestad. (2020). Reaching Natural Growth: Light Quality Effects on Plant Performance in Indoor Growth Facilities. Plants, Vol. 9, No 10,1273.

[16] M. Rehman, S. Ullah. (2017). Light-Emitting Diodes: Whether an Efficient Source of Light for Indoor Plants. Environmental Science and Pollution Research, Vol. 24, No. 32,24743-24752.

[17] N. Liu, F. Ji. (2019). Effects of LED Light Quality on the Growth of Pepper Seedling in Plant Factory. International Journal of Agricultural and Biological Engineering. Vol 12. No 5.,44-50.

[18] W.Gao, D.He. (2020). Effects of Daily Light Integral and LED Spectrum on Growth and Nutritional Quality of Hydroponic Spinach. Agronomy, Vol.10, No.8. 1082.

[19] N. Thanakon. (2012). Control Temperature and Relative Humidity in Greenhouses, Hydroponics. System Control Temperature and Humidity.

[20] N.A.M. Jamil, C. Gomes. (2018). Electrical Stimulation for the Growth of Plants: With Special Attention to the Effects of Nearby Lighting on Mushrooms.Asian J. Microbio, Biotechno, Environ, Sci. Vol.4,12-24.

[21] R. Bi, M. Schlaak.(2010).Alternating Current Electrical Field Effects on Lettuce (Lactuca Sativa) Growing in Hydroponic Culture with and without Cadmium Contamination. Journal of Applied Electrochemistry.Vol.40. No.6, 1217-1223.

[22] W.Chitarra, M. Pugliese. (2013). Effect of Silicates and Electrical Conductivity on Fusarium with Hydroponically Grown Lettuce.555-558.

[23] K. Bonghwan and C. Kyunghan.(2017). Electrical Stimulation and Effects on Plant Growth in Hydroponics.Journal of Engineering and Applied Sciences. Vol. 12, No. 14,4396-4399.

AUTHOR INDEX